The US Ar~
Interagenc,
Historical Pe ~ctives

The Proceedings of the Combat Studies Institute
2008 Military History Symposium

Kendall D. Gott
Managing Editor

Michael G. Brooks
General Editor

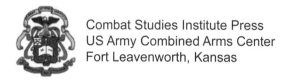

Combat Studies Institute Press
US Army Combined Arms Center
Fort Leavenworth, Kansas

Library of Congress Cataloging-in-Publication Data

Combat Studies Institute Military History Symposium (6th : 2008 : Fort Leavenworth, Kan.)
    The US Army and the interagency process : historical perspectives : the proceedings of the Combat Studies Institute 2008 Military History Symposium / Kendall D. Gott, managing editor; Michael G. Brooks, general editor.
        p. cm.
    Includes bibliographical references.
    ISBN 978-0-9801236-6-1
    1. United States. Army--Organization--History--20th century. 2. United States. Army--Management--History--20th century. 3. Interagency coordination--United States--History--20th century. 4. Military planning--United States--History--20th century. 5. United States--History, Military--20th century. 6. United States--Military policy. 7. National security--United States. I. Gott, Kendall D. II. Brooks, Michael G. III. Title. IV. Title: U.S. Army and the interagency process.

    UA25.C64 2008
    355.6'8--dc22

                                2008048742

First Printing: December 2008

CSI Press publications cover a variety of military history topics. The views expressed in this CSI Press publication are those of the author(s) and not necessarily those of the Department of the Army, or the Department of Defense. A full list of CSI Press publications, many of them available for downloading, can be found at: http://usacac.army.mil/CAC/csi/RandP/CSIpubs.asp

The seal of the Combat Studies Institute authenticates this document as an official publication of the CSI. It is prohibited to use CSI's official seal on any republication of this material without the expressed written permission of the Director of CSI.

Book Design by Michael G. Brooks

Contents

Day 2, Featured Speaker

Day 2, Panel 4: Interagency Case Studies

Day 2, Panel 5: Post-War Germany

Day 2, Panel 6: The Interagency Process in Asia

Day 3, Panel 7: Interagency Process in the United States

Day 3, Panel 8: Military Governments and Courts

Day 3, Featured Speaker

Closing Remarks

Foreword

These proceedings represent the sixth volume to be published in a series generated by the Combat Studies Institute's annual Military History Symposium. These symposia provide a forum for the interchange of ideas on historical topics pertinent to the current doctrinal concerns of the United States Army. Every year, in pursuit of this goal, the Combat Studies Institute brings together a diverse group of military personnel, government historians, and civilian academicians in a forum that promotes the exchange of ideas and information on a pressing topic of national significance. This year's symposium, hosted by the Combat Studies Institute, was held 16-18 September 2008 at Fort Leavenworth, Kansas.

The 2008 symposium's theme, "The US Army and the Interagency Process: Historical Perspectives," was designed to explore the partnership between the US Army and government agencies in attaining national goals and objectives in peace and war within a historical context. The symposium also examined current issues, dilemmas, problems, trends, and practices associated with US Army operations requiring interagency cooperation. In the midst of two wars and Army engagement in numerous other parts of a troubled world, this topic is of tremendous importance to the US Army and the nation.

This year the symposium welcomed as keynote speaker Brigadier General Robert J. Felderman, who provided an overview of the organization and capabilities of the US Northern Command and its ongoing efforts in interagency cooperation. Other featured speakers included Brigadier General (Ret) Mark Kimmit, the Assistant Secretary of State for Political-Military Affairs, and Dr. Richard Stewart from the Center of Military History. This volume also contains the papers and presentations of the seventeen participating panelists, including the question and answer periods following each presentation. The symposium program can be found at Appendix A. The proceedings can also be found on the CSI web site currently located at http://usacac.army.mil/CAC2/CSI/RandPTeam.asp.

These symposia continue to be an important annual event for those students and practitioners of military history who believe that the past has much to offer in the analysis of contemporary military challenges. Every year, both attendees and other recipients of the proceedings have uniformly found them to be of great benefit. It is our earnest hope that readers of this volume will find the experience equally useful. *CSI – The Past is Prologue.*

Dr. William Glenn Robertson
Director, Combat Studies Institute

Day 1—Opening Remarks

Introduction
(Transcript of Presentation)

Lieutenant General William B. Caldwell IV
Commanding General, US Army Combined Arms Center

I want to welcome everybody aboard and tell you we are glad to have you here. We are looking forward to a great three days. I think everybody can see the title—very applicable in the 21st century as we all move forward. The Chief of Staff of our Army keeps talking about the fact that we are in an "era of persistent conflict." In the next couple of years, I think we will all recognize and note that the budgets that the US military has today will only decline, not increase. And when they do, even greater emphasis will be put on everybody collaboratively working together in a coordinated manner as we all move forward. There is no question that in the 21st century anything that we do is going to be done in what we call a JIIM, a Joint Interagency, Intergovernmental, and Multinational kind of an event. We can even take it one step further and say it is going to be comprehensive because it is going to involve all the other entities, too, such as nongovernmental organizations, entities such as NATO (North Atlantic Treaty Organization), and everybody else. We will all be participating in a collaborative manner wherever we find the US military engaged around the world in the future. So today as we move forward, one of the great things about this is that sometimes we say those who do not read history are doomed to repeat it. What I would tell you is that, in fact, there are some great things from history that we do want to repeat. And I am hoping that through these next three days, through these eight different panels, the interactive discussions that will take place, and the off-line discussions that we will be reminded that there are things that we have done well in the past, and we know we have, but that we have just forgotten about, that we will repeat and do well again in the future. So I look forward to this. I think there are some great lessons learned out of the past, historical perspectives that the US Army and the interagency can dialog and talk about and benefit from. I will tell you also that here at Fort Leavenworth there are a couple of other great resources for those that have not been here before or really have not spent any time here. We have our Combined Arms Research Library which we call CARL. Our Combined Arms Research Library is a great resource. You can access it through the Internet. You can spend time over there. It is open to the public. It is not a restricted library, although there are restricted areas, classified areas in there, but the majority of it is wide open to the American public. It just won the 2007 Federal Library of the Year Award. It is that kind of quality library that we have here on this installation. We just received the award last Friday in Washington, DC. So I would encourage you to check it out, if you have not seen it or did not know about it, even if you just want to go by and ask some questions. We also have the Center for Army Lessons Learned here, closely tied in with the Joint Lessons Learned Center at Norfolk in the JFCOM (United States Joint Forces Command) command there. It resides here too. Great lessons learned. And they have an element that just looks at joint interagency lessons learned. It is a huge directorate. So that resource exists here at Fort Leavenworth also. We also have a host of other entities that reside here and are part of this organization. So if you are not aware of those, they exist here and people are more than welcome, and we encourage you to look at them, ask questions about them, or come back if you want at some other point. I want to thank Dr. Glenn Robertson. Glenn is our director for the Combat Studies Institute. He and his team have done a great job over many, many months of getting ready to be where we are today. So Glenn, thank you. We appreciate the hard work and effort that everybody has put into getting us to this point today, and we really look forward to these next three days. Anyway, I just want to officially tell everybody thanks for coming; thanks for being a part of this. This is being televised live throughout Fort Leavenworth. We have a closed-circuit television system that we use here so it allows any of our CGSC (Command and General Staff College) students, the twelve hundred of them that are here daily in school, to be able, in their classrooms, to watch what is happening here if they are interested in a particular panel. It also allows anybody else who works on Fort Leavenworth, outside of the college, to tune in and watch what is happening here. We are also blogging live from this session today. We have three interactive bloggers that will be engaged in blogging so there may be a time when different questions are asked by one or two persons in the back that are in the "blogosphere" that are sharing this with a large audience also. So if you hear other questions coming in, they will probably be coming in through the blogosphere in the back or they will be coming in through some of the classrooms with an instructor passing something down asking a question. So we will take questions that way too. Okay, with that, those are my comments. I just want to say welcome. Take the time to reach out and touch other people. You will see a lot of different uniforms if you are from outside Fort Leavenworth. In that row right there you will see four of our International Liaison Officers. There is another one back over there. They are representing 14 different nations. These officers are assigned here for two to three years. They are an incredible resource. They bring great richness to what we do at this institution and in the school itself where you will find 140 international officers, not as senior as these and not quite as knowledgeable, but they also are here and you will see them around. So there are almost 200 international officers, on a daily basis, that are here at Fort Leavenworth that are part of this whole organization and everything we do here. So if you see them around, please engage them. Ask them what they do. Quiz them. Put them under a little pressure. Do not let them off lightly. But we appreciate you all being here too. Thank you very much.

Day 1—Keynote Speaker

Interagency Coordination
(Transcript of Presentation)

Brigadier General Robert J. Felderman
Deputy Director of Plans, Policy & Strategy
NORAD and USNORTHCOM

Thank you very much. US Northern Command (NORTHCOM) was formed in 2002 based off of the actions of 9-11, 2001. To kick things off since we are talking about interagency, I think it is important for you to have an understanding of who your speaker is today. I like to consider myself one of the most joint military officers at NORAD (North American Aerospace Defense Command) and US Northern Command. I am the son of a retired naval officer. I followed him around for twenty-plus years. I enlisted in the Air Force and worked on F-106s that had a nuclear strike capability, which is one of the capabilities that NORAD used to monitor. I transferred over and was enlisted in the Army, went through Officer Candidate School, was Infantry qualified, attended Armor School and flight school, and then throughout the thirty-plus years in the National Guard I have worked in Infantry, Armor, Aviation, Medical Service, and Cavalry. I have a real estate business that I maintained as a National Guardsman. I stayed in the National Guard because of being a single parent for over six years I realized that was a little bit more important at the time than career. So I have a very successful real estate business that allows me these last almost four years to step up and work at NORAD and US Northern Command. But I also am on some local boards and committees, working down at the city helping write ordinances for real estate, zoning, waste water, and different things along that level. I have worked at the state level with real estate appraisals and real estate boards in licensing and accreditation. I am on the tri-state (Iowa, Illinois, and Wisconsin) American Red Cross Board of Directors. At the national level I have been working at the National Guard Association running resolutions and became certified as a legislative liaison. So I have a broad perspective, but a person is also impacted by whom you have in reach, and my brother-in-law is Director of Mission with USAID (United States Agency for International Development) down in Mozambique most recently, and in Guatemala. So I have a lot of reach out there. All of that comes into play in what I am discussing here today on US Northern Command.

Our primary focus is Homeland Defense. Homeland Defense is the number one mission, but we tend to spend most of our time and I will tell you I just disengaged yesterday from our response from working with Hurricane Ike, and before that Hannah and Gustav. But those are our two primary missions.

As you can see here (Slide 1) the focus of our mission is "anticipates." On the 29th of August as Hurricane Katrina had roared across the tip of Florida in Miami and was working its way up through the Gulf, General Rich Rowe was the Director of Operations. I was running the command center. We were trying to anticipate what would be needed, knowing that this storm was gathering up to huge levels, very similar path to what Gustav did. We tried to anticipate. We tried to look forward and see what was out there. We tried to pull forces in. We reached out to the services and what we got was, "The military does not respond until we are asked for something." FEMA (Federal Emergency Management Agency) and DHS (Department of Homeland Defense) were relatively new, coming together as an organization, and we could not get anything out of them as to an assessment as to what we were going to do. I will tell you that has changed today. It is significantly different and we have pre-scripted mission assignments out there that we have worked. We have coordinated with all of the services. We have the interagency partners that work with us every single day that say, "Here are what our gaps are." Here are the capabilities we do have that we know we will not need, and we reached out, as I will cover a little bit later on, with over 120 entities to do that coordination.

Our area of responsibility (Slide 2) will expect a little bit of a change here with the 2008 Unified Command Plan (UCP), but here today we cover Canada, the United States, Mexico, Bermuda, and Saint Pierre and Miquelon Island, which is a little French territory just off the east coast of Canada. We are going to pick back up the Bahamas, British Virgin Islands, Puerto Rico, and the Virgin Islands and that will give us a bit more expansion into what we now call our third border and that is the Caribbean. We see a significant growth there in our relationship in reaching out to them as we have done with Canada with over 50 years with NORAD. We have a very strong relationship with them at the North American Aerospace Defense Command. They just celebrated their 50th anniversary on May 12th. We also have, and I am very proud to have worked on that team, a bi-national team that will help Canada Command stand up their command, something very similar to our Northern Command and we have developed that relationship where we have a civil assistance plan that was just signed by both nations where if they have a disaster or we have a disaster, that we can offer military support to that Federal response. In fact, for Gustav we had some aircraft and some ships from Canada that were en-route and ready to be shared with us. With Mexico we work with the army SEDENA (Secretaria de la Defensa Nacional) and with the navy SEMAR (Armada de Mexico) on a regular basis. In fact, my boss, General Miller, is down representing General Renuart there right now for a meeting with their senior leaders and he will be headed back up tomorrow. These relationships are expanding on a regular basis. We also have quite a bit of an interagency side where we work with our Department of Homeland Security that reaches out to Canada's Public Safety and then to Mexico's Protección Civil.

This gives you a little bit of an idea of the spectrum that we deal with (Slide 3). Starting over here on your left, we work with national special security events such as the space shuttle. We have a team that goes down and coordinates with them, our Standing Joint Force Headquarters–North and then that mission is being handed off to our Joint Force Air Component Commander, which I will cover in a little bit as we lay out the distribution chart. We also work with the United Nations General Assembly, which begins here this week. I cannot remember the number, but we will work and provide support and assets to the Secret Service as they stand up and protect those folks. We are operating in an area that the government calls a POHA (Period of Heightened Awareness), and we have a period from the last couple of months here through hurricane season and our response, the Republican National Convention, Democratic National Convention, United Nations General Assembly, we have a space shuttle that is getting ready; it was moving out to the pad in the last week or two. We also will be coming into the elections in November, and then the transition into the new government in January, the Super Bowl that we cover every year, and several other different major events where we have thousands of folks in the United States that could be impacted. That will carry forward into July of next year and then from there we start working and coordinating with the 2010 Olympics that we are working with Canada in providing them support in that area. But we also work down here in the middle as you slide across the spectrum, which is the wildfires that we worked last year and the year before significantly in California and the southeast area. We monitor these throughout the country, but we had a huge response to those on the military side. We had immediate response from agreements and memorandums of understanding that we have with the local bases to be able to provide firefighters, either aircraft with buckets of water or flying folks around. We had a much larger response this last year with doing assets that are typically used for surveillance that we did for assessment and aerial views that we were provided that were hyper-spectral for the chemical side to the ultraviolet to be able to show where the fires were starting to pick up and we found that the fire chiefs, being able to send information down through a Blackberry or through a Rover P3 or full-motion video for this ultraviolet, infrared image that they could tell that they had their forces in one area of the nation and they could move and relocate it and be ahead of the fire. They never had the opportunity to do that before. So then everybody knows with the devastation that we have had and the stand up as I talked earlier of Hurricane Katrina, and then rolled into Hurricane Rita where we had First Army east of the Mississippi and Fifth Army west of the Mississippi and realized that we needed to have an Army element to be able to work that whole piece so we now have Fifth Army that works the whole nation for any response. We have 1st Air Force at Tyndall that works the air side and Fleet Forces Command that works the naval asset side, of course, linking into the Coast Guard. We have found that as we build those responses and work it that this becomes almost a day-to-day operation. So while Homeland Defense is our number one mission, the coordination with the interagency, coordination with nations in our area, coordination with the services and providing those assets is where we spend most of our time in this type of response. Now we can also get over into the homeland defense side. One of the areas I work quite a bit is nuclear weapons accidents, but also new NUDETS (Nuclear Detonations) as we work out there. We have two major exercises every year. One that focuses in on Homeland Defense, which is coming up here this November, VIGILANT SHIELD, and then ARDENT SENTRY, which focuses on the consequence management for any kind of a Homeland Defense response. We know that we are going to have a significant civil support response, no matter what that is, whether it is a 10-K NUDET that occurs, which we practiced last year in Indianapolis, or whether it is a pandemic influenza. As we reach out and try to work with all the different agencies that are there, it is very critical to our response to be able to work that whole piece. And this can extend all the way up to dealing with an actual force that attacks the United States.

This is our command and control chart (Slide 4). It gives you an idea. We just made a change. We conducted an organizational mission analysis in the Plans, Policy and Strategy Directorate and we just made a move effective this year. But I would like to start at the top and you can see where it is very important that we have our coordination with the Department of Homeland Security and with the National Guard. For those of us in the military, C^2 always meant command and control, but we have changed that definition of C^2 because we are in support always of others in that while we may have command and control within the service elements, command and control stands for co-ordination and collaboration, and with collaboration that means that you are sitting at the table as even partners. So as we work with the Department of Homeland Security, we have their liaison officers and senior officers that exchange and work with us. As we get into the chart on the Department of Homeland Security and the different agencies I have a huge list of folks that we work with, but with the National Guard it is more than just the National Guard Bureau because that is not a command and control entity. It is reaching out to the states; it is reaching out to the Adjutant Generals that are with each of those 54 states and territories. It is reaching out to 12 of them that are triple- or dual-hatted as their Officer of Emergency Management or their Office of Homeland Security. So we reach out and coordinate and collaborate with those folks giving us our own definition of C^2. We have Joint Task Force Alaska that works the homeland defense and civil support missions up in Alaska. They also are a little bit dual-hatted with the Alaska Command. They work with the Pacific Command out of that area of our world. Then we have Joint Force Headquarters National Capitol Region. You might know them as the Military District of Washington, the Army element. In day-to-day operations they coordinate and work our equities in the National Capitol Region. They are leaning forward and we turn them into a Joint Task Force for different events. For example, the elections, or the inauguration, or when the President gives his State of the Union Speech, they will stand up and operate as our lead element. Here we have the services representatives. Air Force North becomes a Joint Force Air Component Command when we stand them up for a response. Army North, Fifth Army, becomes our Joint Force Land Component, or JFLC. We have Marine Forces North, which works the Marine forces that would respond to an event in support of the Navy or the Army. And then we have the Fleet Forces Command, which today is a supporting command but there have been discussions that they would become a Navy North. Then when we stand them up as we have in the past, they would become a Joint Force Maritime Component Command. Each of those component commanders would also be responsible to be prepared to stand up a Joint Task Force should we need it. For right now, regarding our response into Florida, the required response was not as strong as it could have been. Florida has a significant response force so all we sent in there was 1 of our 10 Defense Coordinating Officers. But had we needed a larger response there or a larger response in Louisiana or in Texas, we could have stood up the Joint Task Force that all of those elements of the services would have coordinated through. We also work through, if the state stands up their own Joint Task Force in a Title 32 capacity or a Federal funded—but under the State's control—Task Force, the Defense Coordinating Officer could coordinate and collaborate with them. We also have some laws on the book that allow us to have a National Guard officer, as in my case, become dual-hatted and to become a Title 10 officer as well. Also we could have a Title 10 Active Duty officer get a commission into the National Guard and then he would have the responsibility of Title 32 National Guard forces or Title 10 Active Duty forces to work that total response. Where that would come into play more than likely for the National Guard is if you had four or five different states, particularly in one state where there is significant response. Let's use Texas as an example. If they did not have the capability that they needed, we could add a lot of forces, a significant number. During Katrina we had 50,000-plus National Guard folks from different states that came down to Louisiana and Texas and over 22,000 Active Duty. About 10,000 of them were boots on the ground. So when you get that large of a force, you need to be able to develop unity of effort, a clear command and control piece of those operations, but all of that is in support of our nation. Then as I said, just recently we moved Joint Task Force North out of El Paso, formerly known as JTF-6. Those folks work the counterterrorism efforts that we have along both of our borders north and south. And then we have Joint Task Force Civil Support. They work the CBRNE (Chemical, Biological, Radiological, Nuclear, High-Yield Explosive) effort that we have if we need to stand up a Joint Task Force. They are in fact today working with Army North at Fort Stewart, running a validation exercise for the first assigned forces that will be given over to Northern Command. On 1 October we have a CBRNE Consequence Management Response Force or CC-MRF. You may have heard of it by different names. We are actually hosting a contest. General Renuart has offered $100 to anyone who comes up with a better name than CCMRF. They are out there being validated today. That is a force that has been built up of about 6,000 folks and right now the plan is to stand up three of those CCMRFs. JTFCS (Joint Task Force Civil Support) would be the element that would work with them on a regular basis. The next two elements as they stand up over the next few years will be a mix of Active, Guard, and Reserve as we work out there and then several of them across the services so it will end up being a joint entity.

We typically run this slide when we are doing a one-on-one with folks (Slide 5). We do it as the build slide. You can see the whole picture, but typically you start right down here in the middle when an event actually happens. In this case you can see that it is depicted as a hurricane. We also discuss it where we have a NUDET or some kind of chemical response. As the hurricane begins you start off with your local responders, that fireman, that police chief, that first response down on the ground. So these are the folks right here as they go out and work that effort. Then as they see the storm or the response is within the national response framework, they see that response is larger than they can handle, then the governor may declare a disaster, either statewide or by different counties. In June in Iowa, my home state, we had 89 out of 99 counties that were declared state disasters, 85 that received Federal assistance. But we have the local responders that are out there working. The governor activated the National Guard to State Active Duty and sent the folks out to respond. Typically for a hurricane high-water vehicles are needed, search and rescue capabilities are required, and planners are needed to be able to help the State and locals put their efforts together. Sometimes it is just having a body to go out and fill sandbags or to direct traffic. The National Guard folks are also allowed to do security, law-enforcement type duties legally when they are in State Active Duty or Title 32 under the governor's control. So they are able to provide that kind of benefit. As they went out during Katrina, they were able to go through the houses, break in the door to make sure if someone was alive or not in the areas that were flooded. We have another valuable resource, EMAC (Emergency Management Assistance Compact). That has really grown over the past few years. It used to be just a few states in one area would help another state within their area—the northeast quadrant, the southeast quadrant, southwest quadrant—all these different compacts out there. Now we have all of the states and territories that have signed on to this. You can have the need for 100 ambulances down in Texas . . . you put out a call and coordinate and they will provide those folks. You can have any assets in that regard that can show up. Where it really has come into play is in the National Guard response. During Katrina, General Blum, Chief of the National Guard, had a video teleconference and had all the Adjutant Generals out there and said, "You know the kind of forces we need. Send us what you can. I need 1,000 folks from each of you. This is how big this is going to be. We need to get down there." Then the paper work caught up. Typically what happens is, for example with Louisiana, Texas was short of aviation assets so they put out the call to and coordinated through the Emergency Management, NEMA (National Emergency Management Association) (we will get into all the different acronyms). It is an entity that helps work that for us and coordinates it. They coordinate through the National Guard Bureau, through the states and all these assets started flowing down into Texas. But as you look at that, as you look at the 26 different hurricane states, which also includes Hawaii, and the capabilities that they have based off of what level of storm that they are responding to they have identified different gaps that are out there and that is where we need to work before a disaster to be able to identify what assets might be needed—medical, aviation, incident awareness and assessment, transportation, search and rescue. So as those folks start getting out there and the Governor asks for a Presidential declaration, that is what is in his request up here, he asks for a Presidential disaster declaration. If the President does declare that, it invokes the Stafford Act. The Stafford Act then allows us to be paid for what we will be responding to in those different states. He reaches out and typically puts into the Department of Homeland Security as the primary Federal agency in that regard, but not always. Everyone knows the I-35 up in St. Paul/Minneapolis that fell in. The Department of Transportation was the primary Federal agency for that response. So we provided Navy divers. Those different requests for assistance become a mission assignment as they pass through the Secretary of Defense for his approval and then work their way down. Then US Northern Command works that response for any of those Title 10 assets. As I mentioned earlier, we have a group of pre-scripted mission assignments that we identified from Katrina, different assets that they needed—communications teams, field housing and sheltering support, clearing debris, incident awareness and assessment, transportation, critical care patient evacuation. So as we identify those we put them on, for lack of a better word, a canned flight plan so that the primary agency working down to the state, with their State Coordinating Officers collaborating with the Defense Coordinating Officer. In the past they were reaching out during Katrina and saying, "I need, because I have been told that the U-2 is the best aircraft to give me a flyover to be able to tell what the roads, the power lines, and the roofs look like so that we know what we have to respond to, whether it is tarps or food, or whether people are stranded." We need to have that U-2 and that is what they asked for. That was a 900 and some thousand dollar request yet we had aircraft with better assets, better capabilities, or equal capabilities that could go out and do it for significantly less. For example, the Civil Air Patrol or the P-3 Orion from the Navy could fly out and take pictures and give us full-motion video. All of those assets we were using so they now ask for a capability and as you work in that capability that allows us to develop it. But within that they still need to have an idea of what we need to know, the five Ws—what, where, when, why, and where are we going to put them. So they now know what they have to give us. What location do you want them in, how long do you want to have them, are you going to provide the support or are we going to provide the support? So we send those entities out there as we work through Northern Command, the Defense Coordinating Officer, the Federal Coordinating Officer in the Joint Field Office and the State Operations Center. Later on I will have some slides that will cover what we did for Gustav regarding interagency coordination, several different agencies, and how that worked into this response. The big thing here on the take-away end is that support has to be requested and that we are always in support of that primary Federal agency as they respond. It does not mean we do not put forces out there to be prepared to deploy. We have the ability to have a group of Tier-1 level, Category One Forces that General Renuart can put out there as the Defense Coordinating Officer, some assessment elements that can go out there and give us some idea of what the situational awareness is. We are not going to send forces forward. We are not going to send boots on the ground into the State of Florida when they know that they have it under control, but we are going to be positioned and ready to go and with a lot of those assets that I will discuss later on.

The next slide (Slide 6) really focuses in on the main part of what you are here for today, the interagency piece. You can see within the military services the folks that we have, but we have over 60 different organizations that work with us from the American Red Cross to the Department of Transportation all the way to the Central Intelligence Agency and the Defense Intelligence Agency, NASA (National Aeronautics and Space Administration). We have over 60 of those folks represented at US Northern Command; 45 of them are full-time. The rest of them are either liaisons and available in the area or we may have a liaison out to work with those folks. As we work with those, that has grown. Secretary Rumsfeld was the one who first put out the Joint Interagency Coordination Group and I will talk about it a little bit later, what its mission and role is, but all these folks work through that Interagency Coordination Team.

The next slide (Slide 7) focuses in . . . you can see that we work in a joint and combined environment, the joint being the different services that are there, the combined when we do a response that includes the different nations. Below the surface is where our success is. That is where we see the piece that we work every single day that we would not be able to do the mission and tasking that we have been given for homeland defense or civil support without those folks.

The next slide (Slide 8) will focus just a little bit on what is interagency. We work within the military. We have our Operations, our J3, we have our Plans, the J5, we have Training and Exercises, J7. Those are directorates; those are functions, so the communications, the logistics, the operations, the plans. Interagency is a process and as that process works through it is important that we reach out and facilitate that. Well, that Interagency Coordination Group does just that. As necessary, they reach out and they focus in on those different areas. So you can see who we reach out to. Most folks do not realize that we work with the tribal nations as necessary. But it starts with the local, tribal, state, and then the Federal side. You can see a couple of them that are up there, the different agencies that we work with on a regular basis—the Department of Homeland Security for a typical Consequence Management type of response for a disaster, the Department of Transportation for something with the highways, Health and Human Services would be more along the lines of pandemic influenza where they might be the primary agency, the Department of Energy for homeland defense with a civil support tag on for a NUDET of some kind. On the nongovernmental agencies it is very important that we work with those—Red Cross, other international humanitarian groups that are out there. We have an entity, and I will bring it up later on, that just handles the donations. The Department of State reaches out and handles donations from other nations, but from local faith-based, nonfaith-based folks that just want to do good. And they start up often before we see something coming. So for a no-notice response, a NUDET, an earthquake, that response starts after it kicks in. But for a hurricane . . . we started working Hurricane Ike as a pressure system off the coast of Africa. I always like to say it started when the butterfly fluttered its wings, but as we came across and had Hannah and Gustav and Ike and were watching them and where they were going, and doing the gap analysis and assessment as to what we do if it was a Category 2, which it finally did even though it had a Category 4 type of impact. What if it hit Florida? What if it hit the Gulf? What if it made a turn and went up and hit in North Carolina or the National Capitol Region? So as we worked those things, these are the folks we have to really reach out and work with. In the private sector we have a lot of different conferences, a lot of different preplanning events, a lot of after-action reviews (AARs) that we do, and we bring folks in just as you are doing now to be able to discuss these types of things. So we work with those. We have one organization, BENS (Business Executives for National Security) that are significant contributors in holding these types of entities out there, but we also focus in on the whole of government and the whole of society. And on here, what is missing are the international partners. Each of these different entities through the Department of State really do reach out and get that piece. During Hurricane Rita when it went into Texas, for the first time Mexico provided us some forces that came across the border and one of my division branch chiefs worked with them, was working out of Mexico, brought this team across with some food and some doctors and folks to be able to provide a response. That was one of the first times. During Katrina, with the levees and that piece, we had the Dutch that flew in with experts to advise us on working with the levees. It does cross the entire spectrum.

Folks say, "Why do you need to have that?" (Slide 9). Well, in the military as we are learning, as your group is doing here today, reaching out and looking at the interagency coordination that needs to take place are very critical. In the United States it works down into a lot of the things that carry all the way back to our Constitution. The states have their rights; the mayors have their rights, and you can see up here the mayor does not necessarily work for the governor and the governor does not necessarily work for the President. We need to be able to have that coordination across the entire spectrum to be able to do that response. The challenges here—there is no one department that can handle all of this, there is not one that has all of the assets, all the capabilities, and so it does require us to reach out and do this. And again, the key here is that we work as an integrated team, that unity of effort and that there is collaboration taking place across it. And the idea here is that everybody wants to do good, but not everyone, or not any one entity, is going to be in command so it does take several different lines of effort to be able to go out and work this piece.

Again, on the why (Slide 10), I actually have my sheets here in front of me because even though I am one of the weak users of acronyms I have to carry it around and look them up on a regular basis. We think in the military we have a lot of acronyms, but it is just like having a translator for several of these different things. There are so many duplicate acronyms out there that are two different meanings so as we work through that, that is what we are looking at, the language piece. Everyone from their foxhole has a different perspective. They think they know what they need, but they do not necessarily know how to make it work and get it coordinated and focused in on a response. So that piece is very important. And then the experiences, as I led off with the different experiences that I have had, I am sure General Caldwell has different experiences, I am sure everyone in the audience has different experiences of where you have come from, and that makes a difference. What we have found is that the military are great planners. When it is time to get in, get the mission completed and get back out, we are able to do that. But we do not look at it from a different perspective. I like to use the definition of what is a disaster and what is a catastrophe? A disaster is something that happens to someone else. A catastrophe is something that happens to you. So as you are there and you are looking at that, your experience shapes what you see. We all wonder why do folks live in "Hurricane Alley?" Why do folks live in earthquake zones in California or the Madrid Fault? Why do folks live where they know they are going to have fires every year? Those experiences, that knowledge is what we reach out to gather and to pick up. The National Response Framework has done wonders in bringing together that entity, the National Incident Management System and the Incident Command System as we work out, that puts a set of rules, a set of guidelines out there for us to follow to be able to coordinate and work those efforts through and then within the interagency they obviously have their own hierarchy that they work through. One of the things that has been very important to me, this has been my fourth hurricane season working at NORTHCOM, and as we go through everybody wants to do great, everybody wants to get out there and get the job done, help those that are suffering, help save lives, and help sustain those lives, and work in that regard, but we in the military typically have come in and we know that we are in charge and we can go in and run it. In this case, we are in support, it is a different role for us so we need to be able to go out and work through that. It also requires significant understanding of each other's capabilities, each other's way that we operate.

This is what the Interagency Coordination Directorate looks like (Slide 11). Mr. Bear McConnell, a very good friend of mine with quite a bit of experience in USAID, State Department, SCS-5 (Senior Executive Service). I understand he is equivalent to a three-star; he leads it. He has a civilian government service executive officer, Jim Castle. He is authorized 6 different military folks and 13 civilians in his directorate, but it is all those other 45 different full-time folks that reach out to a of 60 different agencies. Broken down into four different divisions, the Operations and Training Division, you can see just what that does. They do the training, but they also respond to ongoing operations that are out there, and they work to integrate NORTHCOM personnel into those different agencies. The Emergency Preparedness and Plans Division reaches out to the FEMAs. They reach out to the EPA (Environment Protection Agency), to the Health and Human Services, and then they focus on defense support of civil authorities. They are the ones that will stand up the Crisis Action Team, the Interagency Coordination Group, the Joint Interagency Coordination Group that we have stood up here today, and I will have slides later on that cover a little bit more of where that coordination comes from. Most of us only deal with those two divisions. The Law Enforcement and Security Division is in the bullet with NORAD, for NORAD's aerospace and maritime warning and aerospace control. They also work with our Customs and Border Protection folks. Concepts and Technology Division reaches out with different capabilities that are available. They co-chair the innovation and technology panel. They work with Sandia Labs. They work across the spectrum of different capabilities so that we become aware of what is out there and where we could possibly provide them a linkage into the Defense Department to be able to work that. So Operations, the Preparedness Division works planning, and then obviously focuses in on capabilities, and then the idea is that we have those resident agents that we can reach out to and we will get into some of those a little bit more.

The Interagency Coordination Directorate responds to the commander (Slide 12). As I said, every different agency talks with a different group of acronyms so to be able to provide that interpretation to the commander, be able to provide the staff with the understanding of how that works is invaluable. That is primarily what they do. The Battle Staff works the current fight in the Command Center while the Future Operations Center works what is next. We have the Future Planning Center, plans folks that work for me, and we work or lead the fight and hand it off then when it becomes a current operation. We have the logistics folks to do a support group. We have the communications folks that work that piece. Interagency has a representative to each of those groups so as we work through they can provide us that right person to contact, that right person to reach out to and grab, but it gives us a link into that entire group of 60 folks that are out there. We deal with over 150 different operation centers on a day-to-day basis. And as we do that most people think of Northern Command, most people think of the Department of Homeland Security, most people think of FEMA only when it is time to do that response for a Hannah or a Gustav. They think of us only when it is the Department of Transportation or something and they need that support. But it is a lot more than that. We have 30 to 50 different incidents that we monitor every single day. They do not always turn into something, but it is something we are reaching out and as we work that there is always, everything we do there is a linkage out there to interagency. And more important than that is the fact that we need to reach out in support of those. We are not the lead. So as we work through that, that is where we focus.

Again, the Joint Interagency Coordination Group is a little bit more (Slide 13). Every other week we have speakers come in. They provide us with information. We have anything from the Office of Secure Transportation that comes out and brings equipment and shows us what they actually do provide when they are transporting nuclear assets around the country. We had the Customs and Border Protection folks to come out. They did not bring their horses along with them, but they came out and told us about their mission, where they operated, how they are growing, how they coordinate with the locals and states to be able to be able to go out and do their mission. We bring these folks in on a regular basis and all focused in on a lot of different areas of interest. Not too long ago we had an expert come in and talk about the Madrid Fault so that gives us an idea on where to focus some of our plans in that regard. I discussed the Interagency Coordination Group. That is our Battle Cell, Crisis Action Team that you can kind of think of. They are stood up today. They have been stood up for 24 days and we have been nonstop 24-hour operations in response to following through for Gustav, Hannah, and now Ike. Fortunately, Josephine fell off to the wayside and got blown away. We are still watching one other little pressure system out there. So we are continuing to work. Then as we continue the response we also need to do the transition back to where we hand everything off. They are the ones that are doing that coordination and they provide us, and again I have a set of slides that will kind of cover all the different areas that they give us information so that we know what the . . . the cell towers, how they are standing up so that the emergency responders are doing good. So they reach out and find out what the loop down in Louisiana is that has the ships that come in and pick the oil rigs up. They reach out and provide us with that information so that we can know that we can start transitioning, standing down, or disengaging out there. And we have started doing that. That is a significant part of my directorate, preparing to go in and then also preparing to bring us out.

We could not do it without . . . the next slide here (Slide 14) shows a lot of those different interagency partners. So you can see where it has two asterisks up there that we have a liaison officer that is resident with that and we are continuing to grow those on a regular basis as we take our manpower and put them out there and work through them. I will give you a second there to kind of look through them and see what is out there.

The next slide (Slide 15) gives you an idea of the nontraditional interagency partners that are out here working with us, as I mentioned earlier, Mexico with their Protección Civil and then all the way down to the BENS that I mentioned, the Red Cross, the different governors. There is a big effort right now with the 2010 Olympics, but that is a Canadian response and while the states along the borders can reach out we have to work with those governors and ensure what piece they are working and we help them coordinate that, the National Guard or any other asset that we might be able to.

The next one (Slide 16) shows several different events that we have been doing here this last year or more. You can see here from pandemic influenza tabletop exercises, and we do a little bit more than that. I am part of a pandemic influenza team and we are working with South Korea and Japan knowing that is more than likely where something is going to initiate if it is a pandemic, the NH5N1 strain of virus that could come over so we work with those folks. Well, we have this tabletop exercise here. We have had tabletop exercises that discuss the constitutional law piece of pandemic in responding to germ warfare. We have improvised explosive device conferences. Defense Coordinating Officer, Federal Coordinating Officer conferences, all the way up into weapons of mass destruction conferences with the FBI (Federal Bureau of Investigation) and the border regions. They have started reaching out to interagency in Mexico significantly more for us all the way to California wildfires and humanitarian.

This gives you a couple of more that we are continuing to work (Slide 17). Some of them are ones that occur every year. Obviously, with the National Conventions, support is worked into there. You can see here we have a Transportation Security Conference coming up.

These are the few slides that I was talking about that really gets us into what our Interagency Coordination Directorate does (Slide 18). This is the one where you need the translation to get through some of the acronyms up here. The top one up here is FEMA (Federal Emergency Management Agency) and you can see the different things they work through. The US Geological Service that helps us with imagery that we coordinate that through with our assets that we have for the Incident Awareness Assessment piece. The Customs and Border Protection has a Predator that they have had out there flying and was able to give us some imagery and share it across to all those that are responders. For the fire fighting down in California, we coordinated to have a Global Hawk fly over and focus imagery down to the folks on the ground. Obviously with the Coast Guard working the cutters that are out there, they also worked going through the harbors. Right now Houston's main harbor was just cleared. They had to go out with the vessels to do the side scanning sonar to be able to make sure that the depth and width was there for folks to get into it. Health and Human Services has the Medical Assistance Teams out there. The Army Corps of Engineers, while we think of them as a Department of Defense entity, and they are, but they work Emergency Support Function Number 3 for the Department of Homeland Security and so they are out there working the levees. They are out there doing debris removal where we, NORTHCOM, are only allowed to do debris clearing. So they go out and look at that. They look at the power side so they work with the Department of Energy in working through that piece. But one of the big ones up there, you can see the Joint Task Force Unwatering. When we get flooding the Corps of Engineers are the ones responsible for putting together the plan and monitoring that piece. So with New Orleans, and actually going on right now in Texas, they are looking at what the process will be and how long that will work on the watering side. Fortunately, we have some winds that turned around last night and will be coming out of the north so we are expecting to start drying out pretty well down there. The American Red Cross does the coordination across the different states. And again, this slide was put on the table the 1st of September so these are things that we were working across. The American Red Cross coordinates what is out there and again, the reason we do all of this is to look at any gaps that we would have to fill. The American Red Cross or FEMA have commodities that they need to deliver; I remember in Hurricane Dennis that we were working to get the commodities water, ice, and Meals Ready to Eat (MRE) down to Homestead, putting together the transportation plan working with TRANSCOM (United States Transportation Command) to be able to work that and get those delivered down there. It took four of five different agencies to be able to coordinate and work that.

The next slide (Slide 19), as I mentioned before, the Customs and Border Protection had their Predator that they used. Aidmatrix is a private sector, nongovernmental organization that coordinates donations as they come in. Typically, as we start seeing the Katrinas, the Ikes, the Hannahs, or the Gustavs start and get an idea of where they are going to land, they start coordinating that type of support. They have 50 different field organizations that offer a one-stop system. You probably are not surprised at the number of on-line groups that started up accepting donations, but not a single dime goes toward a response. So we focus in on this piece. I know Secretary Chertoff put out a call to folks to do strategic communications plans to discuss how you go about picking and choosing the right entity to be able to respond with your donations so they work through there. And they help register volunteers. We had some live feed yesterday during FEMA's and DHS's video teleconference. Friday the President walked into the meeting, but typically the last few days Secretary Chertoff has been in there and we had a live feed down to one of the points of distribution. They had folks lined up with their cars for miles each way, but they had just as many people showing up and saying, "I am okay. What can I do to help?" So this type of organization helps work that piece.

This is really what we are all about focusing here on (Slide 20), as I said, the military is there to support. We are there to coordinate and collaborate. We are there to be able to use that interagency group to be able to support unity of effort. We are focused on being part of the team, not running the team and that is a significant challenge for many of us in the military to be able to step up and do that, but then at the end of the day, I can tell you that being there for four years from where we had to fight just to get anyone to pick up the phone to even think about anticipating that support to the folks that are out there, that today we are prepared, that we have done the planning, that we are ready and able to get out there. And a lot of that really focuses in on the practice side so with those exercises that we run from the national level to the tabletops that we run, it is more important to have the interagency perspective, the interagency reach so that they could pick up the phone and knew the right people to talk to to get the answer so that we could determine what asset we needed to get out there, to anticipate and to be able to respond to save lives and mitigate suffering and that is what it is all about. It is a pleasure to be able to present this to you today. I am very fortunate to have worked first in operations and now in the plans, to have been General Renuart's representative for events like this and to be able to carry that message out to you. It was very important to him. I do know that General Webster expected to be your speaker; he was looking forward to coming back here for a visit. Both send their regards and wanted me to pass their thanks on to you for taking this subject on. Thank you very much.

Brigadier General Robert J. Felderman Slide Addendum: Interagency Coordination

Slide 1

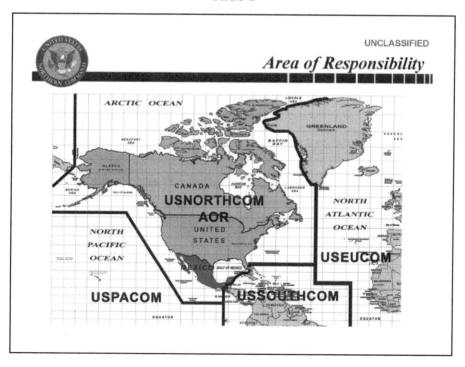

Slide 2

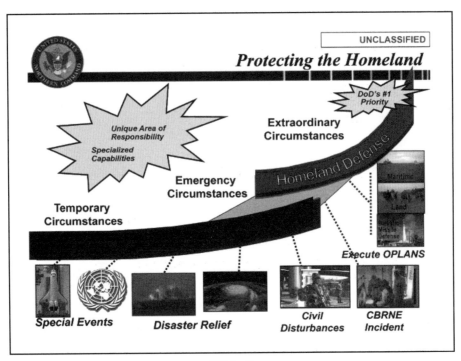

Slide 3

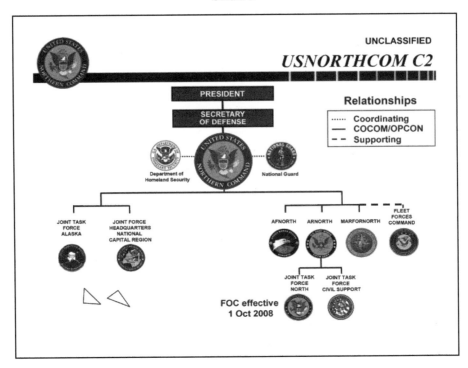

Slide 4

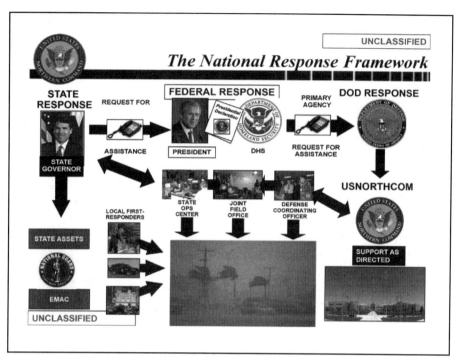

Slide 5

Slide 6

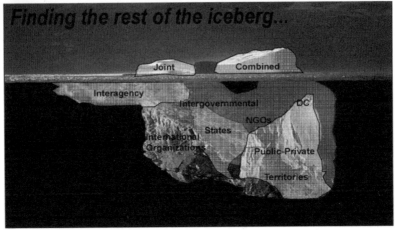

Finding the rest of the iceberg...

Joint Combined

Interagency

Intergovernmental DC

NGOs

States

International
Organizations

Public-Private

Territories

Slide 7

What is Interagency Coordination?

"The Coordination that occurs between agencies of the US
  Government, including the Department of Defense, for the
  purpose of accomplishing an objective" – Joint Publication 3-08

- *We at NORAD and USNORTHCOM include:*
  - US Government, State, Tribal, and Local Agencies
    (e.g. Dept of Homeland Security, State, Transportation,
    Health and Human Services, etc)
  - Non-Governmental Agencies
    (e.g. Red Cross, Humanitarian Int'l Service Group, etc)
  - Private Sector Organizations
    (Academic, business, professional)
- Whole of Government/Whole of Society

"Interagency" is an Adjective, not a Noun

Slide 8

Why Interagency Coordination?

- Many organizations have a common goal, but not a common boss
  - Multiple authorities, jurisdictions, levels of government (Constitutional underpinnings)
    - Mayors don't work for Governors, who don't work for the President
  - Private sector has responsibilities, obligations and capabilities
  - NGOs have their own objectives

- 21st Century security challenges are far too complex for any one department/agency – at any level of government
  - Meeting these challenges requires integration and collaboration among all instruments of US national power
  - Operations inevitably require close cooperation between various organizations with military, political, economic, public safety and other forms of expertise and resources

Slide 9

Why? (cont'd)

- Common goal(s), but not common:
  - Languages
  - Approaches
  - Experiences

- The NRF, NIMS, ICS standardizes some things, but…

- Interagency diversity – differing cultures, hierarchy, biases, and misperceptions makes unity of effort difficult

- In decision-making, non-DOD Departments/Agencies rely more on cooperation and consensus building

- Understanding plans, capabilities and limitations of other stakeholders is key to building your own plan

Slide 10

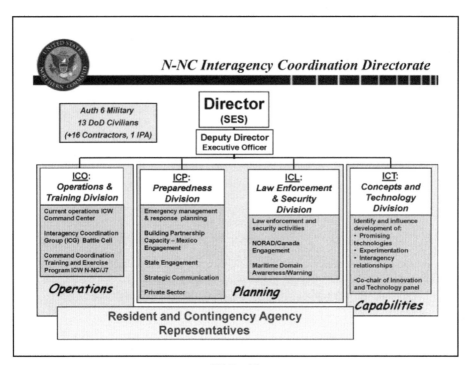

Slide 11

IC Directorate

- **Provide Interagency context to Commander's decision making processes**
- **Provide Interagency perspective to N-NC staff and Department of Defense (DoD) perspective to external Agencies**
- **Anticipate NORAD and USNORTHCOM requests for assistance through National Response Framework (NRF)**
- **Administer Commander's Joint Interagency Coordination Group (JIACG)**
- **Operate the Interagency Coordination Group (ICG) "Battle Cell"**

Slide 12

JIACG and ICG

JIACG – Supports operational planning & initiatives day to day
• Provides interagency situational awareness, Interagency assessments, Interagency reach back (Resident and 'Virtual' membership), and synthesis of Interagency information •   Working Groups formed for issues of interest….   *Law Enforcement, DCO/FCO, Pandemic Flu, Earthquake, Pre-scripted Mission Assignments, Private Sector, etc*

ICG – The Interagency Coordination Group ("Battle Cell")
• Interagency coordination focus point for Agency reps during contingency operations or exercises • Provides the JIACG Assessment to the Commander and staff   • Anticipates gaps/seams that may lead to DOD missions

Slide 13

Civil Interagency Partners

DHS:** Department of Homeland Security
CBP: Customs and Border Protection
FEMA:** Federal Emergency Management Agency
TSA: Transportation Security Administration
FAMS: Federal Air Marshal Service
USCG: US Coast Guard
ICE: Immigration & Customs Enforcement

DHHS:** Department of Health and Human Services
USPHS: US Public Health Service
CDC: ** Center for Disease Control

NASA: National Aeronautics and Space Administration

DOE: Department of Energy
OST: Office of Secure Transportation

USDA: US Department of Agriculture
USFS: US Forest Service

DOS:** Department of State

NOAA: National Oceanographic and Atmospheric Agency

DOT: Department of Transportation

FAA:** Federal Aviation Administration

DOI: Department of Interior
USGS – US Geological Survey

DOJ: Department of Justice
FBI: Federal Bureau of Investigation
ATF: Bureau of Alcohol, Tobacco, Firearms and Explosives
DEA: Drug Enforcement Agency

EPA: Environmental Protection Agency

DNI: Director, National Intelligence

CIA: Central Intelligence Agency

Resident   Contingency
** = DOD LNO at Civil Agency

Slide 14

Slide 15

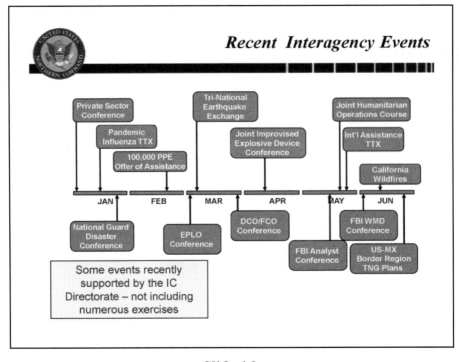

Slide 16

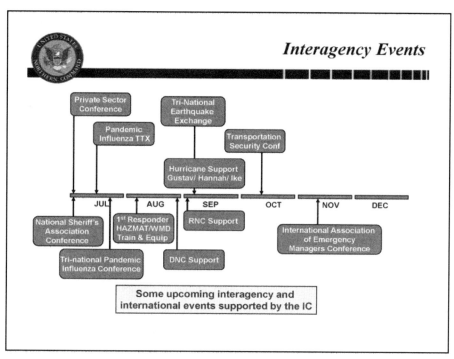

Slide 17

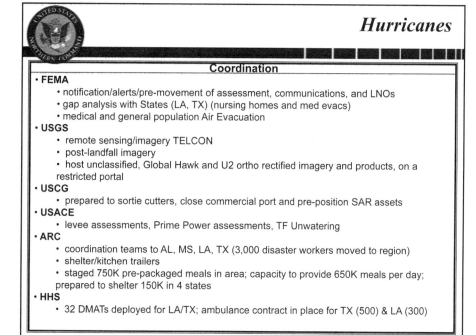

Slide 18

███ ██ ██ ██ ██ ██ ██ ██ ██ ██ ██ ██ ██

Coordination

- **CBP**
  - monitor border operations pre / post landfall
  - Predator UAV available for IAA
- **DOT**
  - MS/LA contraflow operations FEMA
- **PS/NGO**
  - Aidmatrix announced the launch of AL, FL, LA, MS, and TX State portals to manage unsolicited donations
  - actively working with dozens of NGO partners to help faciliate awareness of needs and direc sources to affected areas.
  - Through an established network, N-NC/ICG PS/NGO was able to support two shelters in LA who were in need of critical supplies and resources.
- **DOI/MMS**
  - tracking oil industry off-shore platform evacuations and shut-downs
  - <u>Evacuated</u>: Platforms: 518 of 717; Rigs 86 of 121; <u>Shut in</u>: oil 96.26%; Gas 82.3%
- **DOE**
  - activated Strategic Petroleum Reserve Emergency Operations Center
  - suspended on and offshore ops as 1100 AM EDT 8/30
  - tracked power restoration in LA
- **USDA**
  - LA request for infant formula & baby food funded & delivered

Slide 19

███ ██ ██ ██ ██ ██ ██ ██ ██ ██ ██ ██ ██

- **Unity of effort is easier to say than to accomplish**
  - **Foster an atmosphere of cooperation**
  - **Cannot "over-coordinate"**
  - **Push information – develop distribution lists ahead of time**

- **Be a good teammate, a colleague**

- **Coordination, collaboration, cooperation vs. command and control**

- **At the end of the day, always consider your agency partners in your preparation, planning and response**

Slide 20

Brigadier General Robert J. Felderman
Deputy Director J5, Plans, Policy & Strategy
NORAD and USNORTHCOM

LTG Caldwell

Yes, I will start off. Bob, I have a question for you. I thought your comment about the fact that they are going to dual-hat military officers so they can be Title 10 and Title 32 simultaneously was interesting. Is that accurate?

BG Felderman

Yes, it is.

LTG Caldwell

Is that something we could have always done or something that has been passed and required a legislative type of action from Congress to do and they have approved it? Can you just talk that for a second?

BG Felderman

Yes, I can. It has actually been on the books for quite a while. 525 is the law that goes into that effect. It requires an agreement to be signed between the governor and the President. Both signatures go on the block. We actually looked at it in Katrina, the five different states having five different dual-hatted commanders. The requirement is that you have to be a general of the line. You have to have been trained and been through the certification process and to have been a Joint Task Force Commander. It is authorized to go both ways. Right now I am Active Army, but if I were in a National Guard role then I could be dual-hatted. Then you are the only person, that one person has command and control of the National Guard forces because you are working for the governor and the Adjutant General, and you have control of the Title 10 forces that roll in. We have done it several times. The G-8 Summit. We just did it for the Democratic National Convention in Denver with Tom Mills out of Minnesota, we worked the parallel effort and depending on the event it can go both ways. We actually offered up General Honoré to be given a Title 32 commission in several different states, but particularly in Louisiana and he would have had the command and control of the National

Guard forces and the Title 10 forces. But dual-hatted means he has to respond back to two different leaders, the governor and the President. While we have not yet used that for a crisis response for Consequence Management Response, we are getting closer to that effort when we do more of these. But the law is out there. We do train. We do a course of instruction and certify folks through US Northern Command. We now have Active and Guard folks that have been through the training. About five or six hundred folks have been through it, but it has been going on long enough that as people retire and move on we have a stable of a couple of hundred folks that we can work with right now. It is very tough to be able to go and work through that. You need to have someone who has an understanding of the Active Duty forces. You need someone who understands how to work within the state and someone that can sit on the fence and work both sides.

Audience Member

When you send your liaisons out to these agencies that do not provide them to you, do you send them with communications, computers, all their support equipment or do you expect that they are provided those things when they get to their higher agency that they are supporting?

BG Felderman

Two different ways—on the day-to-day we have those liaisons that are out there, the Department of Homeland Security, we expect them to give them an office, computer, space, access and then we provide that access back. No different than when we send our officers down to Mexico or Canada. We do the same thing. But the Defense Co-ordinating Officers have a team that we send out to each of the FEMA (Federal Emergency Management Agency) regions, all 10 FEMA regions. And we send those folks out with a team of six folks, themselves and six folks. Some of them have a few more. Some have a few less when they are doing transitions. But we send them out with vehicles, communications capabilities, and support capabilities to be able to go out. They have trailers and satellite reach so we can do video teleconferences with them so as they go out and colocate with that Federal coordinating officer at the Joint Field Office and reach out to the principal Federal official or Federal coordinating official. They can bring them in and do a video teleconference right back to Colorado Springs and then from there feed it to wherever it needs to go—the Department of Homeland Security or we have had it where the President has been online and asking direct questions to them. So you can kind of work both packages. That is a good question. Thank you.

Audience Member

We understand or recognize that there is a different perspective on training that a soldier or military person would have versus what an interagency member would have. Their view of training is much different than ours. What successes can you point to in your relationship with the interagencies in building up a training partnership with some of these groups? Clearly it is mutually beneficial to prepare for this before actually executing it; it is in all of our best interests.

BG Felderman

Absolutely. I just recently attended an earthquake exercise in Osaka, Japan—two and a half hours where they had over 5,000 folks responded to this area about the size of three or four football fields that they stepped in and had trained and prepared to be able to go and work it. And I looked at that and I thought to myself and wondered if we could do that here. And I think we have done it to a different level. That is a tactical-level type of response. Spoke last week to New York emergency managers. They have one of the larger response forces and they do these exercises themselves at the state level. Jack Colley who is in Texas every day in his video teleconference talks about the training that they have done at the state and local level and so that is with the National Guard, the Reserve, and then the Active forces, this Defense Coordinating Officer. They work with him on a regular, daily basis. But you get into some areas where we need to expand that training. That is where the tabletop exercises come into play. We do several of those. There is one quarter that is at the presidential level where we have the senior leadership, and we do the national level. ARDENT SENTRY and VIGILANT SHIELD—one is for homeland defense and one is for civil support. So the training piece comes in when you trade those business cards before you get in the foxhole. We also hold pandemic influenza training. FEMA holds exercises with emergency managers. Mr. Gene Pino, a retired Marine, is our SES (Senior Executive Service) that works our Plans Division at NORTHCOM. He reaches out and gets us fit into where we need to be, where we provide folks instruction. It is very long term. We just had the Worldwide Planning Conference that Mr. Pino attended and presented several different opportunities for us to expand into the military side. So as Fifth Army is reaching out and working that, we are doing that in depth. Good question.

Audience Member (Blogosphere)

With regard to cyberspace, Sir, the recent attacks that were preemptory fires for the Russia-Georgia conflict and General Renuart had mentioned that he believes that cyber-domain is one of the biggest threats in the 21st century. Could you comment on the interagency coordination that you have done for that? Who has the rose pinned on them specifically within the interagency?

BG Felderman

Thank you. That is a very good question. We have actually identified several different challenges that are out there, cyber being one of them and the arctic being one of the future. Cyber is relatively new for all of us and working with the Strategic Command is where we expect to have the lead for the military side as they work through that. So they are just at the ground floor now of working that coordination. We have our J6, communications, and then our J2 on the intelligence side has reached out and collaborated with the right folks. Off the top of my head, I do not have the specific agency that they are working with right now, but it is growing every single day. It is moving further up the level, and we do know that after observing what happened with Georgia . . . we know what happened with one other nation where they went in and did basically a cyber attack, a cyber influence impact. We know what we go through every single day, the day-to-day hackers that are out there. As we expand we know that is going to be a continually growing requirement for us at Northern Command to be able to work that. We like to say we have 42 number one priorities, I see that as moving up into one of the top priorities. Great question.

Audience Member

My organization is the Army's experimentation venue for the command and control warfighting function. I noted on one of your slides that within the Interagency Coordination Directorate, I think in the Concepts and Technology Branch, you have an experimentation capability or interest of some type. Might this be an entity that the Command and Control Laboratory here at Fort Leavenworth could interact with and network with to achieve some common experimentation objectives?

BG Felderman

I think you probably hit the nail right on the head with that because we are seeing that reach to be able to do that. Working with the cyber realm, working with nuclear weapons, recapture and recovery, working with the *[inaudible]*, all of that, anything that contributes toward that command and control, that is what these folks do. They do that linkage so absolutely. I will give you a business card afterwards and we will get you hooked up with the right folks.

Dr. Robertson

Sir, with regard to change over time, Katrina was not all that long ago. Some might even call it not history, but more recent events. I tend to take the view that what is in the past, if it happened yesterday, it is a part of history. You had a response to Katrina. You are now having responses to Ike and those that will come after. Could you possibly characterize for us the change over time in that just short number of years, how things have improved and what we are learning from the past?

BG Felderman

It is such huge leaps and bounds from there that every single day it still surprises me. I have a list that I brought with me that I talked about that anticipation piece. That was not part of our mission. That is probably the biggest one. The prescripted mission assignments that we worked. The process to get the Secretary of Defense to approve the use of an asset. We used to present him his book on Thursdays. You would work your way through and if you needed something you got it on Thursdays. There was no such thing as a vocal approval. There were processes that had to be vetted at so many different levels. Now we have it almost instantaneously. We have it . . . it is a defense, a very long acronym, and it is a computer portal page that the Defense Coordinating Officer talking with someone down on the ground that says, "We think we are going to need this." We start then putting together a list of asset potential and start looking at analysis and assessment. First off, if the state has it, and second if the National Guard has it. If they do not, we do have it and work with JFCOM (United States Joint Forces Command) to be able to coordinate that. This is all on a portal page that you can look through and see that the request has happened and it is being worked. We have a signature. It is green. By the time it is signed and agreed, it means it went from the Federal Coordinating Official to the primary agency whether it be . . . in this case the Department of Homeland Security, FEMA and then goes to the SECDEF (Secretary of Defense) and then back down to us. I saw it take 72 hours. It is a matter of, in several cases, from minutes to hours. The anticipation piece where we were not allowed to commit, General Renuart was not allowed to commit any forces. We have a relationship, as I told you, with Fleet Forces Command, with the Navy. They are allowed to, in a training mode, to have a ship start heading down behind a hurricane. Typically, we have had to prepare to do that from Bermuda or the Bahamas. But they will come in behind the storm. We are working the USS *Nassau* [LHA-4] right now that left two days ago from Norfolk and started its way around, swung past Jacksonville, flew out and gave some assets that they needed, capabilities that they needed, swung around, and as of last night was down by Key West ready to go either to Panama City to pick up another capability to deliver over to help with the national interests for the energy or to get into Galveston and provide whatever support they need. That asset now, last night in the video teleconference was determined to send it to Galveston and they could do medical and food because they can generate so many gallons of food and water every day, and then they have the food piece, and they are also a platform because Galveston was devastated that they can then have aircraft come and stage off of there. The video teleconferences that I mentioned, in the past we would first stand up the domestic warning center, not a Command Center, not a Joint Operations Center, and would just monitor things. We now have two to three senior leader video teleconferences and that is typically General Renuart with the representative from OSD (Office of the Secretary of Defense) Joint Staff. Several times it has been the Deputy Secretary of Defense, Gordon England, on that video teleconference with similar partners from FEMA, the Department of Homeland Security, Department of Energy, all those different interagency folks that are there. And then we have three different video teleconferences that has just dropped down to two a day, that we do coordination and we get the information as to where everybody is and while we are at those, we can sit and make decisions and talk and be able to have an asset diverted as we did with the *Nassau*, get another one ready that was out on a training mission. You can get it diverted. You can take an asset we thought maybe needed to do a full-motion video of one area, then we can change it. It is a little bit of a reach into some of the tactical area, but we did not have any of those before, and to get an assessment of what we needed in Katrina took 72 hours. Now we have an idea of what any response is going to need so we can have it ready to go and then the assessment piece can be done because instead of waiting to talk in two or three days, when you found it we are out along side of you, helping you, supporting you to be able to get that, feeding that assessment back. My planners and operators are already coordinating to be able to get that asset ready to go so that 72 hours to a week process to get an asset is minutes to hours, so significantly shorter. That probably is the biggest improvement that I have seen, and just to give you an idea and it is tough to see, these are assets that we had, some on "be prepared to deploy" in that anticipation mode and now, most of them were for search and rescue, and we took about 50 percent. We do not need them now so we started standing them down. But we also, because of that communications, coordination, collaboration where we were starting to stand some of them down, we got a feeling from Texas that it is going to take another day so we stopped and kept those assets going. We would not have had that situational awareness to do that. We did not have that for Katrina for several days into it. Great question.

Panel 1—The Difficulties in Interagency Operations
(Submitted Paper)

The Interagency Process and the Decision
To Intervene in Grenada

by

Edgar F. Raines, Jr., Ph.D.
US Army Center of Military History

In October 1983 at a time when Cold War tensions between the United States and the Soviet Union were at their highest in years, the United States invaded the Caribbean Island of Grenada. This paper examines the workings of the interagency process in the days leading up to the decision by President Ronald W. Reagan to intervene. Interagency process in this instance is considered to be the definition, analysis, coordination, and evaluation of an issue across organizational lines leading to a presidential decision.[1]

The Army as an institution, of course, was not one of the agencies involved in the decision to intervene in Grenada in 1983. Ever since President Eisenhower had signed into law the Defense Reorganization Act of 1958, the Department of Defense had completely superseded the military services in the development of policy relating to issues of war and peace. Individual Army officers often were, on the other hand, key participants in the formulation of policy, because they were members either of the Joint Chiefs of Staff or of the Joint Staff. At the level below policy making, the preparations of Army, Air Force, Navy, and Marine units selected to participate in the incursion were vitally affected by the timing, sequence, and context of the decisions made or avoided in the days leading to the final presidential decision to land troops.[2]

The individuals involved worked in no ivory towers. They made judgments on the fly based on partial and sometimes incorrect information using their own store of personal experiences to weigh the relevance and significance of presumed "facts." Real world events intervened to distract them or spur them to greater efforts. They worked in a technological environment very different from our own, in which personal computers were still a rarity and most of the papers produced in the Pentagon were, in their final versions, the products of word processing pools. As always, the personalities and decision-making styles of key players had an enormous impact on the outcome.

This paper, after briefly discussing the origins of the Grenada crisis, lays out a narrative account of the decision-making—with emphasis on the interagency process—from the first realization on 13 October 1983 that there was a potentially dangerous situation developing in the eastern Caribbean until 24 October when the president made the virtually irrevocable decision to go ahead with the operation. Next there is a brief account of what the US government knew, but didn't know it knew, about the Americans on the island, a matter of some importance because the need to rescue them triggered the intervention. Fourth, the paper considers the impact of President Reagan's style of decision-making and his relationship with his key advisors on the operation. Fifth, it hazards an evaluation of the interagency process during the Grenada operation on the admittedly incomplete information available at present. Finally, it considers what students of the Reagan era don't know yet but need to know to better evaluate the national security apparatus' performance during the Grenada crisis.

The island-nation of Grenada, located in the eastern Caribbean, became a diplomatic and security problem for the US government in March 1979 when a left-wing coup overthrew the authoritarian but western-leaning government of Sir Eric Gairy. The leader of the revolutionary government, Prime Minister Maurice Bishop, kept the island in the British Commonwealth but actively sought closer ties with Cuba and the Soviet Union. At the same time, the United States became the source of all the calamities, both large and small, that befell the island—at least in Bishop's rhetoric. Grenada thus became one small point in the "arc of crisis" that National Security Advisor Zbigniew Brzezinski discerned stretching from Afghanistan through the Horn of Africa and the Caribbean to Central America.[3]

In response, the administration of President James Earl Carter began a modest military build-up in the region. It established a standing joint task force in Miami under US Atlantic Command. The new organization was to monitor Soviet and Cuban activity in the northern Caribbean, draw up various contingency plans, and conduct exercises when needed. At the same time, a second subordinate unified command, also under Atlantic Command, US Antilles Command, performed the same functions for the southern Caribbean. Atlantic Command also staged a series of military exercises as a show of force. None of these initiatives, however, involved any great influx of forces on a permanent basis. With world-wide commitments, the US military was stretched too thin for that.[4]

Initially, the administration of President Ronald W. Reagan did little more than continue the Carter approach of correct but cool relations with the Bishop government. Military exercises continued to demonstrate US interest in the Caribbean. Atlantic Command consolidated its subordinate organizations into one more robust headquarters, US Forces Caribbean, with headquarters in Puerto Rico. Behind the scenes, however, Grenada received much more high-level attention than it ever had before. The Grenadians, with Cuban assistance, had begun building a modern international air terminal capable of servicing the largest trans-Atlantic jets. Its purpose, said the Grenadians, was to revive the local economy by luring tourists to their beaches. In Washington, however, analysts worried that the airport when complete would provide a base for projecting Soviet and Cuban power through the region while serving as a staging area for the Cuban expeditionary forces fighting in Africa. President Reagan made this very clear in a televised address to the nation on 23 March 1983 that was otherwise devoted to unveiling his Strategic Defense Initiative ("Star Wars" in the words of his critics).[5]

Shortly thereafter, the administration adopted a two track policy. On the one hand, it would seek to engage the Grenadians constructively, but if that failed it would bring economic, political, and military pressure to bear to induce better behavior. An offshore medical school owned, operated, and largely attended by Americans, the St. George's University School of Medicine, was the largest source of foreign exchange for the Grenadians. If all else failed, the Reagan administration proposed to encourage that institution to move elsewhere. The first track produced a Bishop visit to Washington and a meeting with National Security Advisor William P. Clark but resulted in no discernable change in Grenadian behavior as far as the administration was concerned. The president's decision to launch track two was sitting on the desk of Clark's successor, Robert C. McFarlane, awaiting implementing instructions, when events on the island rendered it superfluous.[6]

On 12 October 1983 an emergency meeting in Grenada of the Central Committee of the revolutionary party—the New Joint Effort for Welfare, Education, and Liberation Movement Party, most commonly referred to by its acronym as the New JEWEL Party—deposed Bishop and substituted his sometime deputy and finance minister, Bernard Coard, as prime minister. When the party announced the leadership change the next day in the capital of St. George's, a near riot ensued. Coard then withdrew, but he remained very much a power behind the scene, although this was not clear to party outsiders. Just who was in control became, suddenly, ambiguous not only for foreign observers but also for ordinary Grenadians. The central committee, meanwhile, placed Bishop under house arrest. He was held incommunicado.[7]

Events in Grenada, as did those of all countries in the Western Hemisphere south of the United States, fell under the purview of the Restricted Interagency Group of the National Security Council. Upon taking office, the Reagan administration had mandated an elaborate, formal hierarchy of coordinating groups between all the departments and agencies involved in formulating and implementing US national security policy with the State Department at least nominally serving as the lead. At the top of the hierarchy was the National Security Council and the president. The Restricted Interagency Group was at the bottom. Consisting of mid-level officials from agencies involved in national security, it was chaired by the assistant secretary of state for inter-American affairs. It was also a surprisingly influential group, considering the rank of its members in their respective departments.[8]

The Reagan administration considered both the Sandinista government of Nicaragua and the insurgents in neighboring El Salvador as proxies for the Soviets and the Cubans and thus the number one national security threat to the United States in the Western Hemisphere. Over the first almost three years of the administration, the Restricted Interagency Group had served as the primary forum for developing US policy toward Central America. Recently, however, consensus had broken down. Some members of the State Department advocated a diplomatic approach while a loose coalition of National Security Council staffers, officials at the Department of Defense, and the Central Intelligence Agency pursued a military solution. In these circumstances, representatives of the State and Defense Departments were both territorial concerning their prerogatives and suspicious of the intentions of their counterparts.[9]

It was in this inauspicious setting that, as the meeting broke up, Assistant Secretary of Staff Langhorne A. Motley spoke to the representative of the Joint Chief of Staffs, Colonel James W. Connally, US Air Force, Chief of the Western Hemisphere Division of the Plans and Policy Directorate (J5) of the Joint Staff. Grenada had not even been on the group's agenda that day, and Connally's relatively low rank indicated what the Chiefs' nominal representative, the assistant to the chairman of the Joint Chiefs of Staff, Vice Admiral Arthur S. Moreau, Jr., thought of the importance of the topics that were. Ambassador Motley told Connally that there was apparently some unrest on Grenada and that it might become necessary to evacuate US citizens. He asked that the Joint Chiefs review their plans to support such a contingency. Connally thanked him and promised to pass his concern on to his superiors. This apparently was noted, but no further action followed.[10]

That same day, the senior Latin American specialist on the National Security Council staff, Constantine C. Menges, reacted to the same reports of a Grenadian power struggle by proposing military intervention to rescue Americans on the island and restore democracy. Menges had a reputation as an ideologue at odds with the pragmatists who dominated the staff. His proposal appeared so far-fetched that his superiors ignored him for the remainder of the crisis.[11]

Menges's intervention did alert the senior members to the Grenada crisis, but this was more than off-set by the turmoil into which the staff was thrown at 1000 that same day. Without warning, National Security Advisor Clark announced he was stepping down to accept the post of Secretary of the Interior in the president's cabinet. His deputy, Robert C. McFarlane, would succeed him. This development probably slowed the staff's response to the Grenada issue. Not until 2000 did a low-level staffer on the Latin American desk contact an officer in the Joint Operations Directorate (J3) of the Joint Staff. The staffer wanted to know what resources were available if it became necessary on short notice to safeguard the evacuation of Americans there.[12]

This query brought a quick response because the National Security Council was an agency of the Office of the President. At 0800 the next morning, the director of operations on the Joint Staff, Army Lieutenant General Richard L. Prillaman, activated a crisis response cell in the National Command Center to monitor the Grenada situation. The officers assigned to the cell were to assess the situation and prepare possible courses of action. One of them placed a "what if" call to the headquarters of US Atlantic Command at Norfolk, Virginia. Spurred by the inquiry, officers in the operations directorate there began reviewing contingency plans for noncombatant evacuation and a show of force. They also began drafting options specifically related to the situation in Grenada. Initially these efforts were interspersed with the normal business of the directorate and proceeded at a somewhat leisurely pace.[13]

That same day the Restricted Interagency Group held another meeting. This time Grenada did appear on the agenda as a minor item. Ambassador Motley repeated for the entire group much of what he had told Connally the day before. In his view, Bishop's arrest opened the possibility of further radicalization of the revolutionary movement. This development might pose a threat to the safety of the large number of Americans, estimated at 1,000, living on the island. Motley informed the group that the State Department was reviewing its standard evacuation procedures and formally requested that the Joint Chiefs review its plans for such a contingency.[14]

Grenada at this point was still a potential crisis rather than a full-blown one. The 14th was a Friday, and if any of the members of the Restricted Interagency Group missed part of their weekend, it was not because of Grenada. The military staffs, of course, continued working although at reduced levels of activity. Both the planners at the Pentagon and those in Norfolk knew that any noncombatant evacuation involved one major inherent risk—that the intended evacuees could become hostages instead. With this in mind, the planners at Atlantic Command made three assumptions. First, the National Command Authority would make available all the forces listed in the existing Caribbean concept plan, 2360, last updated the preceding March. From this pool of units, the commander of US Atlantic Command would draw whatever appeared appropriate given the situation on the island. Second, neither the Soviet Union nor Cuba would intervene. Third, the bulk of the resident Americans were medical students who lived on the True Blue Campus of the St. George's Medical School, just off the east end of the runway at Point Salines. Based on scanty intelligence, the third assumption would prove to be incorrect.[15]

On Monday, 17 October, Motley convened a special meeting of the Restricted Interagency Group devoted solely to Grenada. This agenda testified to the State Department's rising concern. Colonel Connally's continuance as the Joint Chiefs of Staff representative in attendance, however, indicated the rather less serious view with which his superiors contemplated the situation. At the meeting Motley made a strong case for armed intervention. He began by reviewing the State Department's standard procedures for dealing with a country where American lives might be at risk. The more moderate courses of action involved negotiating with the government in power. Motley rejected them because in his view Bishop's arrest and Coard's resignation meant that no legitimate government remained in Grenada. The only alternative was an evacuation of noncombatants during which the US military would be prepared to use whatever force was necessary to protect them. Motley wanted the Joint Staff to begin planning immediately for such an operation.[16]

Connally dutifully carried Motley's concerns back across the Potomac. The director of the Joint Staff, Army Lieutenant General Jack N. Merritt, found them overblown. In Merritt's opinion, the crisis was "just vibrating"—the situation on the island was neither deteriorating nor improving. Nevertheless, he directed the Joint Staff to join the Atlantic Command staff in developing a full range of options extending from a peaceful evacuation to the use of force. In response to the renewed flurry of planning activity, Atlantic Command activated its own crisis response cell the next day.[17]

On the morning of 19 October, the crisis stopped vibrating. In the Grenadian capital of St. George's, a crowd, estimated at from 3,000 to 4,000, broke into the prime minister's residence and released Bishop, who had been strapped to his bed. Observers reported he appeared dazed, reputedly because he had refused all food and water for several days, fearing that his former comrades might poison him. Bishop and some of his supporters moved to army headquarters at Fort Rupert, where they easily overpowered the guards. Bishop, however, had failed to gauge the depth of Coard's resolve to retain power. Three Soviet-supplied armored personnel carriers rolled up to the fort and without warning fired into the crowd. To prevent a massacre, Bishop surrendered. A firing squad of soldiers loyal to Coard then executed the prime minister and his leading supporters. Over Radio Free Grenada, the Minister of Defence, General Hudson Austin, announced the formation of a 16-man Revolutionary Military Council to govern the island until a new government could be formed. Bishop, he said, had died in the fighting at the fort. During the present emergency, he imposed a 24-hour curfew. Anyone violating it would be shot—words that carried grave import given Bishop's fate.[18]

The first news of disturbances in St. George's prompted a strong response from the US Embassy in Bridgetown, Barbados. Reports of Bishop's rescue by the crowd prompted Ambassador Milan D. Bish to report, long before he learned of Bishop's death, that conditions on Grenada "posed an imminent danger" to US citizens resident there. In Washington, his assessment prompted yet another meeting of the Restricted Interagency Group. This time Admiral Moreau attended in person. Although junior in the bureaucratic pecking order to Ambassador Motley, he was in many ways the most influential person in the room. A nuclear submariner noted for his keen intelligence, Moreau was very close to the chairman of the Joint Chiefs of Staff, Army General John W. Vessey. Vessey, a low key unflappable soldier with a droll sense of humor and a military career that stretched back to before World War II, was a favorite with both the secretary of defense and the president. Moreau's ideas thus had a way of reaching the very highest echelon of government. It made him a force with which to reckon.[19]

All the members of the interagency group agreed with Bish's analysis and further concurred that the military needed to begin immediately to plan for a noncombatant evacuation. Moreau stated that the Joint Chiefs understood the situation and would instruct the relevant commands to monitor events on the island. At the same time, he emphasized that only the vice president in the Special Situation Group, which was the National Security Council when presided over by the vice president, or the National

Planning Group, which was the National Security Council when presided over by the president, could direct the Joint Chiefs to prepare invasion plans. Moreau did not add to this statement the phrase "not the State Department," but that implication was clear from the context of his remarks.[20]

Later that same day, following an intelligence briefing, the Joint Chiefs of Staff held its first formal meeting on the situation in Grenada. The Chiefs decided to dispatch a warning order for a possible evacuation of Americans to US Atlantic Command. Issued at 2347 local time on 19 October that instruction directed the commander of Atlantic Command, Admiral Wesley L. McDonald, to prepare by dawn of the next day an estimate of the courses of action available to protect and evacuate US and designated foreign citizens. The Chiefs envisioned a three- to five-day operation. Possible scenarios included a show of force, seizure of evacuation points, combat operations to defend the evacuation, and postevacuation peacekeeping. This list encompassed a range of political objectives that extended from minimal involvement in the internal affairs of the island nation to creation of a posthostilities democracy. The amount of combat power envisioned for each increased in line with the scale of the objectives. US Readiness Command, as a supporting command that might have to provide forces to Atlantic Command out of the strategic reserve, received an information copy of this warning order. Shortly thereafter, the Joint Staff gave the operation the code name URGENT FURY.[21]

In the early hours of 20 October, Admiral McDonald replied to the Joint Chiefs of Staff warning order with his estimate of the situation. Later that morning, General Vessey flew to Norfolk where McDonald's staff briefed him. Atlantic Command's plans envisioned both the possibility of an uncontested evacuation and a contested one. There was about a battalion's worth of Cuban construction workers on the island, all of whom had military training. If the Grenadians and Cubans permitted a peaceful evacuation, McDonald outlined two possible courses of action: the use of either chartered commercial passenger planes or Military Airlift Command aircraft to fly the evacuees out of the country. The use of military aircraft would include stationing a small security detachment at the departure airfield. Either approach would require extensive negotiations with the Grenadians.[22]

If the Grenadians refused to permit a peaceful evacuation, McDonald proposed four options involving military forces. The first was a show of force with warships from the Atlantic Fleet. Each of the other three involved landing about one battalion's worth of ground troops. The first used the 22d Marine Amphibious Unit supported by the aircraft carrier USS *Independence* and its accompanying battle group. Some of these forces were already at sea en route to the eastern Mediterranean; others were preparing to embark for that destination. All could reach the eastern Caribbean in five days. The second substituted another Marine amphibious unit, but that force would require nine days to arrive on station. Finally, McDonald might use an Army airborne battalion. Depending on the unit the US Readiness Command provided, it might be ready for action even before the 22d Marine Amphibious Unit.[23]

McDonald stated very clearly that he preferred to use Marines for both the assault force and any stay-behind occupation force. The second Marine option, however, dropped out of the discussion very early because of the unit's late availability. The first, employing the 22d Marine Amphibious Unit, became what the Joint Chiefs referred to as Atlantic Command's "small option." Vessey focused on the third option involving an Army airborne battalion. He proposed that McDonald use Army Rangers in the initial assault force because they specialized in seizing airfields. The potential hostage situation also suggested to him that this operation required the special skills possessed by the Pentagon's hostage rescue specialists—the special operations units controlled by the Joint Special Operations Command at Fort Bragg. Vessey envisioned elements of the 82d Airborne Division at Fort Bragg serving strictly as the occupation force. The Joint Chiefs referred to this course of action, which General Vessey clearly preferred, as Atlantic Command's "large option."[24]

Vessey's hands-on style of revising Atlantic Command's plans suggested that he was less than impressed with that headquarters' product. Ever since the creation of the command almost 40 years earlier, its primary mission had been to keep the sea lanes open to Europe and to transport large numbers of ground forces there in the event of war with the Soviet Union. Given this mission, the headquarters had been a single-service "blue water" organization throughout its history. It had no permanently assigned Army or Air Force units. The general's concern that Admiral McDonald and his staff lacked the knowledge and skills to conduct a ground operation was thus no reflection on those individuals. It was the product of the Unified Command Plan and the assumptions upon which it rested.[25]

Upon his return to Washington, Vessey took the first of two key actions that day. He directed the Joint Staff to analyze the impact of executing Atlantic Command's large option on the world balance of forces between the United States and the Soviet Union. Throughout preparations for URGENT FURY, the Joint Chiefs would have as their primary concern keeping any conflict localized—and that meant deterring any Soviet or Cuban inclination to intervene. They proposed to accomplish that goal by massing enough land- and sea-based air power to dissuade the Soviets and Cubans from attempting to reinforce the Grenadian People's Revolutionary Army and the Cuban construction workers.[26]

While Vessey reviewed McDonald's plans, news of General Austin's 24-hour shoot-on-sight curfew prompted the newly appointed National Security Advisor, Mr. McFarlane, to decide that the crisis required White House oversight. The National Security Council replaced the State Department as the lead agency in the decision-making process. With the shift came a change in the name of the coordinating group. At McFarlane's direction his deputy, Rear Admiral John M. Poindexter, convened a Crisis Pre-Planning Group at 0800 on 20 October. Institutional representation remained the same as in the Restricted Interagency Group but with an expanded number of often more senior representatives.[27]

As the attendees discussed the deteriorating situation, a member of the National Security Council Staff, Lieutenant Colonel Oliver L. North, US Marine Corps, mentioned that the USS *Independence* battle group and the 22d Marine Amphibious Unit had recently sailed for the Eastern Mediterranean. Ambassador Motley wanted the Defense Department to divert them to the Eastern Caribbean until the crisis eased. Admiral Moreau refused to entertain the idea short of a written presidential order. In the end, Poindexter's committee urged that the Special Situation Group, a committee of the most senior policy makers chaired by Vice President George H. W. Bush, assume responsibility for managing the crisis. The president accepted this recommendation.[28]

At 1645 that same day, just as the vice president prepared to enter the first meeting of the Special Situation Group, a staff member handed him a copy of a message from Ambassador Bish reporting that Barbadian Prime Minister J. M. G. M. "Tom" Adams had requested American assistance in overthrowing the Austin junta. The vice president had just returned from a four-day visit to Jamaica, where he had received an in-depth analysis of the crisis from the Reagan administration's favorite Caribbean leader and free-market advocate, Prime Minister Edward Seaga of Jamaica. Seaga believed that the coup and Bishop's murder posed a threat, by example, for all the democratic governments in the Caribbean. If the Austin-Coard clique remained in place, every adventurer in the area would have a working model for how to take power. Bush led off the meeting by reading the Bish cable and then summarized his own conversations with Seaga. "These people" he concluded, "are asking us to do something."[29]

Detained because he was testifying on Capitol Hill, Secretary of State Shultz entered midway in the meeting and outlined the State Department's plans to evacuate American citizens. He also noted that US forces would probably have to protect the evacuation. If that became necessary, Shultz advocated disarming the Grenadian armed forces as a safety measure. This was only one step short of outright regime change.[30]

General Vessey briefed the attendees on the risks of using force and the possibilities of Soviet or Cuban intervention. The Joint Chiefs ". . . were determined," he said, "to make sure that [Fidel] Castro got the message that interference was not an option for him and that the message was clear and early." If the president decided to intervene, they wanted to send a very clear message to Cuba: "Hands off!" A representative from the Defense Intelligence Agency informed the group that the People's Revolutionary Army would oppose any evacuation but that force was militarily ineffective. On the other hand, he added, the Cubans and Soviets simply lacked the means to intervene in sufficient strength to affect the outcome.[31]

The Special Situation Group anticipated that conditions on Grenada would continue to deteriorate and that at some point events would compel the president to rescue the Americans on the island. As a result, Bush and his associates decided that the Joint Chiefs should prepare a detailed operational plan for this contingency and directed McFarlane to begin drafting a decision directive covering such a circumstance for

Reagan's signature. The committee also recommended the immediate diversion of the *Independence* Battle Group and the Marines. Shortly thereafter, Secretary of Defense Caspar W. Weinberger ordered the change of course for these ships without waiting for the White House to issue the order.[32]

Late that evening, about 2100, General Vessey contacted the commander of the Joint Special Operations Command, Major General Richard A. Scholtes, on a secure line and informed him that military intervention in Grenada was possible. He directed Scholtes to develop a plan and to come to Washington and brief him early the next morning. Specifically, Vessey wanted to know what targets Scholtes considered essential and how in general terms he would envision the operation taking place. At that time, Scholtes assumed that his men would be working directly for the chairman as they had done often in the past.[33]

On Friday, 21 October, President Reagan formally directed the Department of Defense to continue contingency planning, enjoined the State Department to contact allies and regional governments to determine both their assessment of the situation and their willingness to participate in a multilateral intervention if one became necessary, and confirmed the diversion of naval and marine elements to the Eastern Caribbean. Early that same day the director of operations on the Joint Staff, General Prillaman, telephoned General Scholtes and indicated that the Secretary of Defense and the Joint Chiefs had decided to make Atlantic Command the supported command. Prillaman directed Scholtes to brief Admiral McDonald on the concept of operations and the outline plan Scholtes' staff had developed the night before.[34]

At this point three different military commands were preparing for an invasion of Grenada. Because Admiral McDonald could not be certain that the Grenadians would wait until the special operations forces or the Marines were ready, he had the 82d Airborne Division preparing to seize both the airport under construction at Point Salines and the only operational field on the island at Pearls. The Joint Special Operations Command had its planners working on the same objectives along with several others. Even though both organizations were located on the same post, rigid compartmentalization prevented either from having more than a very general idea of what the other was doing. Finally the officers of the 22d Marine Amphibious Unit were looking at the same objectives. Neither the Joint Chiefs nor Admiral McDonald deemed coordination between these groups necessary or even useful because they were developing competing plans for the operation. Still the distances involved and the need to feed these organizations information from Washington led General Vessey to direct that everyone involved take special security precautions. These directions came too late to prevent first CBS News and then the *Associated Press* from breaking the story of the diversion of the *Independence* Battle Group, a fact confirmed by a "Defense Department official" during a briefing of Pentagon correspondents the evening of the 21st. The intended destination of Amphibious Squadron Four made the front pages of both the *Washington Post* and the *New York Times* the next day. Whatever the origins of the report, whether an unintended "leak" or calculated indiscretion designed to influence the president's decision, it produced intense concern in the White House about operational security.[35]

Meanwhile, the Reagan administration attempted to maintain a public facade of "business as usual." On Friday afternoon, President Reagan and Secretary Shultz and their wives departed for a scheduled weekend of golf at the Augusta National Golf Course. Because the situation in Grenada appeared to be heading toward a crisis, the new national security advisor, Mr. McFarlane, joined the party at the last moment. That evening, also conforming to his previously announced itinerary, General Vessey left to deliver a speech in Chicago. At the same time, "two senior presidential aides" accompanying the president to Georgia assured the press pool that the *Independence* Battle Group was in the area only to protect Americans and that no invasion was contemplated.[36]

After Vessey departed for Chicago, the acting chairman, Admiral James Watkins, the new Chief of Naval Operations, attended a second meeting of the Crisis Pre-Planning Group. New intelligence reports suggested that the Cubans as well as the Grenadians might resist. The Cubans, according to one inaccurate report, might have introduced 240 combat troops onto the island when the freighter *Vietnam Heroica* docked at St. George's on 6 October. At this point senior leaders were inclined to believe that Cuban machinations lay behind the Coard coup. The intelligence reports hardened the consensus that the United States would have to use military force to protect the evacuation and might have to disarm all Grenadians and Cubans, even those well removed from the evacuation point.[37]

For both decision makers and planners, the available information about Grenada was seriously flawed. The figures on Grenadian and Cuban defenders given to the Joint Chiefs of Staff, for example, represented an overestimate on the order of 190 percent. In the same way, Bishop and his supporters were more closely associated with Cuba than Coard and his faction. The intelligence was the best available, but it rested largely upon inferences rather than hard data.[38]

This lack of accurate and up-to-date information was the product of major structural problems in US intelligence agencies and misguided policies in the local American embassy. The United States had drastically cut back its intelligence assets in the wake of the Vietnam War. In an attempt to economize, for example, the Department of Defense failed to assign a defense attaché to the American embassy in Barbados during most of the period following the Grenadian Revolution. Finally, in 1982 Lieutenant Colonel Lawrence N. Reiman, US Army, opened a one-man shop. As the Grenada crisis began, intelligence assets in the region were skeletal at best. All the agencies involved had to play catch-up.[39]

So did the State Department. From the US embassy in Bridgetown, Barbados, Ambassador Bish, a Nebraska businessman with no previous experience in government, had quite reasonably concluded that the Bishop regime was communist shortly after taking up his post. Rather than seeking to obtain more information about what was transpiring on Grenada, however, he had arbitrarily directed his staff to drastically reduce even routine visits to the island. (This meant that Reiman, for example, concentrated on establishing contacts with friendly forces in the region.) The reporting from Bridgetown on events in Grenada during the crisis thus consisted of a composite of interviews with American citizens recently there, summaries of local press reports, transcripts of Radio Free Grenada broadcasts, and whatever information friendly governments with better sources chose to pass along to the embassy. Policy makers in Washington consequently received little if any special insight into the events or psychology of the key figures in Grenada that high quality diplomatic reporting could have provided. Even more important, because the policy makers in Washington were unaware of Bish's embargo on Grenada visits, they assumed that the embassy reports were much more solidly based on first person observation than they were.[40]

Reaction in the eastern Caribbean to the diversion of the *Independence* Battle Group was emphatic. The same day as the Department of Defense announcement, the heads of government in the region met in Barbados. They unanimously agreed to intervene in Grenada to restore order and, because the forces at their disposal were minuscule, to request the assistance of both Barbados and Jamaica. The prime ministers delegated the chairman of their organization, Prime Minister Eugenia Charles of Dominica, a 53 year-old woman of keen intellect and forceful personality, to approach Great Britain and the United States for additional forces. They also prepared to make their case to the larger Caribbean community for still more local assistance. Attempts the following day to enlist other nations from the Caribbean proved largely unsuccessful.[41]

In Barbados the American embassy struggled without avail to find a peaceful solution to the problem of US citizens on the island. The whole thrust of the post coup diplomatic offensive by the members of the Organization of Eastern Caribbean States was to isolate the Austin regime. Ambassador Charles A. Gillespie, Motley's new deputy, interrupted an orientation tour of the Caribbean to provide Bish with on-the-scene advice. Neither he nor Bish wanted to undercut this promising diplomatic development, but, at the same time, they had to start a dialogue with Austin, Coard, or whoever was in charge to ensure a peaceful evacuation. In short, the Americans had to negotiate without seeming to negotiate.[42]

Initially, Gillespie and Bish used the vice chancellor of the St. George's University School of Medicine, Dr. Geoffrey Bourne, as a go-between. They conducted conversations via teletype with the school, and then Bourne conveyed their views to representatives of the Military Council. Sometimes Austin personally came down to the campus. After much discussion they convinced the Grenadians to allow a consular party headed by Kenneth Kurze, an officer in the Bridgetown embassy, to visit the island to check on the resident Americans. Kurze and one other embassy official landed at Pearls Airport on 22 October.[43]

Kurze confirmed that the Americans on the island were unharmed, but his efforts to negotiate a resolution to the crisis floundered upon Grenadian intransigence. While General Austin proclaimed the Revolutionary Military Council's readiness to allow all foreign nationals to depart peacefully, the Council's negotiator in this matter, Major Leon Cornwall, found "technical objections" to each course of action proposed. When the Cunard Lines, for example, offered one of their ships to evacuate free of charge all foreign nationals who wished to depart, Cornwall denied the vessel docking privileges and said that the Grenadian Army would fire on it if it entered Grenadian waters. (Grenadian antiaircraft guns did fire on the ship when it appeared on the horizon.) Cornwall's behavior convinced the senior officers at the Bridgetown embassy that the Grenadians were already attempting to use the Americans on the island as bargaining chips.[44]

One officer of the Central Intelligence Agency entered Grenada during this period of limited access. Ms. Linda Flohr spent over two days dodging Grenadian Army patrols while reporting on the situation via clandestine radio. The Grenadians, she noted, had confined the students to their dormitories and had posted sentries to keep them in and everyone else out. In her view the students were already hostages. She urged an immediate invasion.[45]

So, too, on the 22d did the Governor General of Grenada, Sir Paul Scoon, regarded by the regional governments as the only legitimate source of authority remaining on the island. Scoon confided to a British official resident on the island that he desired an intervention to overthrow the Austin clique. The State Department learned of the appeal early the next morning.[46]

During the night of 21–22 October, Prime Minister Eugenia Charles of Dominica, acting in her capacity as chair of the Organization of Eastern Caribbean States, formally petitioned the United States to intervene in Grenada. US Ambassador Bish in Bridgetown immediately forwarded her oral request to the State Department. On 22 October, this message precipitated a very early morning meeting of the Special Security Group of the National Security Council and a call to McFarlane. Shortly afterwards, the president, dressed in slippers and robe, met with McFarlane and Secretary Shultz in the living room of the Eisenhower Cottage at Augusta National where he and Mrs. Reagan were staying. They briefed him on the latest developments. The president's reaction, recalled Shultz, was emphatic: The US had to respond positively to such a plea from small "democratic neighbors." And then there was the danger posed to resident Americans. Reagan telephoned Washington at 0558 and spoke with Vice President Bush, who had been chairing the Special Security Group to develop options and recommend a course of action for the president. Next Reagan spoke with Secretary of Defense Weinberger, who had also participated in the meeting.[47]

The president made no irrevocable decisions either then or at an 0900 teleconference with all the senior members of his National Security team. It was, as one scholar has observed, at best a 75 percent decision, but it certainly gave a military intervention decided impetus. Everyone now agreed that there was no longer any possibility of a peaceful evacuation of the American residents in Grenada, and Reagan ordered the Joint Chiefs of Staff to draw up plans to seize the country.[48]

Back from Chicago and anticipating such a request, General Vessey, had plans already in hand. Since his meeting with McDonald, the Joint Staff, working with the US Atlantic Command, had prepared two force packages for the operation that were remarkably similar in size. One, originally Atlantic Command's small option, consisted of a Marine Battalion Landing Team with Navy SEALS attached; the other, the so-called large option, included two battalions of US Army Rangers and a contingent of special operations forces from the Joint Special Operations Command. Each package numbered about 1,800 men and could be reinforced by two or more airborne infantry battalions from the 82d Airborne Division. The Chiefs anticipated, however, that the airborne units would function primarily as occupation troops in either scenario.[49]

Two of the president's counselors, both veterans of fighting in the Pacific during World War II, Secretary Shultz and Secretary Weinberger, expressed concern that these elements were too light for the mission. Weinberger insisted that the United States should apply overwhelming force to minimize casualties. He was determined to avoid the sort of mistakes that had led to the costly failure in 1980 to rescue American hostages held in Iran, the DESERT ONE disaster, and so he told the Joint Chiefs to double whatever strength the theater commander considered adequate. On his own, using a similar rationale, Shultz advised Reagan to double the number of troops the Joint Chiefs recommended. At this time, the president did not appear to make any final determination on the matter.[50]

The president did decide to send a special envoy, Ambassador Francis J. McNeil, a career foreign service officer, to the meeting of the Organization of Eastern Caribbean States to gauge just how committed the heads of government were to intervention and to obtain their request in writing. Reagan wanted an independent evaluation of the situation "before making a 'go/no go' decision." As the Assistant Secretary of State for Inter-American Affairs, Ambassador Motley, remarked to McNeil, "It isn't everyday that we get a request like this."[51]

As these more detailed preparations began, Reagan, Shultz, and McFarlane debated whether Reagan should remain in Georgia and adhere to his schedule. They decided ultimately that an early return to Washington would lead to intense press speculation that might precipitate hostage taking on the island. While the president continued his golf, Shultz and McFarlane monitored the Grenada situation using a satellite telephone to call Washington. Even the efforts of an emotionally disturbed gunman who crashed the security fence at Augusta National in a Dodge pickup truck and barricaded himself in the pro shop with five hostages—including two members of the White House Staff—failed to shake the president's resolve to maintain a facade of normalcy. The

Secret Service did insist that the president leave the course while the pro shop crisis gradually moved toward a peaceful resolution.[52]

Reagan did not allow the excitement of the day to divert him from his Caribbean concerns. On the evening of the 22d, shortly before 1700, the National Security Planning Group, the highest level of the National Security Council presided over by the president (in this instance using a secure telephone), formally directed the Joint Chiefs to dispatch an execute order to the responsible theater commander. The order authorized, but did not require, Admiral McDonald to combine the troops in both options. Earlier in the day using a secure phone, General Vessey had suggested this possibility to McDonald. The chairman had told McDonald that the chiefs thought he needed to beef up the landing force to "intimidate" the Cubans. In the order, the Joint Chiefs estimated that the earliest possible time they could stage the operation was Tuesday, 25 October. They told the admiral to use that date as a target.[53]

On Sunday, 23 October, Admiral McDonald accompanied by General Scholtes flew to Washington to brief the Joint Chiefs of Staff on his concept of operations. At this meeting the Joint Chiefs decided to consolidate the two options. The chiefs planned a nighttime assault by Rangers and special operations forces using night vision devices against the Point Salines airport and various military and political objectives in and around the capital of St. George's. The Marines would land at dawn near the town of Pearls, which had the one fully operational airfield on the island. After he returned to Norfolk, McDonald chose the commander of the US Second Fleet, Vice Admiral Joseph Metcalf III, to command the overall Grenada operation. In the process, he set aside US Forces Caribbean, the headquarters responsible for the region. He did not think it was robust enough to conduct an actual operation.[54]

Another early morning telephone call, this time at 0239 on Sunday, 23 October, informed the president of the bombing of the Marine Corps barracks in Beirut, Lebanon. The news convinced all concerned that Reagan needed to return to Washington immediately. Throughout the day, as information accumulated, the scope of the disaster became clearer: 241 American servicemen were dead and another 70 wounded. Much of official Washington was in shock. Ambassador McNeil, who had arrived in Washington early on Sunday, thought that the Beirut crisis would abort the whole Grenada enterprise. Nevertheless, after a briefing on the situation in the eastern Caribbean at the State Department, he and a representative of the Joint Chiefs of Staff, Major General George B. Crist, US Marine Corps, left for Barbados that afternoon. Crist, whose normal post was vice director of the Joint Staff, was to make arrangements for military participation by the Caribbean governments.[55]

Upon arrival in Washington, the president and his advisors embarked on a round of almost non-stop National Security Council meetings that alternated between Lebanon and Grenada. As Grenada appeared to be progressing without problems, General Vessey concentrated on Lebanon and delegated most of the Grenada briefings to

Admiral Moreau. The ghastly news from Beirut dampened everyone's spirits. At one point the president hung his head in his hands and wondered aloud if his administration would suffer the same fate as that of President Carter, undone by the hostage crisis in Iran. His advisors believed that a Grenada operation would only detract from his popularity. General Vessey, who like his colleagues on the Joint Chiefs did not think an intervention was necessary at this point, ventured that with the 1984 presidential election only a year away, perhaps Reagan should call off the invasion. The president shot back that he intended to consider this operation strictly on its merits.[56]

The president might have temporarily lost his ebullience but not his resolve. The key issue for him was that American citizens were at risk. As soon as he heard that hostages were involved, he made up his mind that he would use military force if necessary. He carefully refrained from telling anyone of that decision, however, because he intended to keep his options open until the very last minute. Periodically he asked Vessey if he had made any decision that had irreversibly committed him to a military operation. Vessey always assured him that he had not reached that point.[57]

That evening, 23 October, in the White House residence, after his advisors had departed, the president signed a National Security Decision Directive for the invasion of Grenada. "The Secretary of Defense and the Chairman of the Joint Chiefs of Staff . . . will land US and allied Caribbean military forces in order to take control of Grenada, no later than dawn Tuesday, October 25, 1983." Reagan carefully stipulated that the State Department would not notify the Organization of Eastern Caribbean States of his decision until after 1800 on 24 October. The president did not inform his principal advisors of his action until sometime the next day. The explanation for this reluctance appears to have been more personal than political. He was very much aware that a decision to intervene would cost lives, and he refused to make that determination until he absolutely had to.[58]

The news from the Caribbean continued to pressure the president toward military action. Ambassador McNeil arrived in Barbados on 23 October and immediately went into a meeting with the Caribbean heads of state that lasted almost three hours. Probing their rationale for intervention, he found their advocacy thoroughly grounded in the realities of the situation and concluded that they were deeply committed to action as the only way to preserve democracy in the area. Following instructions, he gave the prime ministers no hint as to what his recommendation would be.[59]

McNeil also spent some time reviewing all locally available intelligence on Grenada with particular emphasis on the medical students. He concluded that they were not hostages yet but that this well might be the Grenadians' next step. He believed the situation was deteriorating daily and was dangerous. With the fate of his colleagues in Tehran, Iran, during the 1979 takeover of the American embassy very much in mind, he recommended that the president should order immediate military intervention. His one qualification was that it had to be quick, before surprise was lost.[60]

The Organization of Eastern Caribbean States heads of government joined by Prime Ministers Adams of Barbados and Seaga of Jamaica knew nothing of these developments and frankly doubted American resolve. Ambassador McNeil had brought a list of State Department concerns about the repercussions of American military action. Based on the discussion of these points, the group drafted a formal request for American intervention. The chairperson of the Organization of Eastern Caribbean States, Prime Minister Charles of Dominica, signed it on the evening of 23 October. She declined to forward it to Washington, however, until she received a "final positive US decision . . . ."[61]

Although the president signed the National Security Decision Directive for the intervention in Grenada on the evening of 23 October, he remained less than totally committed to the operation. Shortly after 1200 on 24 October, he met with the Joint Chiefs of Staff to ask each member individually to give his personal assessment of the plan. Did he agree with it? Was something more needed? The chairman and the chiefs of service were unanimous on two points. First, they preferred a negotiated, peaceful evacuation of the students to armed intervention. Second, if the situation required intervention, they were satisfied with the plan and the forces committed to its execution. The meeting broke up with the president reassured about the plan but with his option of whether to execute still open.[62]

Listening to this exchange, Secretary of Defense Caspar W. Weinberger became convinced that Reagan had concluded to invade the island barring a last-minute diplomatic breakthrough. That afternoon, the president confirmed that this was his decision and gave the secretary the signed National Security Directive. Weinberger immediately returned to the Pentagon. With a military operation looking ever more likely, he delegated to General Vessey, as chairman of the Joint Chiefs of Staff, full power to conduct the operation in the secretary's name. This decision reflected both the president's and the secretary's confidence in Vessey's professional abilities and their affinity for his low–key operating style. With this decision, Vessey gained more control over a major American military operation than had any uniformed officer since the Korean War.[63]

That evening Weinberger, Vessey, and the chiefs went to the White House for a meeting with the president, the other members of the National Security Council, and the House and Senate leadership. At that time President Reagan asked Secretary of State George P. Shultz to describe the situation in Grenada for the group. General Vessey followed with a briefing on the rescue plan. In the discussion that followed, the Speaker of the House, Representative Thomas P. ("Tip") O'Neill, and the Senate Minority Leader, Senator Robert C. Byrd, expressed their unhappiness with the idea of military intervention but could offer no alternative. The president observed that it looked as if it had to be done and that he would do it.[64]

As the meeting broke up, Reagan took Vessey aside and asked him what his decision times were. When did he need to decide to launch the operation? What was the latest time at which he could abort? Vessey told him that if he wanted to stage the operation the next day, 25 October, he had to make the decision immediately. Planes, ships, and troops were already deploying to launch positions. Vessey said that the latest time for an abort would be shortly before 0500 when US aircraft would first enter Grenadian airspace. With that information in hand, the president said: "Go."[65]

Just at that moment the new national security advisor, McFarlane, walked over and told Reagan that he was activating the White House situation room. The president could come there at any time during the night and receive a briefing on the latest information from Grenada. Reagan turned to Vessey and asked what he intended to do that evening. The general responded that first he was going to make a call to the Pentagon to set the operation in motion. Once he had made that call, there was nothing further he or the president could do that evening, unless the president decided to call off the operation. That being the case, Vessey said he intended to go home and go to bed. In the morning, once the troops were on the island, he might be able to do something to assist them. He wanted to be well rested and alert when that time came. President Reagan replied that he intended to do the same.[66]

At the time he issued his "go" order, President Reagan and the other decision makers and planners in Washington, Norfolk, and Fort Bragg believed that almost all the Americans on Grenada could be found on the True Blue Campus of the St. George's University School of Medicine. This assumption was incorrect because it overlooked the even larger Grand Anse Campus located just south of St. George's on the west coast of the island and the sizable collection of students and tourists living near Prickly Point, a peninsula east of True Blue. The misapprehension was due to the failure by Atlantic Command to double-check the information about the Americans' location.

Their whereabouts was hardly a state secret. Because US Forces Caribbean had the mission of conducting operations in the Caribbean, intelligence officers at Key West had already developed detailed information on the location of Americans living on Grenada as a precautionary measure in case their evacuation became necessary at some future date. When Admiral McDonald decided not to use this headquarters to direct Operation URGENT FURY, however, he cut these intelligence assets out. Even before that, when the Joint Chiefs first raised the possibility of evacuating Americans and other foreign nationals from the island, McDonald's senior intelligence officer should have requested all the information that US Forces Caribbean had compiled and distributed it to the participating headquarters, but there is no evidence that he ever attempted to do so.[67]

Despite Ambassador Bish's prohibition on routine visits to Grenada, his staff at Bridgetown also knew the general location of the students. On 20 October 1983 Bish outlined to the State Department where the students lived, but he never sent an information copy to the national command center, so the message never reached military intelligence. A copy did eventually come to rest in the records of the State Department's Grenada Working Group. The military liaison officer from the Pentagon with this group should have passed on the information, but there is no indication of when the working group received the message. It may have arrived after the fact or with a mass of other cables. Although Bish devoted considerable space to describing the students' domiciles, his subject line only referred to their attitudes about evacuation. Given the short time available, members of the working group could easily have overlooked this buried information.[68]

Intelligence agencies also overlooked two other obvious sources of information about the students. Most of the Americans at the medical school received federally guaranteed student loans. The Department of Education, which mailed the checks, had the street addresses of these students, including those who lived off-campus. Because the department was reviewing the school's accreditation, it had prepared a chart that listed all the students by name with their addresses alongside. The administrative office of the school had even more information, but no one went to Bay Shore, Long Island, to collect this material.[69]

At the very last moment, the Defense Intelligence Agency learned of the Grand Anse campus quite by accident. In a casual conversation, one of the secretaries working with the agency mentioned that she had a brother attending the medical school in Grenada, who lived at the Grand Anse Campus. She had recently visited him there and even had photographs. She had nothing to do with tracking events on Grenada, but the person with whom she was talking did and immediately recognized the significance of her remarks. A flurry of research confirmed the existence of a second campus. At 1800 on 22 October an intelligence analyst from the Joint Special Operations Command picked up a package containing the new material and flew with it to Fort Bragg, arriving there late that evening. At the same time, the Defense Intelligence Agency dispatched similar packages to three other intelligence offices, including that of the Atlantic Command's director of intelligence. In the end, for reasons that remain obscure, no plans were changed, and the information never reached the units preparing to invade the island.[70]

*   *   *

Accounts of the Reagan administration often comment on the president's unusually passive style of decision-making coupled with a general lack of curiosity about a wide range of policy areas. At times his advisors were reduced to reading his body language to determine whether he approved or disapproved of a particular option. Historians have suggested that his age and his philosophy of focusing his time and energy on a few big issues on which he had well developed opinions contributed to this style of decision-making. Granting the truth of both these observations and mixing in a personal predisposition to allow an issue to ripen before taking action—Calvin Coolidge was, after all, one of the president's personal favorites among his predecessors—there still remains room for a political explanation for his behavior. Like all administrations, the Reagan team represented a coalition of constituencies in power. It was an assemblage of movement conservatives, who saw the federal government as the cause of most problems in America; business conservatives, who espoused less government regulation; social conservatives, who had abortion and homosexuality at the center of their concerns; and national security conservatives, who saw the United States failing to keep up militarily in its world-wide competition with the Soviet Union. None of these groups agreed entirely with the others, and they competed with one another to set the administration's agenda. The coalition was new—previous Republican administrations had not reflected this particular mix of groups—and some of its members had never even been in power before. All these factors contributed to making the political consensus buttressing the administration fragile with Reagan the one figure around whom all could rally. In this setting, the president's adoption of a passive posture allowed his supporters in their own minds to endow him with their own hopes and desires. If he ultimately adopted a position at variance with their own, they could always blame it on bad advisors and recite the mantra "Let Reagan be Reagan."[71]

The Grenada crisis, of course, fell within the general area of national security—a topic on which Reagan had decided and long-standing opinions. At all the big decision points, he expressed his views in a clear and straight-forward way, except, that is, until he had to sign the presidential directive actually launching the forces. Then he temporized. He may have concluded that it would be easier for some of his supporters to accept a negotiated end to the crisis—should one become possible—if they did not know he had already signed an execute order than would be the case if they knew he had drawn it back at the last minute to accommodate a deal. This analysis is based not on the president's words but on his actions.

While Reagan gave every indication that he meant exactly what he said about his own motives—that his highest priority in addressing the crisis was securing the safe return of the Americans and any other foreign nationals who wanted to leave—this was almost certainly not the stance of everyone involved in the decision-making. Advisors who approached the crisis from a geo-political perspective saw the Americans on the island as a convenient pretext for a demonstration of American power and purpose that would cheaply send a message to the Soviets and the Cubans about the danger of meddling in a US sphere of influence. For these counselors, the ultimate fate of the potential hostages counted for little against this larger purpose. In the end, Reagan's coyness did not in any way retard military preparations, which were developing on a track parallel to the policy making, but it did allow him to be certain in his own mind that he was making the right decision for the right reason.[72]

By law and the president's inclination, Reagan's two senior advisors during the Grenada crisis were the Secretary of State and the Secretary of Defense. Secretaries Shultz and Weinberger shared several traits—both were Californians, had served in World War II, had first come to national prominence in the administration of President Richard M. Nixon, and were strong-minded men with considerable bureaucratic skill.

Unlike Shultz, however, Weinberger had originally opposed Reagan in California politics, and some of the original Reaganites still considered him tarnished by a liberal patina. As a consequence, Weinberger tended to restrict his advice to the president to issues that he considered to have major import and in which he had developed personal expertise. He was also careful not to allow any of his subordinates to get publically to his right. He believed that his main responsibility was to manage the defense buildup, ensuring that the administration retained the political support necessary to sustain the effort. Generally, he regarded Third World military adventures as unnecessary expenditures of the political capital he needed for that larger project. Shultz, lacking Weinberger's history, was more uninhibited in his advice. He believed in diplomacy backed by military power. At the time when Grenada became an issue, he and Weinberger were engaged in a dispute over whether military power could usefully support US diplomacy in the Middle East. Neither man had any previous personal involvement with the eastern Caribbean. Both came to the Grenada issue late in the process and, when presented with the available evidence, recommended intervention. Their primary contribution to the operation had to do not with whether to go but with the size of the force.[73]

Two other senior advisors, Vice President Bush and General Vessey took diametrically opposed views as to the wisdom of the Grenada operation. Bush, based on his recent Caribbean trip, was decidedly in favor of the invasion. Scholarly research on the first president Bush is just beginning and most of it has focused on his four years in the White House. The impression conveyed by the existing literature is that before his meeting with Prime Minister Seaga he had very little first-hand knowledge of the eastern Caribbean. This assumption needs testing by in-depth research. At the same time, his world view, particularly in regard to the portion of the Western Hemisphere south of the United States, might help explain his readiness to act upon Prime Minister Charles' appeal. In contrast to the vice president, Vessey was consistent in saying that the Chiefs preferred a negotiated settlement. By implication that meant that in the Chiefs' view all the military rationales used to justify seizing the airfield at Point Salines, premised upon the outbreak of war with the Soviets in Europe, did not suffice because the possibility of such a conflict was remote. Vessey, who normally handled himself with aplomb in the higher counsels of government, made one misstep when he advanced a political rather than a military reason for not intervening—but this may simply represent how well he fit into the relaxed and convivial air that Reagan generated among his senior officials. There is something to be said for military advisors not becoming too comfortable in such surroundings.

The interagency process was designed to provide a thorough airing of issues at low levels so that those that survived to rise to senior levels would receive a thorough vetting from many points of view. In that light, several questions arise concerning the process's performance during the Grenada crisis. How well did the system work? How thoroughly did lower level participants consider a range of options before senior policy makers became engaged? To what extent did the time compression affect the decision-making?

The Restricted Interagency Group became aware of a potential problem on 13 October. Eleven days later, the president ordered US forces to intervene. In such a time frame there was a tendency for events to overtake careful deliberation of alternatives. This compression of time between the first flickering awareness of the problem and the use of force suggests that Grenada was not a fair test of the interagency process. Any system of decision making—good, bad, or indifferent—might be overwhelmed under similar circumstances. Moreover, for participants the sense of that a succession of events cascading into a torrent was heightened by the unfamiliar speed with which they learned of occurrences. Grenada was one of the first crises in which the State Department used tactical satellite radios developed by the Army Signal Corps to communicate between its Washington headquarters and its embassies overseas. The new technology meant that policy makers in Washington could dispense with the elaborate memoranda with which they (or at least their subordinates) had analyzed past crises. This shift in procedures had at least four consequences. First, it made the record of decision-making much sparser and complicated the job of anyone seeking to render any definitive judgment upon what happened. Second, it meant that policy makers lost that "second look" which the discipline of writing imposes by requiring a writer to consider the inner logic of his or her subject. Participants relied on the "first look" alone—reading the messages as they arrived and reacting ad hoc to their contents. Third, the speed with which information arrived accelerated decision cycles increasing the psychological pressures that accumulate on the individuals involved when an organization is called upon to process and act upon information received in greater quantity and with greater frequency than anticipated. Finally, the increasing speed of the decision cycle meant that whatever information or misinformation that was immediately at hand became relevant with little time for checking—or at least that was how the individuals in the process acted.[74]

All of the above was certainly true, but to halt the analysis at this point would give the national security bureaucracy too much of a free pass. Grenada was a subject in only four Restricted Interagency Group meetings spread over five working days if the Motley-Connally exchange at the end of the 13 October meeting is included in the total. Ambassador Motley thus used the first two sessions to simply broadcast to key agencies the fact that a potential problem existed on the island. Over the weekend, 15–16 October, at a time when the situation on the ground did not appreciably change, he concluded that a full-blown crisis existed and that the only solution was US military intervention and the restoration of democratic government on the island. It would be interesting to know how, why, and when the ambassador reached these conclusions. Bishop's assassination implicitly validated his approach, but this event occurred two days later. On the surface, at least, it appears that Motley thought he knew the answer before the question was asked.

There is no evidence that the Restricted Interagency Group ever debated a range of options with discussion of the advantages and disadvantages of each approach. The record, such as it exists, of the meetings consists of the recollections of participants.

No notes, minutes, or memoranda of record have surfaced to date. Once Motley called for intervention, the debate appears to have polarized into a binary exchange for and against intervention. Even that description may suggest too much coherence in an exchange between two feuding bureaucracies. Motley argued for intervention. No one, it appears, argued against it on its merits. Connally was too low ranking to do anything more than carry the mail, while Moreau slow pitched the process by appearing to drag his feet. (Actually, more military planning was going on than he cared to share with Motley.) The admiral balked at the appearance of State Department direction, although that was the official purpose of instituting the formal interagency process.

Of course, Moreau and Connally faced a dilemma. As uniformed members of the military, they were responsible for carrying out policy, not making it. Deciding to intervene in Grenada was definitely a policy decision, a responsibility of the State Department, the senior civilian leaders in the Department of Defense, and the president. The military's role in the government thus precluded Connally, Moreau, and at a higher level, Vessey from open debate on the merits of an intervention. They were supposed to deal with the question of how, not whether. Vessey's suggestion that the president defer landing because of domestic political considerations—totally inappropriate given the traditions of US civil-military relations even if induced by the warm fellow feeling that Reagan induced among his senior advisors—does suggest the depth of the Joint Chiefs of Staff's opposition to Operation URGENT FURY.

The student of the Grenada decision-making process is left with the disquieting conclusion that the issue appears to have received its most thoughtful consideration when the president took his own counsel. This observation speaks volumes for the president but is also an indictment of the advisory process supporting his decision-making. The feckless actions of the post-Bishop Grenadian leaders and their minions justified the decision that Reagan ultimately reached, but much of that behavior occurred after rather than before Motley made his recommendation.

Serious, in-depth research into the Reagan years has just begun, so it may well be that future discoveries of personal notes or other records contemporaneous to events or prepared immediately afterwards will clarify some of the puzzles sketched above. While this account is only a first cut at the available evidence, it does suggest several fruitful lines of inquiry. One approach would be a series of in-depth studies of the roles of particular individuals. Political scientist Robert J. Beck has already provided a model for this kind of research with his excellent article focusing on the activities of Ambassador McNeil. Similar accounts on Ambassador Motley, Vice President Bush, General Vessey, Admiral Moreau, Secretary Weinberger, Secretary Shultz, and President Reagan that considered not only each man's participation in the crisis but his previous experience with, interest in, and knowledge of the Caribbean region in general and Grenada in particular might substantially alter the conclusions presented in this paper.[75]

The proposals above are micro studies, but there is also room for broader ranging works. The collection, analysis, distribution, and use by decision-makers of intelligence about Grenada, utilizing the records of all the intelligence agencies involved, would greatly clarify a great many issues. Possibly in the days of our great-grandchildren, the records will become available for such an inquiry. Equally distant but equally useful would be accounts based on archival research in Cuba and Russia that would analyze Soviet and Cuban intentions and actions with regard to Grenada from at least the point at which the island gained its independence but with emphasis on Bishop's years in power and the final crisis of the regime. A monograph grounded in primary materials that focused on the US Department of State's role in the crisis would make clear many things currently out of focus. Finally, someone needs to write a history of Grenada based upon archival and manuscript sources that firmly sets the Grenadian revolution and its denouement in the context of local traditions.

Notes

1. The views expressed in this paper are those of the author and do not reflect the official policy or position of the Department of the Army, the Department of Defense, or the US government. There is no official definition of the "interagency process." The Department of Defense defines interagency coordination as "within the context of Department of Defense involvement, the coordination that occurs between elements of Department of Defense and engaged US government agencies for the purpose of achieving an objective." *US, Joint Chiefs of Staff, Department of Defense Dictionary of Military and Associated Terms*, 12 April 2001 (As Amended Through 30 May 2008), *Joint Pub 1-02* (Washington, DC: Joint Chiefs of Staff, 2008), p. 273.

2. Department of Defense Reorganization Act of 1958, 6 Aug 58 (72 Stat. 514), in Alice C. Cole et al., *The Department of Defense: Documents on Establishment and Organization, 1944–1978* (Washington, DC: Office of the Secretary of Defense Historical Office, 1978), 188–231.

3. For background, see Sandra W. Meditz and Dennis M. Hanratty, "Islands of the Commonwealth Caribbean: A Regional Study," *DA Pam 550–33* (Washington, DC: Headquarters, Department of the Army, 1989), pp. 349–52; Tony Thorndike, *Grenada: Politics, Economics, and Society, Marxist Regimes Series*, Bogdan Szajkowski, ed. (Boulder, Colo.: Lynne Rienner, 1985), pp. xvii–xx; and Frederic L. Pryor, *Revolutionary Grenada: A Study in Political Economy* (New York: Praeger, 1986), pp. 201–06. On Communist influence, see Jiri and Virginia Valenta, "Leninism in Grenada," *Problems of Communism* 33 (Jul–Aug 84): 2–4. The "mood setter" for US-Grenadian relations was in an early Bishop speech: Maurice Bishop, "In Nobody's Backyard," in *Maurice Bishop Speaks: The Grenada Revolution and Its Overthrow, 1979–83*, Bruce Marcus and Michael Taber, eds. (New York: Pathfinder Press, 1983), pp. 26–31. On Grenadian views of US destabilization, see Chris Searle, *Grenada: The Struggle Against Destabilization* (New York: W. W. Norton, 1983), pp. 53, 68–70. Bishop explicitly endorsed these accusations in Kwando M. Kinshasa, "Prime Minister Maurice Bishop: Before the Storm," *Black Scholar* 15 (Jan–Feb 1984): 41–59, the transcript of an August 1983 interview with Bishop, see especially pages 47, 50–51.

4. Burton I. Kauffman, *The Presidency of James Earl Carter, Jr., American Presidency Series*, Donald R. McCoy et al., eds. (Lawrence: University Press of Kansas, 1993), pp. 151–166; Robert A. Pastor, *Whirlpool: US Foreign Policy Toward Latin America and the Caribbean* (Princeton, N.J.: Princeton University Press, 1992), pp. 42–64; Ronald H. Cole et al., *The History of the Unified Command Plan, 1946–1993* (Washington, DC: Joint History Office, 1995), pp. 70–71.

5. Speech, Ronald Reagan, 23 Mar 83, sub: Address to the Nation on Defense and National Security, in *Public Papers of the President of the United States: Ronald Reagan, 1983*, 2 vols. (Washington, DC: Government Printing Office, 1984), 2: 440. On the strategic importance of Grenada, see Issues Paper, [Interagency Core Group], [18 May 83], sub: Grenada, Chairman, Joint Chiefs of Staff (CJCS) Files (Vessey), 502B (NSC Memos), Ronald Reagan Presidential Library (RRPL) Simi Valley, California.

6. Issue Paper, Interagency Core Group], [18 May 83]; Memo, Charles Hill, Executive Secretary, Interagency Core Group, for William P. Clarke, National Security Advisor, 18 May 83, sub: Grenada; Memo, William P. Clark, National Security Advisor, for George P. Shultz, Secretary of State, 15 Jun 83, both in CJCS Files (Vessey), 502B (NSC Memos), RRPL.; Memo, William P. Clark, National Security Advisor, for President Ronald Reagan, 4 Oct 83, sub: Grenada; National Security Decision Directive (NSDD) 105, President Ronald Reagan, 4 Oct 83, sub: Eastern Caribbean Security Policy; both in NSDD 105, Box 91,291, OAPNSA, RRPL.

7. Minutes, CC, NJM Party, 14–16 Sep 83, sub: Extraordinary Meeting of the CC of NJM, in Michael Ledeen and Herbert Romerstein, eds., *Grenada Documents: An Overview and Selection* (Washington, DC: Department of State and Department of Defense, 1984), p.112. See also "The Alienation of Leninist Group Therapy: Extraordinary General Meeting of Full Members of the NJM," *Caribbean Review* 12 (Fall 1983): 14–15, 48–58; Notes, no date, sub: CC Meetings; Resolution, PRA Branch, NJM Party, 12 Oct 83; Ltr, Nazim Burke to Maurice Bishop, 11 Oct 83; all in US, Departments of State and Defense, Individual Documents Released (IDR), Nos. 000136, 100103, 100270, Historians files, US Army Center of Military History (CMH), Washington, DC Frank J. Prial, "Grenada Curtails Communications to the Outside," *New York Times* (16 Oct 83): Sect. I, 22; R. S. Hopkin, *Grenada Topples the Balance in West Indian History: A Day To Day Report of the Grenada Episode* (Grenada: Mr. Merlin, 1984), p. 8, a journal kept by one of the protestors. Thorndike, Grenada, pp. 153–56. Ltr, Vincent Noel to CC, 17 Oct 83, in Paul Seabury and McDougall, eds., *The Grenada Papers* (San Francisco, Calif.: Institute for Contemporary Studies Press, 1984), pp. 329–339, on the emotional climate.

8. National Security Defense Directive 2, President Ronald W. Reagan, 12 Jan 82, sub: National Security Council Structure (Ms., Armed Forces Staff College Library, Norfolk, Va.).

9. William M. LeoGrande, *Our Own Backyard: The United States in Central America* (Chapel Hill: University of North Carolina Press, 1998), pp. 72–146, 188–96, 356–57, provides a detailed account of the Reagan administration's Central American policy.

10. Interv, Ronald H. Cole with Connally, 25 Jul 84. Msg, Amb Milan D. Bish, Bridgetown, Barbados, to Sec. State, Washington, DC, 141508Z Oct 83, sub: Report of a Marxist Coup by DPM Bernard Corad, 83 Bridgetown 06209, State Dept. See also, Stephen E. Flynn, "Grenada as a 'Reactive' and a 'Proactive' Crisis: Models of Crisis Decision-Making" (Ph.D. diss., Fletcher School of Law and Diplomacy, 1991), pp. 103–05; and Ronald H. Cole, *Operation URGENT FURY: The Planning and Execution of Joint Operations in Grenada, 12 October–*

2 November 1983 (Washington, DC: Joint History Office, 1997), pp. 11–12. Cole's account provides an excellent description of the crisis from the perspective of the Joint Chiefs of Staff.

11. Constantine C. Menges, *Inside the National Security Council: The True Story of the Making and Unmaking of Reagan's Foreign Policy* (New York: Simon and Schuster, 1988), pp. 60–62. The judgment about Menge's subsequent role in the operation is in Robert J. Beck, *The Grenada Invasion: Politics, Law, and Foreign Policy Decisionmaking* (Boulder, Colo.: Westview Press, 1993), pp. 120–21, note 56.

12. Robert C. McFarlane and Zofia Smardz, *Special Trust* (New York: Cadell and Davies, 1994), pp. 258–60, describes the impact of Judge Clarke's departure. On the approach to the Joint Chiefs of Staff, see: Chrono, 140000Z [Oct 83], JCS, URGENT FURY Miscellaneous Scenario Events List, Planning/Execution Systems, p. 84, Historian's Files CMH; Cole, Operation URGENT FURY, p. 12.

13. Chrono, 141200Z [Oct 83], JCS, URGENT FURY (hereafter UF) Miscellaneous Scenario Events List, Planning/Execution Systems, p. 84, Historians' files, CMH. See also Cole, Operation URGENT FURY, p. 12.

14. Flynn, "Grenada as a 'Reactive' and 'Proactive' Crisis," p. 105, provides the only available account of this meeting.

15. Chrono, Atlantic Command (hereafter LANTCOM), sub: Operation (hereafter Op.) URGENT FURY, encl in Ltr, Admiral Wesley L. McDonald to General John W. Vessey, Jr., 6 Feb 84, sub: Op. URGENT FURY Report, in Historians' files, CMH; Msg, Amb Milan D. Bish, Bridgetown, Barbados, to Sec State, Washington, DC, 200739Z Oct 83, sub: Grenada: Attitudes of the Grenadian Medical School Toward the Possible Evacuation of Their Students/Staff, DOS. Msg, Admiral Wesley L. McDonald, CINCLANT, Norfolk, Va., to JCS, Washington, DC, 200616Z Oct 83, sub: Commander's Estimate of the Situation, Grenada Evacuation, encl in Memo, LTG Jack N. Merritt, Director, Joint Staff, for Directors and Heads of Agencies, OJCS, 30 Jan 84, Joint History Office (JHO), Washington, DC; Interv, Major Bruce R. Pirnie with Tony Nelson, DIA, 9 Dec 85, Historians' files, CMH.

16. Statement, Langhorne A. Motley, Assistant Secretary for Inter-American Affairs, before the House Armed Services Committee, 24 Jan 84, sub: The Decision To Assist Grenada, in US, Department of State, Bureau of Public Affairs, *Current Policy*, No. 541 (24 Jan 84): 1. Flynn, "Grenada as a 'Reactive' and 'Proactive' Crisis," pp. 105–07, provides the most detailed account of the interagency meeting.

17. Interv, Cole with LTG Jack N. Merritt, USA, 21 Jun 84, JHO; JCS, Chrono, 180000Z [Oct 83], URGENT FURY Miscellaneous Scenario Events List, Command and Control—Task Organization, p. 1, Historians' files, CMH; Chrono, LANTCOM, sub: URGENT FURY Op.

18. "Leader of Grenada Is Reported Killed by Troops," *New York Times* (20 Oct 83): A1, A3; V.S. Naipaul, "An Island Betrayed," *Harper's* 268 (Mar 84): 61–72; Alister Hughes, "Island Bloodshed 'Started with Army Rockets,' Journalist Says," *Washington Post* (30 Oct 83): A11; Transcript, Tape Recording by Alister Hughes, Market Square, St. George's, Grenada (GJ), 19 Oct 83, in US, Congress, Senate, Congressional Record, 27 Mar 84 (Washington, DC: Government Printing Office, 1984), pp. S 3278–S 3279; Transcript, Broadcast by BG Hudson Austin over Radio Free Grenada, 19 Oct 83, approximately 2230, in Ibid., S 3729–S 3780. Statement, [BG Hudson, Austin], 19 Oct 83, sub: Rev. Soldiers and Men of the PRA, IDR, No.

000091, CMH; "Military Council Says It Now Rules Grenada," *New York Times* (21 Oct 83): A8. Mark Adkin, *URGENT FURY: The Battle for Grenada* (Lexington, Mass.: D. C. Heath, 1989), pp. 47–81, provides the fullest reconstruction of these events.

19. Msgs, Amb Milan D. Bish, Bridgetown, Barbados, to Sec State, Washington, DC, 192356Z Oct 83, sub: Planning for Possible Emergency Evacuation of AMCITS, 83 Bridgetown 06387; 200409Z, sub: Grenada: Bishop Dead; Army Takes Over Fully; Imperialism Warned To Keep Hands Off, 83 Bridgetown 06388; both in DOS. Richard Halloran, "Reagan as Military Commander," *New York Times Magazine* (15 Jan 84): 61; Richard Halloran, "A Commanding Voice for the Military, *New York Times Magazine* (15 Jul 84): 18–25, 52. For Moreau's background, Memo, 1 Jul 86, sub: Transcript of Naval Service for Vice Admiral Arthur Stanley Moreau, Jr., US Navy, Naval Historical Center (hereafter NHC) Archives, Washington, DC. On the Vessey-Moreau relationship, see Roy Gutman, *Banana Diplomacy: The Making of American Policy in Nicaragua, 1981–1987* (New York: Simon and Schuster, 1988), p. 138. Interv, Cole, with Connally, 25 Jul 84.

20. Intervs, Cole, with Connally, 25 Jul 84; with V Adm Arthur S. Moreau, Jr., Assistant to the CJCS, 12 Jul 84, JHO; Flynn, "Grenada as a 'Reactive' and 'Proactive' Crisis," pp. 107–08.

21. Msg, Gen John W. Vessey, CJCS, to USCINCLANT, Norfolk, Va., USCINCMAC, Scott AFB, Ill., and USCINCRED, MacDill AFB, Flor., 200347Z Oct 83, sub: Warning Order—Grenada NEO, attached to Grenada Timeline, JHO Archives; Chrono, JCS, UF Miscellaneous Scenario Events List, Command and Control—Task Organization, p. 1; Chrono, LANTCOM, sub: Op. UF; Historians' files, CMH; Cole, Operation URGENT FURY, pp. 13–14.

22. Msg, McDonald, to JCS, 200616Z Oct 83; Cover Sheet, LTC Daniel E. Staber, Office of the Deputy Chief of Staff for Operations and Plans (hereafter ODCSOPS), Department of the Army (hereafter DA), for DCSOPS, 21 Oct 83, sub: Grenada Evacuation (UF), JHO.

23. Ibid.

24. Ibid.; Interv, Cole with General John W. Vessey, chairman, Joint Chiefs of Staff, 25 Mar 87; Ltr, GEN John W. Vessey, US Army (Ret.), to BG David A. Armstrong, US Army (Ret.), Director, Joint History, no date, JHO Archives, Washington, DC; Cole, *Operation URGENT FURY*, pp. 14–16; Pirnie, *URGENT FURY*, pp. 63–66.

25. Unclassified portions of Anon., Annual Historical Report for Calendar Year 1983, Commander in Chief, US Atlantic Command (Norfolk, Va.: US Atlantic Command, 1984), pp. vii–ix; Interv, John T. Mason, Jr., with Admiral Charles K. Duncan, USN, Ret., 3 Nov 76, US Joint Forces Command Historical Office, Norfolk, Va.; Memo, [Dr Leo P. Hirrel, Command Historian, US Joint Forces Command], for author, [22 May 07], Historians files, CMH.

26. Cover Sheet, Staberfor DCSOPS, 21 Oct 83; Interv, Cole with General John W. Vessey, chairman, Joint Chiefs of Staff, 25 Mar 87; Ltr, GEN John W. Vessey, US Army (Ret.), to BG David A. Armstrong, US Army (Ret.), Director, Joint History, no date, JHO Archives, Washington, DC; Cole, *Operation URGENT FURY*, pp. 14–16; Pirnie, *URGENT FURY*, pp. 63–66.

27. Menges, *Inside the National Security Council*, p. 68; Cole, *Operation URGENT FURY*, p. 16; Don Oberdorfer, "Reagan Sought To End Cuban 'Intervention,'" *Washington Post*, 6 Nov 83, A1, A21.

28. Interview Cole with Moreau, 12 Jul 84; Agenda, 20 Oct 83, sub: CPPG Meeting, NSC

Crisis Management Center (hereafter CMC) Records, file "Grenada I (9), Box 90, 931, RRPL; George P. Shultz, *Turmoil and Triumph*, (New York: Scribner Publishing 1993), pp. 326–27.

29. Msg, Amb Milan D. Bish, Bridgetown, Barbados, to Sec State, 201945Z Oct 83, sub: Barbadian PM Tom Adams Pleas for US Intervention in Grenada: Believes Leadership of the Region Would Strongly Support and Fully Associate with US, 83 Bridgetown 06430, DOS; Flynn, "Grenada as a Reactive' and 'Proactive' Crisis," p. 111, has the fullest account of the Vice President's role in this meeting.

30. Statement, George P. Shultz, 25 Oct 83, sub: Secretary Shultz's News Conference, in US, Department of State, Bulletin: The Official Monthly Record of United States Foreign Policy, 83 (Dec 83): 69; Cole, *Operation URGENT FURY*, p. 16.

31. The quotation is from Ltr, Vessey to Armstrong, no date; Info Paper, Dan Landers, Office of the Deputy Chief of Staff for Operations and Plans, Army Staff, 21 Oct 83, sub: UF; Memo, COL Peter Cummings, Assistant Deputy Director for Current Intelligence, DIA, [c. 20 Oct 83], sub: Evaluation of the Threat, all in JHO. Cole, *Operation URGENT FURY*, pp. 16–18.

32. Caspar W. Weinberger, *Fighting for Peace: Seven Critical Years in the Pentagon* (New York: Warner Books, 1991), p. 109; Statement, Shultz, 25 Oct 83, *Bulletin*, p. 69; Interv, Cole with Moreau, 12 Jul 84.

33. Interv, Lawrence A. Yates with MG Richard A. Scholtes, 4 Mar 99, Historians files, CMH.

34. National Security Decision Directive (hereafter NSDD) 110, Ronald Reagan, President, 21 Oct 83, sub: Grenada: Contingency Planning, in NSDD, 1–250, RRPL, Box 1; Interv, Yates with Scholtes, 4 Mar 99.

35. Edward Cody, "Cuba Condemns Grenada Coup, Will Review Tie," *Washington Post* (22 Oct 83): A1, A12; *New York Times* (22 Oct 83): B. Drummond Ayres, Jr., "US Marines Diverted to Grenada in Event Americans Face Danger," *New York Times* (22 Oct 83): Sect. I, 1, 12.; Interv, Cole with GEN John W. Vessey, Chairman JCS, 25 Mar 87, JHO.

36. R. Drummond Ayres, Jr., "US Marines Diverted to Grenada in Event Americans Face Danger," *New York Times* (22 Oct 83): Sect. I: 1, 5; Fred Hiatt, "US Says Situation Still Unclear as Naval Force Nears Grenada," *Washington Post* (23 Oct 83): A24; Statement, Langhorne A. Motley, Assistant Sec. State for International Affairs, before the HASC, 24 January 1984, sub: "The Decision To Assist Grenada," in US, Department of State, Current Policy No. 541 (24 Jan 84): 2. On the trip, see: *Ronald W. Reagan, An American Life* (New York: Simon and Schuster, 1990), p. 449; McFarlane and Smardz, Special Trust, p. 261.

37. Cole, *Operation URGENT FURY*, p. 19.

38. Log, [Grenadian GS], 24 Oct 83, sub: Strength of the Armed Forces in Terms of Permanent, Reserve, and Party Comrades, GDC, MF 005213, NARA.

39. For an overview of US intelligence history, see: Christopher M. Andrew, *For the President's Eyes Only : Secret Intelligence and the American Presidency from Washington to Bush* (New York: Harper Collins Publishers, 1995), pp. 350–502. Robert M. Gates, *From the Shadows: The Ultimate Insider's Story of Five Presidents and How They Won the Cold War* (New York: Touchstone, 1996), pp. 56–63, is perceptive on the weakening of the CIA in the 1970s. On the situation at Bridgetown, see Msgs, Amb Frank V. Ortiz, Bridgetown, Barbados, to Sec. State, 151010 Mar 79, sub: Query re Gairy's Activities, 79 Bridgetown 0924; 191444 Mar 79, sub: Grenada: US Presence on the Island, 79 Bridgetown 1013; both in State Dept. Interv,

COL Walter M. Loendorf with LTG James A. Williams, USA, Ret., 14 Dec 1990, Senior Officer Oral History Program, US Army Military History Institute, Carlisle, Pa. (hereafter MHI). General Williams was Director of the Defense Intelligence Agency at the time of the Grenada operation.

40. This evaluation of the quality of diplomatic reporting is based on the author's reading of all the Bridgetown cables pertaining to Grenada between 12 and 25 October 1983. See also Shultz, Triumph and Tragedy, p. 327. Sally A. Shelton, "Comments," in *Grenada and Soviet/ Cuban Policy: Internal Crisis and US/OECS Intervention*, Jiri Valenta and Herbert J. Ellison, eds. (Boulder, Colo.: Westview Press, 1986), pp. 236–37. Shelton was US Ambassador to Bridgetown between Frank V. Ortiz and Bish. Interv, Loendorf with Williams, 14 Dec 1990.

41. Edward Cody, "Caribbean Nations Discuss Response to Violence in Grenada, *Washington Post* (23 Oct 83): A24; Frank J. Prial, "American Envoys Going To Grenada," *New York Times* (23 Oct 83): Sect. I, 1, 13. Statement, Motley before the HASC, 24 Jan 84, p. 2. On the call to Austin, see Lewis, Grenada, p. 95.

42. Shultz, Turmoil and Triumph, p. 333–34. Msgs, Amb, Milan D. Bish, Bridgetown, Barbados, to Sec. State, Washington, DC, 211558 Oct 83, sub: Embassy Officers' Travel, 83 Bridgetown 06476; 222219Z, sub: More Discussion with Eastern Caribbean Leaders, Unity Prevails, 83 Bridgetown 06544; both in DOS.

43. Ibid.; Msgs, Amb, Milan D. Bish, Bridgetown, Barbados, to Sec. State, Washington, DC, 212046Z, sub: Apparent Invitation from General Hudson Austin to Send USG REP to Discuss the Situation, 83 Bridgetown 06490; 212354Z Oct 83, 83 Bridgetown 06500; Second Telex/ Phone Contact Oct 22 with Medical School Authorities, 83 Bridgetown 06524; all in DOS.

44. Msg, Amb Milan D. Bish, Bridgetown, Barbados, to Sec. State, Washington, DC, 230442Z, sub: Travel of Political Counselor and Vice Counsel to Grenada, 83 Bridgetown 06548; 241740Z Oct 83, sub: Possible Evacuation of American Citizens; Strong Media Interest as Well as Concern by Parents, 83 Bridgetown 06597; both in DOS; Statement, Motley, 24 Jan 84, *Current Policy*, p. 2.

45. Duane R. Clarridge and Digby Diehl, *A Spy for All Seasons: My Life in the CIA* (New York: Scribner, 1997), pp. 250–51. Clarridge was chief of the Latin American Division in the Directorate of Operations.

46. Msg, Amb Milan D. Bish,, Bridgetown, Barbados, to Sec. State, Washington, DC, 241659Z Oct 83, sub: Grenada: Meeting with Seaga and Adams, 24 Oct, Morning, 83 Bridgetown 06594, DOS; Mark Adkin, *URGENT FURY: The Battle for Grenada* (Lexington, Mass.: DC Heath, 1989), pp. 97–99, has the fullest account of the Scoon appeal although it is wrong in one key detail. Ambassador William J. McNeil did not carry a draft letter for Scoon's signature to Barbados on 23 October. Robert J. Beck, "The 'McNeil Mission' and the Decision to Invade Grenada," *Naval War College Review* 44 (Spring 91): 103.

47. Msg, Ambassador Milan D. Bish, American Embassy Bridgetown, Barbados, to Secretary of State, 220735Z Oct 83, sub: Organization of Eastern Caribbean States Officially, Formally Resolves Unanimously To Intervene by Force if Necessary, On Grenada and Pleads for US Assistance, 83 Bridgetown 06514, Department of State; Oberdorfer, "Reagan Sought To End Cuban 'Intervention,'" p. A–21; PDD, 22 Oct 83; *Reagan, An American Life*, pp. 449–52. Shultz and McFarlane agree as to the major decisions in this meeting but differ about some of the details. Shultz, T*urmoil and Triumph*, pp. 329–30; McFarlane and Smardz, *Special Trust,*

pp. 261–62. See also *Ronald Reagan, The Reagan Diaries*, Douglas Brinkley, ed. (New York: HarperCollins, 2007), pp. 188–89. In his diary entries, the president was more certain than his subsequent actions would seem to indicate.

48. Cole, *Operation URGENT FURY*, p. 23. The scholar mentioned in the text is Robert J. Beck, "The 'McNeil Mission' and the Decision to Invade Grenada," *Naval War College Review* 44 (Spring 91): 101.

49. Cole, *Operation URGENT FURY*, pp. 23–26.

50. Caspar W. Weinberger, *Fighting for Peace*, pp. 108–12; Intervs, Caspar W. Weinberger with Alfred Goldberg, Maurice Matloff, and Stuart Rochester, 12 Jan 88; with Alfred Goldberg and Maurice Matloff, 21 Jun 88; both in Office of the Secretary of Defense (hereafter OSD) History Office, Washington, DC; Shultz, *Turmoil and Triumph*, p. 329.

51. Msg, George P. Shultz, Sec. State, to Am. Emb., Bridgetown, 231833Z Oct 83, sub: Instructions for Dealing with Caribbean Friends, 83 State 302418, DOS. Robert J. Beck, "The 'McNeil Mission' and the Decision to Invade Grenada," *Naval War College Review* 44 (Spring 1991): 93–112, the "go/no go" quotation is from p. 99. Francis J. McNeil, *War and Peace in Central America* (New York: Charles Scribner's Sons, 1988), p. 173, is the source for Motley's observation.

52. Shultz has a concise discussion of the debate over whether to stay in Augusta. Shultz, *Turmoil and Triumph*, pp. 329–30. Bernard Gwertzman, "Steps to the Invasion: No More 'Paper Tiger,'" *New York Times* (30 Oct 83): 1, 20; *Reagan, American Life*, p. 451; Francis X. Clines, "Reagan Unhurt as Armed Man Takes Hostages," *New York Times* (21 Oct 83), pp. A1, A28.

53. Intervs, Frasche with Akers, 22 Nov 83; Cole with Vessey, 25 Mar 87; Cole, Operation URGENT FURY, pp. 26–27.

54. Intervs, Yates with Scholtes, 4 Mar 99; Cole with Vessey, 25 Mar 87; Matt Schudel, "Joseph Metcalf: Led Grenada Invasion," *Washington Post* (11 Mar 07): C7. Chrono, 16 Apr 84, sub: Chronology [Operation URGENT FURY] in Department of the Navy, *Operation URGENT FURY Lessons Learned: Discussion, Conclusions, and Recommendations* (Washington, DC: Department of the Navy, 1984), p. III–3, gives the time of Metcalf's selection.

55. President's Daily Diary, 23 Oct 83; E–Mail, Danny J. Crawford, Head, Reference Section, History and Museums Division, Headquarters, US Marine Corps (hereafter cited as HQMC), to author, 18 Nov 2003, 0938, sub: Beirut Casualties, 1983; Paper, Hqs, USMC, Division of Public Affairs, 2 Dec 88, sub: General George B. Crist; both in Historian's files, US Army Center of Military History, Washington, DC (hereafter CMH). Beck, "McNeil Mission," p. 99; Interv, Dr Ronald H. Cole, JCS History Office, with Maj Gen George B. Crist, USMC, 16 Feb 84, Grenada Files, Joint History Office (hereafter JHO) Archives, Washington, DC.

56. Interv, Cole with Vessey, 25 Mar 87; Memo, Oliver L. North for John M. Poindexter, 9 Jan 85, sub: Revised Presidential Correspondence, Executive Secretariat, National Security Council (hereafter NSC), "Grenada, Vol IV" file, RRPL, Box 91, 370; Oberdorfer, "Reagan Sought To End Cuban 'Intervention,'" A21. Clarridge and Diehl, *A Spy for All Seasons*, p. 254, is the source for Vessey's suggestion.

57. Memo, North for Poindexter, 9 Jan 85; Oberdorfer, "Reagan Sought To End Cuban 'Intervention,'" A21; Weinberger, *Fighting for Peace*, p. Ltr, Ronald Reagan, President, to Greg-

ory W. Sanford, 8 Jan 85, Executive Secretariat, NSC, "Grenada, Vol IV" file, RRPL, Box 91, 370.

58. Memo, North for Poindexter, 9 Jan 85. The quotation is from NSDD 110A, Ronald Reagan, President, 23 Oct 83, sub: Response To Caribbean Governments' Request To Restore Democracy on Grenada, NSDD 1–250, Box 1, RRPL.

59. Msg, Amb Milan D. Bish, Bridgetown, Barbados, to Sec. State, Washington, DC, 252203Z Oct 83, sub: Uncleared, Informal Minutes of Meeting between Ambassadors Bish and McNeil with West Indian Heads of Government To Discuss the Grenada Situation, 83 Bridgetown 06654, DOS; Beck, "McNeil Mission," pp. 100–109.

60. McNeil, *War and Peace in Central America*, pp. 174–75; Beck, "The McNeil Mission and the Decision to Invade Grenada," pp. 100–01. Prime Minister Charles laid out the Organization of Eastern Caribbean States' rationale in Statement, Eugenia Charles, [c. 25 Oct 83], DOS, Accession 59–99–0424 (10/13).

61. Msg, Amb Milan D. Bish, Bridgetown, Barbados, to Sec. State, Washington, DC, Grenada: Text of Final Draft OECS Invitation, 83 Bridgetown 06574, DOS; Ltr, Eugenia Charles, Chairman, OECS, to Ronald Reagan, President of the US, 23 Oct 83; DOS, Accession No. 59–97–0323 (4/6).

62. Cole, *Operation URGENT FURY*, p. 39.

63. Msg, Milan D. Bish, American Embassy, Brigdetown, Barbados, to Secretary of State, Washington, DC 241659Z Oct 83, sub: Grenada: Meeting with Seaga and Adams, 24 Oct, Morning, Msg, McDonald to Cavazos et al., 24 Oct 83, 0007Z; Interv, Cole with Vessey, 25 Mar 87; Halloran, "A Commanding Voice for the Military," p. 18. Cole, Operation URGENT FURY, pp. 28–40, provides a much more detailed discussion of the events leading up to these decisions.

64. Interv, Cole with Vessey, 25 Mar 87.

65. Ibid.

66. Ibid.

67. Interv, author with Daly, 30 Jul 86.

68. Msg, Amb Milan D. Bish, Bridgetown, Barbados, to Sec State, Washington, DC, 200739Z Oct 83, sub: Grenada: Attitudes of the Grenadian Medical School Toward the Possible Evacuation of Their Students/Staff, DOS.

69. Interv, author with Joseph M. Hardman, Chief, College Eligibility Unit, Department of Education (DOE), 31 Jan 84, Historians' files, CMH; Ltr, Joseph M. Hardman, DOE, to Charles R. Modica, Chancellor, St. George's University School of Medicine, Bay Shore, NY, 22 Jul 83, Roger W. Fountaine Files, "Grenada–1983 (1)," RRPL, Box 90,118.

70. Interv, Major Bruce R. Pirnie with Tony Nelson, DIA, 9 Dec 85, Historians' files, CMH.

71. Lou Cannon, *President Reagan: The Role of a Lifetime* (New York: Simon and Schuster, 1991), pp. 120–42, 172–205, is especially perceptive and even handed. See also Sean Wilentz, *The Age of Reagan: A History, 1974–2008* (New York: Harper, 2008), pp. 3, 137–39. Alan Greenspan provides a revealing anecdote in *The Age of Turbulence: Adventures in a New World* (New York: Penguin Press, 2007), pp. 87–88.

72. Clarridge and Diehl, *A Spy for All Seasons*, pp. 207–55, is revealing in this context.

73. Shultz, *Turmoil and Triumph*, pp. 8, 26; Caspar W. Weinberger with Gretchen Roberts, *In the Arena: A Memoir of the 20th Century* (Washington, DC : Regnery, 2001), pp. 147–153. On

Weinberger and the crisis in Central America, see Leo Grand, *Our Own Backyard*, pp. 140–42, 152, 202, 209, 316–17.

74. Dennis E. Showalter, "Even Generals Wet Their Pants: The First Three Weeks in East Prussia, August 1914," *War and Society* 2 (September 1984): 60–86, provides the classic account of such a situation. This is not to say that the Reagan administration at any level broke down the way the German system did, but that there was an unaccustomed strain for all the participants.

75. Robert J. Beck, "The 'McNeil Mission' and the Decision to Invade Grenada," *Naval War College Review* 44 (Spring 91): 93–112.

Panel 1—The Difficulties in Interagency Operations

(Submitted Paper)

The Interagency Non-Process in Panama: Crisis, Intervention, and Post-Conflict Reconstruction[1]

by

John T. Fishel, Ph.D.
National Defense University

Introduction

Nineteen years have passed since Operation JUST CAUSE was executed in Panama. Twenty-one years ago, the Panama crisis began when General Manuel Noriega forcibly retired his only remaining rival for the leadership of the Panama Defense Forces (PDF) and, therefore, the nation. Because that rival would not go quietly, events were set in motion that resulted in US intervention. With respect to the interagency process—or non-process—one has to consider both time and place.

The time was June 1987. The landmark Goldwater-Nichols Department of Defense Reorganization Act had passed the Congress only the previous year. Its effects were still to be felt. The Washington interagency process was still changing its structure with every new administration. Not until George H. W. Bush took office on 20 January 1989 would today's familiar structure of a Principals' Committee, a Deputies' Committee, and a series of subordinate committees or working groups be put in place. Instead, the interagency committee structure was very much in flux with roles of key players in the National Security Council staff very much works in progress.

The place was the Republic of Panama, born in a rebellion against Colombia in 1903 and midwifed by President Theodore Roosevelt. Teddy built the Panama Canal and, according to the Hay-Bunau Varilla Treaty, occupied a ten mile swath of Panamanian territory on either side of the Canal where the US could "act as if it were sovereign . . . in perpetuity" or, at least until a new treaty was negotiated. That happened in 1977 and it went into effect in 1979. In 1987 the US found itself with three separate, largely independent US government institutions located in Panama; two of which were in the former Canal Zone where the US had acted as if it were sovereign for 76 years. Those institutions were: the American Embassy, located in Downtown Panama City, the US Southern Command, located at Quarry Heights and other "Defense Sites" in the former Zone, and the Panama Canal Commission, located in the Canal Administration Building on the slope of Ancon Hill in the former Zone.

US Institutions in Panama

The American Embassy in Panama, like all American Embassies, contains all the elements of the US government that operate in Panama and all work under the direction of the Ambassador. In Panama, in the 1980s, these included the standard Embassy sections (Political, Economic, Consular, Security, USAID, USIS, and others), a CIA Station, a Defense Attache Office, and with regional responsibilities, the Drug Enforcement Administration, and the USMC Latin America US Embassy Guards command and control office. Leading the Embassy in 1987 was Ambassador Arthur Davis, a political appointee who had, however, acquitted himself well as Ambassador to Paraguay. His Deputy Chief of Mission (DCM), John Maisto, was a career Foreign Service Officer who would later go on to several Ambassadorships and the post of Senior Director for Latin America on the National Security Council Staff. Neither Davis nor Maisto had particularly good reputations within the Embassy or with the other US institutions in Panama.[2]

The United States Southern Command (USSOUTHCOM) was the second US government institution located in Panama until the late 1990s. It consisted of the SOUTHCOM commander and staff, located at the Quarry Heights Defense Site, halfway up Ancon Hill, the several service components: Army South operating out of Fort Clayton and other Pacific side sites and several sites on the Atlantic side, Southern Air Force (SOUTHAF) operating out of Howard AFB and Albrook Air Force Station on the Pacific side, Navy South (NAVSO) and the Marines (MARSOUTH) at Rodman Naval Station on the Pacific side, and the sub-unified command, Special Operations Command South (SOCSOUTH), at first located on Quarry Heights and later on Albrook. As part of the senior staff, under the Commander-in-Chief (CINC), General Fred F. Woerner, Jr., was a senior Foreign Service Officer (FSO) called the POLAD, a senior CIA officer called the Regional Affairs Officer, the SOUTHCOM Treaty Affairs Directorate, and the several Staff Directorates. As CINC, General Woerner was responsible for all US military activities in his Area of Operational Responsibility (AOR) which ran from the Mexico-Guatemala border south to cover all of Central and South America plus the seas out to 200 miles from land. Panama, of course, falls within the SOUTHCOM AOR.

The last major US government institution in Panama was the Panama Canal Commission's Administration. The Administrator was former SOUTHCOM CINC, LTG (ret.) Dennis P. McAuliffe while the Deputy Administrator was a Panamanian, Fernando Manfredo. The PCC administration was responsible for running the Canal and conducting business with the Panamanian government regarding the Canal.

Neither SOUTHCOM nor the PCC Administration felt any obligation to coordinate with the Embassy when conducting its core business. For SOUTHCOM, core business included all activity in the AOR except that which dealt directly with Panama, except those activities expressly stated in the Panama Canal treaties as being between the US military and the PDF (which was nearly everything related to the treaties). This meant, as well, that there would be more coordination between SOUTHCOM and the PCC than with the Embassy.

Stovepipes to Washington

As Ambassador David Passage tells it, every agency working in an American Embassy communicates by stovepipe with its Washington DC headquarters.[3] Even though the CIA Station, for example, is supposed to be subordinate to the Ambassador, the Station Chief receives orders directly from CIA Headquarters—some of which are not shared with the Ambassador. Nor is CIA alone in this. Each agency has both communication and command channels to Washington that are simply not shared with the Ambassador. In Panama, this problem was compounded by the existence of independent entities such as SOUTHCOM and the PCC Administration. The CINC worked for the Secretary of Defense and normally communicated through the Chairman of the Joint Chiefs of Staff. Staff sections communicated with their counterparts in the Joint Staff. Service Components communicated with their Service Chiefs. For example, the Army South commander, Major General Bernard Loeffke, was in regular communication with Army Chief of Staff, General Carl Vuono. Similarly, the PCC Administration communicated with the Commission in Washington. So, each institution, and nearly every component within those institutions, had its own lines to somewhere in Washington both for normal communication and to receive direction. Lateral communication—that is, between components and institutions—was haphazard, at best. Indeed, interagency coordination in Panama was only a sometime thing.

Issues & Efforts Requiring Interagency Coordination During the Crisis

In these circumstances, there were any number of events that produced issues that clearly begged for interagency coordination, both in Washington and on the ground in Panama. These issues will be addressed in this section.

Indicting Noriega

The Panama crisis began with Noriega's firing of his second in command, Colonel Roberto Diaz Herrera, in June 1987. Diaz Herrera struck back with allegations of drug trafficking by Noriega as well as far-fetched accusations that he had been responsible for the death of Omar Torrijos in 1981. These allegations provoked the opposition National Civic Crusade (NCC) to mount street demonstrations that Noriega suppressed ruthlessly. In response, the Reagan Administration approved some minor economic sanctions including withholding US military assistance to the PDF and cancelled the annual combined defense of the Canal exercise, KINDLE LIBERTY. For the most part, however, the conflict was internal to Panama with the US supporting the NCC's goals. At the same time, both the CIA and DEA continued their cooperative relationships with Noriega and the PDF. The DEA chief in Panama, Alfredo Duncan, was reported to have acknowledged that Noriega had some "questionable" drug trafficker associations but he also said that Noriega's cooperation was of much greater value.[4] Nevertheless, in the fall of 1987 two US Attorneys in Florida—one in Miami and the other in Tampa—began Grand Jury investigations of Noriega's involvement in drug trafficking.[5] There is no evidence that these Grand Jury investigations were coordinated with, or even known to, the US Attorney General, Edwin Meese. There is solid evidence that the investigations were totally unknown to the State Department.

On 5 February 1988, while General Woerner was visiting Assistant Secretary of State Elliott Abrams, the latter received a phone call. As Woerner recalls it, "Abrams turned white as a sheet." When he hung up the phone, he turned to Woerner and said, "They just indicted Noriega!" The General was completely convinced that Abrams was as clueless as he was.[6] As a result, Woerner immediately directed his staff to begin planning for operations in Panama where the PDF was the enemy.

Economic Sanctions: The "Battles" of the Electric Bills

As the Panama crisis deepened throughout the spring of 1988, the US government chose to keep tightening the economic screws on Noriega through more sanctions. Interagency battles in Washington resulted in policy decisions for ever greater sanctions.[7] The sanctions did raise the pressure on the PDF and, on 16 March 1988, a group of officers attempted a coup. It failed. On 8 April the Reagan Administration invoked the International Emergency Economic Powers Act of 1977 to put Noriega under new and greater sanctions. The Presidential Executive Order prohibited "American citizens and companies operating in Panama from paying any taxes or fees owed the Panamanian government."[8] This sanctions policy was opposed by American businessmen in Panama as well as by the US Treasury Department which put up bureaucratic obstacles to its implementation.[9]

Nevertheless, the American Embassy was fully supportive and immediately decided not to pay the electric bills on the apartments it rented for all US government personnel assigned to the Embassy. In addition to those official Americans, there were any number of US citizens who worked for SOUTHCOM and its components as well as the PCC who lived in apartments and houses, rented or owned, in Panama City. These Americans also owed the Panamanian government for their electricity. Unlike the Embassy, these Americans simply ignored the order.[10]

The Embassy, on the other hand, chose to ignore the electric bill when it came.[11] The Ambassador, DCM, and Administrative officer apparently believed that there was no way the government electric company would cut off so many individual apartments rented to the Embassy for assigned personnel. When the due date came and no payment was forthcoming, the electric company issued a warning—pay by 21 April or electricity would be cut off. 21 April 1988 was a Friday. It came and went. Electricity was not cut off so there were celebrations by the Embassy leaders; nothing would happen until

Monday. Early Saturday morning, however, electric company trucks were observed outside the luxury apartments in Punta Paitilla where Embassy personnel were housed; the crews were removing electric boxes cutting power to the appropriate individual apartments!

The Administrative officer moved quickly to get all the Embassy personnel rooms in the best hotel in Panama, the Marriott. Then, he began work on a longer term solution. He found an apartment hotel (Aparthotel) that would rent small suites and he paid a month's rent up front. Personnel were notified and went to check the place out—they were appalled. Personnel from agencies other than State Department complained to their parent agencies and were told that they could remain in the Marriott. Then, the State Department personnel objected to the double standard and the Embassy backed down and paid for rooms at the Marriott for a month! At that point, the Embassy and the US government capitulated and agreed to pay the electric company, figuring that it would take some time to restore the electricity to the apartments. Of course, Noriega's electric company restored the electricity within 24 hours and the US government was stuck with bills for two hotels—already paid—and the sanctioned electric bill, now paid! Clearly, this was evidence of a failed interagency process, at best, and no process, at worst.

Planning for Intervention

When General Woerner returned to SOUTHCOM from his meeting with Elliott Abrams where they were stunned with the news that Noriega had been indicted, he directed his Operations Directorate (J3) to begin planning for action against the PDF. At the same time, he requested from the Chairman of the Joint Chiefs of Staff (CJCS) direction to begin the same planning. On 28 February he received the order to plan and to do so using the Crisis Action Planning process of the Joint Operational Planning System (JOPS), then in effect. This gave the planning process a sense of urgency that the Deliberate Planning process did not have. The J3 planners produced a four phase plan which they briefed to General Woerner on the first Saturday in March. When the briefing ended, Woerner asked where Phase V—the postconflict phase was. Phase V planning was initiated the following day by the Civil Affairs section of the Policy, Strategy, and Programs Directorate (J5).[12] The resulting five phase plan was code-named, ELABORATE MAZE. The only coordination done was internal to DOD—especially between SOUTHCOM and its subordinate components and SOUTHCOM and the Joint Staff.

By summer, the Joint Staff decided that the five phase plan should be divided into a series of separate individual plans that Woerner directed be capable of execution sequentially, simultaneously, or independently, or any combination of these—as he had done for the phases of ELABORATE MAZE. Collectively, the plans were referred to as THE PRAYERBOOK; it was made up of POST TIME—the buildup of forces; KLONDIKE KEY—a noncombatant evacuation operation or NEO; BLUE SPOON—defensive and offensive operations (Phases III and IV of ELABORATE MAZE); and KRYSTAL BALL (later changed to BLIND LOGIC)—postconflict operations. Of these, only KLONDIKE KEY was coordinated and shared with the Embassy.[13] This was because Embassy personnel were among those Americans who were to be evacuated in the event of a NEO.

On 18 May 1989, General Woerner was briefed on the status of BLIND LOGIC. This briefing had become necessary due to a reorganization of the SOUTHCOM staff which had transferred the Plans Division from the J3 to the J5 and the Civil Affairs section from the J5 to the J3. Discussions among staff officers had resulted in the proposal to leave BLIND LOGIC in the J5 under the Chief of the Policy and Strategy Division where it would be revised and updated.[14] Upon reviewing the plan, it immediately became clear that BLIND LOGIC was, in the words used at the time, "treading all over State Department turf." As a result, he requested from his superior, one of three Deputy J5s, permission to brief BLIND LOGIC to the Embassy Political Counselor and discuss its implications with him. In his opinion, this was a "no brainer"—not only did the Political Counselor have a need to know but he was also a personal acquaintance. The answer was a surprise; he was told, "Not only no, but hell no!" and informed that the plan was held exclusively within JCS channels. The planner persisted and extracted permission from his boss to visit the Political Counselor, sound him out on some of the issues, but always talking "around the plan."[15] One positive result of the conversation was that the planners incorporated the Embassy assumption that the PDF would be disbanded. What they did not know, and, therefore, could not include was the intention to inaugurate as President and Vice Presidents the winners of the elections of early May 1989, an intention that vitiated a principal planning assumption for postconflict, short term military government.

Fissures, the Kozak Mission, and Fissures II

Assistant Secretary of State Michael Kozak began a series of shuttle trips to Panama in mid-March 1988, shortly after the coup attempt on 16 March. Kozak had a long history of involvement with Panama dating back to the successful Panama Canal treaty negotiations in 1977 during the Presidency of Jimmy Carter. Indeed, Mike Kozak had developed a good relationship with then Panamanian dictator, General Omar Torrijos, and had come to know then Colonel Manuel Noriega. As a result, Kozak became the point man for the Reagan Administration's diplomatic efforts to get Noriega to resign and go away, thereby ending the crisis. On his first visit, there was no effort to coordinate in any way with SOUTHCOM. Indeed, the command was kept completely in the dark. On his second trip, however, in mid April, SOUTHCOM had been sent a classified message with US policy goals for Panama that had originated with White House Chief of Staff Howard Baker.[16] Those policy goals were: protect US citizens, defend the Canal and US installations, make certain the US was in the best possible position to deal with a post Noriega government, and the removal of Noriega from power.[17]

With this guidance, such as it was, General Woerner convened a series of inter-agency workshops at his headquarters in Quarry Heights. Attending were people from the Embassy, CIA, State, and SOUTHCOM staff. The purpose was to develop an interagency strategy that would assist US efforts to achieve those goals in the event that the Kozak-Noriega negotiations failed, as seemed likely.[18] The discussions at Quarry Heights produced a degree of consensus among the participants on actions to take that would most likely wean the PDF officer corps away from Noriega in the interest of saving their institution. General Woerner and his staff worked the agreed upon actions into a plan that the General called Fissures. When Kozak's negotiations with Noriega failed, as expected, Woerner sent the Fissures plan through JCS channels with the admonition that it needed to be executed as a fully coordinated effort. If it were to be executed in pieces, it would fail. When orders to execute were finally received, they came in exactly the piecemeal manner that General Woerner feared.[19] Indeed, because those orders came from JCS, they were military only—there is no indication that the JCS had made any effort to coordinate the Fissures plan outside the Joint Staff, let alone the civilian side of DOD or its interagency partners.

A little over a year later, General Woerner decided to try again with his Fissures concept. Revising the Fissures plan to take account of the changes of the past year, he forwarded Fissures II to Washington. Fissures II suffered the same fate as its predecessor, no interagency involvement and direction to implement individual pieces.[20] Apparently, neither Washington, under both Reagan and George H. W. Bush, nor the interagency players in Panama were able or willing to work together at the decision maker level to produce a coordinated strategy to address the disparate issues of the Panama crisis.

An Issue of Perception—Drugs

The indictment of General Manuel Noriega in February 1988 for drug trafficking changed the Panama crisis from being primarily internal to one between the US and Panama. But it also revealed deep rifts within the US government over Noriega's and the PDF's roles in the illicit drug trade. Interestingly, these rifts were not entirely closed until approximately the end of Operation JUST CAUSE on 31 January 1990—41 days after it had begun. The positions of the most relevant US agencies with respect to Noriega's drug dealing are detailed below.

The State Department position from the perspective of the American Embassy in Panama was that Noriega was intimately involved with the Colombian drug cartels. However, this was merely one more count in the Embassy indictment of the General. In addition, the Embassy was convinced that the entire senior leadership of the PDF was dirty, and it was not far off the mark. Still, drugs were seen as a tool to move the Reagan Administration and its successor toward active opposition to Noriega's continued dominance of Panama's government and politics. The drug issue was instrumental as far as the Embassy was concerned.

For the CIA Station, the drug issue was something that got in the way of their real business. CIA was concerned with Cuban and Nicaraguan involvement with the insurgencies in Central America. Noriega was obviously a less than sterling individual but he had been a paid source for many years. From the time he became G2 of the Panamanian *Guardia Nacional* through to his current tenure as Commander of the PDF, he was head of a Liaison Service. CIA funding went to the PDF G2, not personally to Noriega. Clearly, nobody in the Station believed for a moment that Noriega did not take his cut (and give out cuts to other PDF leaders) but he was not directly on the payroll any longer and had not been since soon after the coup in 1968 that brought General Torrijos to power. That said, the CIA valued Noriega's cooperation in providing both information and assistance in conducting other intelligence related operations from Panama. What, then, was a little drug trafficking when there were bigger fish to fry?

Within the regional DEA office in Panama there was considerable debate and disagreement regarding Noriega. The DEA chief, Alfredo "Freddie" Duncan, had expressed the opinion that despite his culpability in drug trafficking, Noriega was worth far more to the US as a source of information on the Colombian cartels' activities than in jail as a convicted trafficker. Moreover, Duncan remained of this opinion long after Noriega had been indicted. Others in the office were not so sure as their boss. While they all agreed that Noriega was dirty and that he cooperated with DEA, these agents believed that the General was basically feeding them information about those cartels that were not cooperating with him. Thus, he was playing DEA (and, not coincidentally, Duncan) for fools.

SOUTHCOM, despite having been tasked during the Reagan Administration with the conduct of an interagency counterdrug mission in Bolivia—Operation BLAST FURNACE (1986)—and by the Bush Administration with the monitoring and interdiction of drug smuggling from Colombia to the US, generally took the position that addressing Noriega's and the PDF's roles in the drug trade was not its job. The SOUTHCOM staff was well aware that Noriega was up to his ears in the drug trade but SOUTHCOM's mission did not involve what Noriega was doing on that front. Rather, SOUTHCOM had to address the PDF as, first, an ally in the defense of the Canal and the support it gave to the command's activities in other parts of Central America, and, then, as a potential threat to the Canal, US defense facilities and personnel, and the US civilian community in Panama. With all this, as long as Noriega was not directly engaged in drug smuggling, his other drug trade activities simply were not part of the command's portfolio.

Finally, the PCC Administration was focused exclusively on the effective and neutral operation of the Canal. What Noriega did, or did not do, with regard to the drug trade just was not the business of the PCC. Its attitude was, "Hey, leave us out of this!" In short, finding common ground, or even common perceptions, regarding the drug trade among and within the US government agencies in Panama was a losing proposition.

Operations JUST CAUSE and PROMOTE LIBERTY

Operation JUST CAUSE began at 2345 hours on 19 December 1989. Operation PROMOTE LIBERTY began at 1000 hours on 20 December 1989. Even before JUST CAUSE was launched, an event of critical importance took place at Quarters 25 on Fort Clayton, a US Defense Site. That was the swearing in as President of Panama, Guillermo Endara, his First Vice President, Ricardo Arias Calderon, and the Second Vice President, Guillermo "Billy" Ford, by a Panamanian Justice of the Peace. This event trampled on a primary assumption of the postconflict plan, BLIND LOGIC, which had assumed that the CINC would be the military governor of Panama for a period of about 30 days following a US intervention. Clearly, the decision to inaugurate the new government had been made sometime prior to the decision to execute; clearly, the State Department was aware. Who in the military knew prior to the evening of 19 December is an open question. Even more intriguing is the question of who in SOUTHCOM knew prior to 17 December when the decision to execute BLUE SPOON/JUST CAUSE was made by President Bush. As of 15 December, nobody with responsibility for BLIND LOGIC had any idea that there would be a new Panamanian government sworn in before the operation began.[21] As was true throughout the crisis, there was a dearth of interagency coordination even on the eve of launching an invasion.

The immediate question for President Endara and his Vice Presidents was how they were supposed to begin governing the country. While they had won the 7 May 1989 elections, Noriega had stopped the official count and annulled the vote. The last gasp for the winners was the street demonstration on 10 May where all three had been attacked by Noriega's thugs in the Dignity Battalions and the PDF. Endara and Ford had been injured while Ford's bodyguard was killed. From that moment on, these three had no expectation of governing Panama. Then, nearly 10 months later, they were informed that they would be sworn in and were expected to begin governing Panama the next day. Early on the morning of 20 December, the government of Panama opened for business in the Legislative Assembly building. It was constituted by three men, one the President and two Vice Presidents. Advising the government was the new DCM/Charge d'Affaires at the American Embassy, John Bushnell. At 1000 hours, General Thurman told his J5, Air Force Brigadier General Benard Gann, to "Get down to the Legislative Assembly and keep Mr. Bushnell out of trouble."[22] That, in effect, was the order to execute OPORD BLIND LOGIC as Operation PROMOTE LIBERTY.

One of the most important parts of Operation PROMOTE LIBERTY was providing security to the Panamanian public. This was achieved by reconstituting the Panama National Police (PNP) and its sister institutions of the Panama Public Force. To advise and assist the new police force in becoming organized, trained, and operational, Army Major General Marc Cisneros, Commander of US Army South, established the US Forces Liaison Group (USFLG). The FLG was treated by the Charge d'Affaires, Mr. Bushnell, as a member of the Country Team at the Embassy. Early on, Bushnell requested the members of the FLG to stop wearing their uniforms and to work in civilian clothes in order to influence the PNP's self perception that it was to be a civilian, not a military, police force. Overnight, the uniforms disappeared at the offices of the FLG. At another meeting of the Country Team, Mr. Bushnell asked the FLG to determine whether there were Reserve Component soldiers who were police officers in their civilian jobs who could be assigned to Panama. The answer was positive, and what were called the "RC Cops" began arriving in short order where they were teamed with Army Special Forces as advisor/monitors in the PNP precincts. Clearly, Bushnell's decision to treat the FLG as a member of the Country Team had the effect of significantly improving interagency coordination in Panama during the first six months of Operation PROMOTE LIBERTY.

In February 1990, the US Congress passed the Emergency Assistance to Democracy in Panama Act. This legislation gave the interagency State/Justice Department International Criminal Investigative Training Assistance Program (ICITAP) full responsibility for training the PNP and other police entities. ICITAP arrived in Panama with three USG employees, its Director (a senior FBI Agent), his deputy (also an FBI Agent), and the State Department representative on its Directorate. All its "worker bees" were former FBI Agents (to include retired Assistant Directors of the FBI) now employed by ICITAP's contractor, Miranda Associates. The ICITAP Director had been an Army Special Forces officer as a young man and had worked with a junior officer in the Panamanian *Guardia Nacional*, who now was the Director of the PNP. In anticipation of a smooth hand off to ICITAP, the FLG established an office for the ICITAP Director to share with the Chief of the FLG. Since the nominal chief of the FLG was General Cisneros, for practical purposes, the office would belong to ICITAP.

From the moment they arrived, ICITAP tried to have as little to do with the FLG as possible. They never occupied the office prepared for them, preferring to commandeer other space to do their business when it was necessary to be at PNP headquarters. Otherwise, ICITAP operated out of the Marriott Hotel.[23] At the hotel, they held classes for some of the PNP in classical lecture format with interpretation—only one of the Miranda contractors was a Spanish speaker. They also launched plans to establish a model police precinct; the model precinct never got off the ground. Meanwhile, it was essential to have the new police take to the streets. The FLG found them interim uniforms, ordered new khaki police uniforms through the Army/Air Force Exchange System, arranged for a purchase of trucks from US military stores, and developed and implemented a 20-hour course that was given to every member of the PNP by April of 1990. Not until July would ICITAP have a training course in place. From February until July, ICITAP was represented in Panama by its Director for two weeks and then by his deputy for two weeks. Only in July did ICITAP bring permanent government staff down and begin their training program. After six months, the FLG was finally able to hand over police training responsibilities to ICITAP. The conclusion to be drawn from these episodes is that where there was the will, on both sides, to make something work —as there was with the FLG working with the Embassy—then interagency coordination and cooperation could be effective. Where the will to make it work was one sided —despite good will at the worker level—there was little or no interagency coordination or effective systemic cooperation.

Is There Any Evidence That We Learned Anything In Panama?

The status of interagency coordination during the Panama crisis and Operations JUST CAUSE and PROMOTE LIBERTY was practically nonexistent. The process in Washington was very much in the same state of flux that it had been in since the passage of the National Security Act in 1947. Only in 1989 did the new Bush Administration begin to institutionalize what has become the current NSC system of interagency coordination. During the last parts of the Panama crisis, that system was just beginning to take hold. In Panama, by contrast, there was even less of a system of interagency cooperation and coordination—as this paper makes clear. In short, the glass (as they say) was very nearly empty.

Two decades later, depending on one's perspective, the glass is only half empty or half full. The Washington process has been relatively well institutionalized. However, the process in the field still has a long way to go. Although mechanisms, often "work arounds," have been developed to address issues of interagency concern, they still depend, far too much, on the personal chemistry of the principals. If it is good, as it was in the case of General David Petraeus and Ambassador Ryan Crocker in Iraq, it can be very, very good. If, on the other hand, it is bad, as it was in the case of Lieutenant General Ricardo Sanchez and Coalition Provisional Authority (CPA) Administrator, Ambassador L. Paul "Jerry" Bremer, also in Iraq, then it is horrid! While we have clearly learned from the Panama experience, we still have much more to learn.

Notes

1. The author wishes to acknowledge here an intellectual debt to two friends and colleagues, Gabriel Marcella and Larry Yates. Gabriel, as International Affairs Advisor to USCINCSO, General Fred F. Woerner, played a major, if largely unsung, role in these events. His writing, both alone and with General Woerner, has given me the benefit of insight from levels above my position. Gabriel has been generous with both his advice and friendship. Larry Yates' recently published official history of the Panama crisis is likely to be the definitive work on the subject. Suffice that I have benefited greatly from Larry's knowledge and insights over the last 19 years where we have shared data and experiences—at least once over pizza which he brought to an interview.

2. Author's conversations with several individuals assigned to the Embassy at the time and with senior staff at USSOUTHCOM.

3. Ambassador David Passage has described this process in a number of lectures given on regular occasions to the Command & General Staff College at Fort Leavenworth, KS, throughout the 1990s.

4. Author's conversations with several DEA agents assigned to Panama in 1987 and 1988.

5. One of those US Attorneys was Dexter Lehtinen, husband of then Florida Legislator and now US Congresswoman, Ileana Ros-Lehtinen.

6. Author's interview with General Fred F. Woerner, Jr., Boston MA,

7. See Lawrence A. Yates, *The US Military Intervention in Panama: Origins, Planning, and Crisis Management June 1987–December 1989* (Washington, DC, 2008: Center for Military History), pp. 40-43.

8. Yates, p. 63.

9. Ibid.

10. The author writes from personal experience; he was not going to be without electricity for cooking, hot water, or air conditioning with daytime temperatures in the 90s.

11. The source for the paragraphs that follow is the author's observations at the time and contemporaneous conversations with friends in DEA and the USMC who were stationed at the Embassy and lived in Embassy rented housing.

12. See the author's, *The Fog of Peace: Planning and Executing the Restoration of Panama*, (Carlisle, PA, 1992: Strategic Studies Institute).

13. Several years later, the author confirmed this with Ambassador John Maisto who had been DCM in Panama at the time.

14. The author was the Chief of Policy and Strategy at the time; his civilian education and experience made him particularly well qualified to take charge of the plan.

15. Later, in an interview with the former Political Counselor, Mike Polt, the author asked him if he recalled their strange conversation prior to JUST CAUSE. When Polt said he did, the author told him that this (postconflict planning) was what it had been all about!

16. Yates, p. 98.

17. Ibid.

18. Ibid.

19. Yates, 99-100, Woerner interview.

20. Yates, 192-195, Woerner interview.

21. The author can make that statement with absolute certainty because up until 15 December 1989, he was responsible for OPORD BLIND LOGIC. There is no evidence that any changes were made before the afternoon of 17 December because the review of the plan began on the morning of that day.

22. The official version of this statement, which the author published in his monograph, *The Fog of Peace* (Carlisle, PA: 1992, SSI) read "support Mr. Bushnell." The reason was to avoid embarrassment to all concerned, but nearly 20 years later, it is hard to imagine any lingering embarrassment for the individuals involved. General Thurman long since has passed away and General Gann retired. So, the time has come to tell the story exactly as it was without the gloss.

23. This is not to say that good relations were not established with members of ICITAP's contractor team. Indeed, they were, but relations between the two organizations per se were anything but harmonious. Author's participant observation.

Panel 1—The Difficulties in Interagency Operations
(Submitted Paper)

Interagency Actions and the US Intervention in Lebanon, 1958

by

Lawrence A. Yates, Ph.D.
US Army Center of Military History

In mid-July, 1958, President Dwight D. Eisenhower ordered nearly 15,000 US combat troops into Lebanon in an effort to prevent that country from falling to what Washington perceived as radical Arab elements operating in the Middle East. What I would like to do this morning is examine certain policies and decisions affecting the intervention in Lebanon from the perspective of interagency processes within the executive branch of the US government, starting with the National Security Council (NSC).

In the American experience prior to World War II, what later came to be categorized as national security issues had been addressed mainly by the President working with the State, War, and Navy departments; or with various advisers, specialists, and friends. This interaction occurred in cabinet sessions, personal meetings, and interdepartmental correspondence and communications. The scope of the Second World War, the complex issues it left in its wake, and the great power responsibilities it thrust upon the United States caused several high-ranking officials to advocate supplementing these methods with a formal, statutory organization that would advise the President on national security matters and promote interagency coordination and cooperation on a systematic and ongoing basis, in a way the State-War-Navy Coordinating Committee had sought to do during the last year of the war.[1]

The result was the National Security Act of 1947, which established the National Security Council, together with the Central Intelligence Agency, the National Military Establishment (later renamed the Department of Defense), and the short-lived National Security Resources Board. After amendments made to the act in 1949, NSC membership included the President and Vice President, the Secretaries of Defense and State, and the NSRB (later the Office of Defense Mobilization) director. Also attending meetings as advisers were the director of Central Intelligence and the chairman of the Joints Chiefs of Staff (JCS), with Eisenhower adding the Secretary of the Treasury, the Mutual Security Administration director, and others, when needed, to the list. The legislation establishing the NSC called on it to "advise the President with respect to the integration of domestic, foreign, and military policies relating to the national security so as to enable the military services and the other departments and agencies of the Government to cooperate more effectively in matters involving the national security."

In this capacity, the NSC was to "assess and appraise the objectives, commitments and risks of the United States in relation to our actual and potential military power."[2]

Under Harry Truman, the NSC produced coordinated policy papers (for example, the landmark NSC-68 of 1950 that charted a new course for America's containment of "international communism"), but the President himself rarely attended its meetings until the Korean War compelled him to seek the council's advice. In contrast, his successor, President Eisenhower, in the words of one historian, "came closer to implementing the NSC as it was originally conceived than any of the Presidents who followed him." The staff process embedded in the council's operations readily appealed to the former general's military mindset, even though he believed the organization had become "moribund" under Truman's stewardship.[3]

To revitalize the NSC, Eisenhower made lawyer and businessman Robert Cutler his special assistant for national security (the forerunner of the national security adviser position created by President John F. Kennedy). Cutler addressed his assignment by developing "Policy Hill," a process that would permit the NSC to perform its advisory mission in a systematic and efficient way. At the bottom of the "hill" were officials in each of the participating departments and agencies, writing policy recommendations that were then forwarded to a Planning Board that generally included departmental assistant secretaries and officials of equivalent rank from other agencies, with the group chaired by the NSC's executive secretary. The board tried (though not always successfully) to resolve interagency differences before moving the revised papers up the hill to the NSC, which, during Eisenhower's second term, generally met once a week around 0900. NSC members and others present would discuss and debate the papers, with the President—who chaired well over 300 of the 346 NSC meetings during his two terms—then approving a policy or authorizing follow-on actions he wanted taken. His decisions then moved back down Policy Hill to departmental deputies and assistant secretaries on the Operations Coordinating Board, a bureaucratic umbrella for over forty interagency working groups based on countries, regions, and subjects. The job of the OCB was to ensure that NSC decisions were properly coordinated and implemented at all levels of government. While the board never lived up to Ike's expectations in this respect, it did serve as another venue for interagency communication and the sharing of information.[4]

This, in general, was the NSC setup in effect during the Middle East crises of the mid- to late-1950s, and during the Lebanon crisis of 1958, in particular. Time constraints and the focus of this paper do not permit an in-depth assessment of these crises,[5] save to say that, from the administration's perspective, the ramifications of each could be felt at three levels: international, regional, and local. The Cold War, which by the 1950s was defined in general terms as the struggle between the "free world" and "international communism," provided the international context for the crises. From 1945 to 1954, this ideological and geopolitical struggle had been largely confined to Europe and Asia. In the mid-fifties, the Middle East entered the picture, as the Soviet Union sought to extend its influence into the area, first, by taking the Arabs' side in their conflict with Israel and, second, by offering military and economic aid on a selective basis. At the regional level, the principal Middle Eastern recipient of Soviet largess was the Egyptian government of President Gamal Abdel Nasser. In the few years since 1952, when he had been one among the group of "Free Officers" that had overthrown the pro-Western government of Egypt's King Farouk, Nasser had become the country's strongman and a popular and vocal proponent of Pan-Arab Nationalism, a position saturated with anti-Western, anti-colonialist sentiment. Some early efforts by the Eisenhower administration to work with Nasser had proved productive, but were soon overshadowed by the Suez war of 1956 and Nasser's efforts to undermine the region's remaining pro-Western governments in Saudi Arabia, Iraq, Jordan, and Lebanon. Neither Eisenhower nor his Secretary of State, John Foster Dulles, believed Nasser to be a communist, but both saw him as a destabilizing influence in an area of strategic importance to the United States, and as a conduit for Soviet penetration of that area.

Of the governments Nasser had targeted, some were already unstable, as was the case with Lebanon in 1957 and 1958. There, a tenuous political structure had fractured along confessional, geographical, and family lines. In accordance with the National Pact of 1943, the Lebanese president was a Maronite Christian, the prime minister a Sunni, and the president of the parliament a Shiite. Demographic changes over fifteen years, president Camille Chamoun's use of electoral fraud in 1957 to deny his rival *zuama* (strongmen) the seats in parliament essential to maintaining their patronage systems, his openly pro-West positions, and his attempt to amend Lebanon's constitution to allow him a second term in office combined to produce a volatile situation that exploded into open violence in May 1958. Sides in the fighting generally—but not entirely—reflected the country's Muslim-Christian divide. While Eisenhower and Dulles had qualms about Chamoun, the fact that many of his opponents embraced Nasser's Pan-Arabism concerned them more. To prevent a radical, anti-Western Arab regime from taking power in Beirut, the administration was prepared to use military force, but only as a last resort.

Against this backdrop of ominous change and upheaval in the Middle East, Eisenhower turned to the National Security Council in February 1957 and, again, in July to review US policy in the region. In January 1958, he approved the coordinated recommendations presented by the NSC's interagency Planning Board. Designated NSC 5801/1, the policy paper, which reflected Washington's perception of the new realities in the Middle East, set forth four US objectives: (1) maintaining the availability of resources, strategic positions, and passage rights of the Near East, and denying these to Soviets; (2) maintaining stable and friendly governments in the region; (3) achieving an early resolution of Arab-Israeli dispute; and (4) limiting Soviet influence. With respect to Nasserism, the United States would seek to avoid confrontation while hoping to guide "revolutionary and nationalist pressures" into channels not hostile to the West. In the process, pro-Western governments might fall, but a "neutralist orientation" by Arab states would be acceptable, provided that it was "reasonably balanced" by relations with the West.[6]

At the outset of this policy review, the NSC had called on the Pentagon to assess the military implications of the US position in the Middle East. Specifically, the Joint Chiefs were directed to report on the status of planning for three scenarios: global war, another Arab-Israeli war, and other contingencies. In June 1957, the Joint Chiefs responded that it was impossible to plan for the latter scenario—there were just too many possibilities. In general, however, they declared that small, mobile, nuclear capable forces from Europe could handle contingencies in the Middle East, short of a major war, requiring the introduction of US troops.[7]

At one point in the Middle East review process, the President demonstrated how he could use an NSC meeting to modify the Policy Hill process to suit his needs. The meeting in question took place on 18 July 1957, with Ike expressing concern that the NSC Planning Board was inappropriately getting involved in contingency planning. To elevate that effort to a higher interagency level, he instructed the Secretaries of Defense and State, the chairman of the JCS, the CIA director (Foster Dulles' brother, Allen), and Cutler to meet to discuss the range of possible US military operations in the Middle East. Soon thereafter, this group of principal advisers met and considered six courses of action, from deterrence to all out intervention. On 8 August, the NSC reviewed the group's findings, then passed them to the Planning Board for incorporation in subsequent drafts of what would become NSC 5801/1. It is possible that the meeting of the small group from State, the Pentagon, and the CIA also opened the door to further interagency cooperation on contingency planning, such as when Dulles queried the relatively new JCS chairman, Air Force General Nathan Twining on 17 October 1957, as to what forces the United States could put into Lebanon or Jordan in 24, 48, and 72 hours, respectively, for the purpose of establishing "the authority of the friendly local government and to help maintain order." Twining's response provided Dulles with all the information the Secretary had requested.[8]

The following May and June, as the political crisis in Lebanon escalated into armed conflict and president Chamoun made his initial request for US military intervention, CIA Director Allen Dulles included updates about the worsening situation in his regular briefings to NSC attendees. But the NSC was not designed to deal with crisis management, as opposed to policy issues and military planning. Thus, as the conflict became more acute, almost all of the interagency activities and operational decisions related to it took place outside the NSC in more traditional modes that included a myriad of personal meetings, interdepartmental memoranda, cables, and telephone calls. This limited utility of the NSC in a crisis was no better illustrated than by the decision itself to intervene in Lebanon.

On 14 July, a group of "radical" army officers in Iraq overthrew the pro-Western government there, in the process killing the royal family and the prime minister. As word of the bloody coup reached Washington early that morning, Eisenhower conferred repeatedly with the Dulles brothers, General Twining, and others over the telephone. The weekly NSC meeting was already on the President's schedule for 9:45 that day, with the main item on the agenda being a discussion of some NSC-directed

Civil Defense studies. Over the phone, Foster Dulles made the obvious observation to Eisenhower: the Iraqi coup and the possibility that it would trigger the downfall of pro-Western governments in Lebanon and Jordan were "more important" issues than the topics before the NSC. Ike agreed, but rather than alter the council's agenda or cancel the meeting, he simply sat through the Civil Defense reports until Dulles arrived at the White House from the State Department, where he had been meeting with State, Pentagon, and CIA officials. At that point, the President adjourned the NSC and convened a meeting with his principal national security advisers in the Oval Office. That was the forum in which he would make one of the more critical foreign policy decisions of his second term.[9]

During the Oval Office session, the Dulles brothers, General Twining, and others briefed the President. From the outset, all agreed on the need to accede to Chamoun's renewed request to send US troops into Lebanon. As more than one person noted, US interests in the region would suffer more from inaction than from intervention, even though the President's key advisers acknowledged that they did not know precisely, or even generally, what US forces would need to do to stabilize Lebanon, or when and under what circumstances the troops would be withdrawn. That notwithstanding, Eisenhower made his decision, and the next afternoon, Lebanon time, the first US Marines began landing south of Beirut. A few days later, Army units from Germany would join them.[10]

In the weeks that followed, Allen Dulles continued to include Lebanon updates in his briefings that opened each NSC meeting. On 24 July, in the first council meeting after the intervention, both Dulles brothers gave detailed reports, which resulted in extensive discussion and a directive for the Planning Board to prepare "a list of relevant policy issues arising out of the present situation in the Near East, together with arguments for and against taking various possible courses of action."[11] Still, despite this use of the NSC as a forum for discussing the crisis, the critical interagency actions taken to effect an acceptable resolution to the intervention continued to take place outside the council.

In retrospect, one of the most important decisions during the intervention was Eisenhower's approval of Foster Dulles' recommendation, made during an Oval Office meeting, to send diplomatic trouble-shooter Robert Murphy to Lebanon. Ironically, both the President and the Secretary of State believed it would be a short visit, about one week, designed to patch up some serious problems caused by the perceived failure of US personnel on the ground to implement a State Department-Department of Defense agreement reached in Washington the previous month. The problem surfaced when the lieutenant colonel commanding the first Marines to land south of Beirut confronted not Lebanese rebels but two uniformed US defense attachés from the American embassy who conveyed to him Ambassador Robert McClintock's instructions for the troops to get back aboard their ships and disembark at the port of Beirut instead. The Marine officer refused the order, citing that the ambassador was not in his chain of command. Unfortunately, the US military commander for the operation, Admiral

James A. Holloway, Jr., was still en route to Lebanon and without the communications needed to resolve the problem before his arrival. When he did arrive the next day, he and the ambassador crossed paths near a road to Beirut, where the Marines were about to engage in what would have been a politically disastrous firefight with the Lebanese army. In a meeting to break the impasse and defuse the situation, Holloway talked tough to the Lebanese commander, but ended up accepting the more conciliatory suggestions made by the ambassador. For McClintock, this incident reconfirmed his belief that the military was not going to honor the aforementioned agreement that, according to the ambassador, gave him the final word on any military movements that could have political consequences—meaning virtually all military movements. McClintock had already complained to Washington, and Dulles responded by dispatching Murphy to "establish better relations *between our own military and diplomatic people*, Lebanese military, and the Lebanese government."[12]

By the time Murphy arrived in Beirut, he found that McClintock and Holloway were working well together. The "interagency" crisis in the field had been resolved. Then, as Murphy's "visit" turned into a full-scale diplomatic effort to end the crisis—an effort backed by US and Lebanese military force and the support of the administration in Washington—the harmonious relationship that these three US representatives on the scene developed over several weeks proved essential in facilitating the process. It was a highly coordinated undertaking: McClintock and Murphy kept Chamoun in line, while Holloway made sure that the actions of American troops ringing Beirut did not undermine negotiations but rather serve to further them. Murphy also traveled throughout the country, reassuring key rebel leaders that the United States was not trying to prop up Chamoun, a pledge that he redeemed by helping to engineer the election of the Lebanese army commander as Chamoun's successor, thus effectively reconciling most of the warring parties in the country's internal conflict. Toward the end of his mission Murphy visited the new leaders in Baghdad, and Nasser in Egypt, finding that the former posed no imminent threat to US interests and that the latter sought to play a constructive role after having failed to convince the Soviet Union to intervene militarily on his behalf. In October, the last of the US forces withdrew from Lebanon, leaving behind a relatively stable situation that would prevail for almost twenty years.[13]

According to one historian, President Eisenhower and his staff insisted throughout his administration that "the President made his most critical national security policy decisions through the NSC." The truth, she goes on to note, is that "We now know that Eisenhower's NSC was just one part of a multifaceted foreign policy process."[14] As that multifaceted process relates to the topic of this paper, it seems to have worked very well. While the NSC chaired by the President was instrumental in developing a sophisticated if ethnocentric policy toward the volatile Middle East that sought to employ resources from the various governmental agencies and departments associated with national security, outside the Policy Hill framework a plethora of interagency activity also produced positive results, at least so far as keeping Arab nationalism at bay and maintaining a fairly stable, pro-Western government in Lebanon. The question, which must be answered briefly here, is "Was this success the result of interagency procedures that Eisenhower enacted? Or was it the result of other, less tangible factors specific to the time and the place?" The answer, as one might expect from a historian, is both.

The Policy Hill process, in which the President regularly approved policy and authorized actions in the presence of his key national security advisers, not only made the NSC responsive to Ike's needs in foreign policy, it also facilitated interagency activity in more traditional arenas. In the case of the Middle East crises, especially during 1957-1958, the NSC setup worked well, even though it fell far short of eliminating all interagency disputes—a goal no one in authority considered remotely attainable anyway. In Washington, for example, the JCS believed the United States should be doing more to help end the Arab-Israeli dispute than John Foster Dulles' State Department was prepared to execute. On the separate issue of contingency planning for an intervention in Lebanon, military planners tried to impress upon the State Department the need to arrange overflight rights with those countries within whose air space US troop transports would have to fly. State refused, arguing that any such approaches to these governments before the President actually made a decision to intervene could compromise any operation and create political problems within several of the affected governments.[15] Yet, these and other points of contention that remained unresolved on 14 July seem minor given the interagency consensus that surrounded the key issues, especially the far-reaching decision to intervene.

That consensus, in its most general expression, was not formed by processes and procedures that allowed for interagency discussion and debate. Nor was it fortuitous. Rather it represented the thinking of most American citizens at the time on the subject of the Cold War. The confrontation with international communism appeared as a zero-sum game, in which, to avoid any significant setback, the United States needed to respond to virtually every Sino-Soviet threat with whatever means required, even nuclear war, if need be. Thus Eisenhower could order US troops into Lebanon with no idea of what they would do once there, and no idea of when and under what circumstances they would be withdrawn; and he could do so with little criticism, save for some partisan political fallout and the reservations of some officials like CNO Admiral Arleigh Burke, who questioned the wisdom of plans calling for placing US troops in the Arab Middle East, and UN ambassador Henry Cabot Lodge, Jr., who, following an informal interagency meeting at Foster Dulles' home on 22 June, questioned the Secretary, his boss, several times as to what the troops' mission would be and when they would be withdrawn, all the while offering Dulles reminders on the limited utility of military intervention. (Once Eisenhower decided to intervene, Lodge was on board, presenting the administration's case at the United Nations.)[16] The Vietnam War would shatter the rock-solid Cold War consensus of nearly twenty years, but, at the time Eisenhower acted in Lebanon, that development was still a decade in the future. In 1958, to repeat, the Cold War consensus itself facilitated interagency cooperation, coordination, and agreement.

Related to the Cold War consensus was how the personalities of the men concerned tended to minimize interagency infighting or inaction. To begin with, John Foster Dulles was a strong Secretary of State—the last until Henry Kissinger during President Richard Nixon's second term—and he had no intention of letting any other administration official usurp his role as the President's principal foreign policy adviser. Dulles had his detractors in the administration—MSA Director Harold Stassen, for one—but they were in no position to challenge the Secretary, who had Eisenhower's full backing. The fact that Foster's younger brother ran the CIA facilitated interaction between the two organizations they headed, while, in the case of Lebanon, the JCS chairman, General Twining, was as much an advocate of intervention as the President and his principal civilian advisers. As for the special assistant for national security, Robert Cutler, his job was to manage the NSC agenda and direct the council's discussions, not to formulate policy or offer operational advice. (The role of a strong national security adviser in the White House capable of eclipsing the Secretary of State and other officials would be crafted by Eisenhower's successor.) Given this setup, only General Andrew Goodpaster, Eisenhower's staff secretary and defense liaison officer, as well as close friend, was in a position to ignite an interagency turf battle, especially with the State Department, given his easy access to the President, but Goodpaster performed his duties in such a way that allowed him to assert his influence without running afoul of Foster Dulles.

In conclusion, then, interagency activities in effect during the Middle East crises of 1957-1958, of which I have only sketched the bare bones here, worked well for President Eisenhower, providing him with sound policy guidance and permitting him to introduce US armed forces into Lebanon in a timely way. After a rocky start, the success of that intervention owed much to diplomats and military officers working together, both on the scene and in Washington, to realize a peaceful outcome based on diplomacy backed by military force, but also diplomacy in which virtually all parties to the conflict felt they had achieved their objectives. Yet, if interagency cooperation was essential to resolving the Lebanese conflict, there were also reminders—the near shoot-out between the Lebanese army and the newly arrived US Marines being but one example—of the role also played in the outcome by what former Secretary of State Dean Acheson termed, in referring to the resolution of the Cuban Missile Crisis, as "pure dumb luck." A discussion of that observation, however, must be deferred to another time. Thank you.

Notes

1. Anna Kasten Nelson, "National Security Council," in *Encyclopedia of American Foreign Policy*, edited by Alexander DeConde, Fredrik Logevall, and Richard Dean Burns (2d ed., New York: Scribner, 2002); "History of the National Security Council, 1947-1997," at http://www.whitehouse.gov/nsc/history.html.

2. The text of the National Security Act of 1947 can be found at http://www.intelligence.gov/0-natsecact_1947.shtml#s101.

3. The quoted sentence is from Nelson, "National Security Council." See also the section on the Eisenhower administration in the "History of the National Security Council" cited above.

4. For Cutler's own account of his service as special assistant for national security, see Robert Cutler, *No Time for Rest* (Boston: Little, Brown, and Co., 1966), pp. 293-365. See also, Nelson, "National Security Council;" "History of the National Security Council;" Anna K. Nelson, "The Importance of Foreign Policy Process: Eisenhower and the National Security Council," in *Eisenhower: A Centernary Assessment*, edited by Günter Bischoff and Stephen E. Ambrose (Baton Rouge: Louisiana State University Press, 1995), pp. 111-25; John Prados, *Keepers of the Keys: A History of the National Security Council from Truman to Bush* (New York: William Morrow and Co., 1991).

5. For a discussion of the Middle East and Lebanon crises, see William B. Quandt, "Lebanon, 1958, and Jordan, 1970," in *Force without War*, edited by Barry M. Blechman and Stephen S. Kaplan (Washington, DC: The Brookings Institution, 1978), pp. 222-88; Lawrence A. Yates, "The US Military in Lebanon, 1958: Success Without a Plan," in *Turning Victory Into Success: Military Operations After the Campaign*, edited by Brian De Toy (Fort Leavenworth: Combat Studies Institute Press, [2004]), pp. 123-33; and Salim Yaqub, *Containing Arab Nationalism: The Eisenhower Doctrine and the Middle East* (Chapel Hill: University of North Carolina Press, 2004).

6. For the complete text of NSC 5801/1, see US Department of State, *Foreign Relations of the United States* [hereafter FRUS], *1958-1960,* Vol. XII, *Near East Region; Iraq; Iran; Arabian Peninsula* (Washington, DC: Government Printing Office, 1993), pp. 17-32. For the NSC meeting at which NSC 5801/1 was approved on 24 January 1958, the JCS had submitted much of their input through the Planning Board. A JCS memo recommending "decisive political and diplomatic action to solve the Arab-Israeli dispute" was submitted directly to the NSC. There, the Secretary of State opposed it in favor of a vaguer statement, which the President approved.

7. Robert J. Watson, *History of the Office of the Secretary of Defense*, Vol. IV, *Into the Missile Age, 1956-1960* (Washington, DC: Historical Office of the Secretary of Defense, 1997), p. 205.

8. Ibid., pp. 205-6; Declassified and sanitized manuscript for JCS official history for 1957-1960, Chapter VIII, "The Eisenhower Doctrine and Middle East Plans," no date; Dulles' 17 Oct 57 memorandum and Twining's response are contained in author's notes for Lebanon lesson in A645: US Interventions since 1945. The Planning Board's near excursion into contingency planning came as a result of reviewing the JCS response to an NSC request for information from the Pentagon related to the council's consideration of policy toward Iran.

9. Cutler, *No Time for Rest*, pp. 362-65; FRUS, *1958-1960,* Vol. XI, *Lebanon and Jordan* (Washington, DC: Government Printing Office, 1992), p. 211.

10. FRUS, *1958-1960,* Vol. XI, *Lebanon and Jordan* (Washington, DC: Government Printing Office, 1992), pp. 211-15.

11. Ibid., 382-86. The Planning Board paper was considered at the NSC meeting on 31 July. Prados, *Keepers of the Keys,* p. 85

12. Italics mine. For an account of the Marine-Embassy standoff and the near firefight with the Lebanese army the next day, see Yates, "The US Military in Lebanon, 1958," pp. 128-29. On Murphy's mission, see FRUS, *Lebanon and Jordan,* p. 256; Dwight D. Eisenhower, *The White House Years: Waging Peace, 1957-1961* (Doubleday book club edition; Garden City, NY: Doubleday & Co., 1965), p. 279. Another reason for dispatching Murphy seems to have been some doubts harbored by Foster Dulles and Vice President Richard Nixon concerning McClintock's luke-warm response to the intervention and his interference in the Marine landing. In a phone conversation with the Secretary, Nixon "indicated the need for people who can stand up better than that." He then suggested sending Murphy to Lebanon in order to address "the political problems involved in the crisis." FRUS, *Lebanon and Jordan,* pp. 251-52, n3. Prados, *Keepers of the Keys,* p. 83, includes Allen Dulles among those in Washington who had doubts about McClintock.

13. On the Murphy mission, see Robert D. Murphy, *Diplomat Among Warriors* (Garden City, NY: Doubleday & Co., 1964), pp. 394-418; FRUS, *Lebanon and Jordan,* passim.

14. The quotations are from Nelson, "The Importance of Foreign Policy Process," pp. 112, 113. Prados, *Keepers of the Keys,* pp. 81-85, also examines the limited role of the NSC in decisions surrounding the actual intervention in Lebanon.

15. On the dispute over overflight rights, see Watson, *History of the Office of the Secretary of Defense,* p. 209.

16. On Lodge's concerns, see FRUS, *Lebanon and Jordan,* pp. 168-69; Memorandum, Telephone call [to Secretary Dulles] from Amb Lodge, June 18, 1958, Dwight D. Eisenhower Library, Abilene, Kansas.

Panel 1—The Difficulties in Interagency Operations
Question and Answers
(Transcript of Presentation)

Edgar F. Raines, Jr., Ph.D.
John F. Fishel, Ph.D.
Lawrence A. Yates, Ph.D.

Moderated by Richard Stewart, Ph.D.

Audience Member

I was struck by your closing comments about the glass being totally empty at the beginning of JUST CAUSE because later on this afternoon we are going to have a panel that features a paper, actually two papers, one of which talks about the CORDS (Civil Operations and Rural Development Support) Program in Vietnam. So there were some things that were learned in the 1960s. And I suppose Larry would also add that there were some things learned in the 1950s, so what occurred do you think between 1973 with our exit from Vietnam and then what happens in the 1980s? Do those people move on, is the knowledge forgotten? I am curious as to what your perspectives are.

Dr. Fishel

Can I take that for starters? I have wanted to write an article that steals from a Bob Dylan line that says, "When will we ever learn? Oh, when will we ever learn?" The thing is, I think we have learned something. We do remember CORDS. We do remember the lessons. Some of the guys at the Joint Center for International Security Force Assistance here on post keep making a point that it took us far less time in Iraq to learn what to do right than it took us in Vietnam and some of the other places. I think the point I was trying to make was that in Panama itself, on the ground in a non-war situation but pre-war, we were not really doing much coordination. The only piece of planning that we really coordinated with the Embassy on was the Non-Combat Evacuation Plan (NEO) which directly involved them. I talked with John Maisto a couple of years later. We were on a panel together and I took him to the airport and we were chatting. I asked him specifically, "What did you know?" "Well, I knew about the NEO." That is all he had been told because we would not talk and we could not talk, but that is why I said there were some drops at the bottom of the glass and that was just our experience there. I hope that answers at least some of it.

Dr. Stewart

We do tend to learn occasionally but then empty the glass out every few years just to make sure we can start all over again.

Dr. Yates

Given the dates you mentioned, one, the consensus I talk about is gone that you had during the 1950s and 1960s. The Cold War consensus ends with Vietnam. Someone in foreign affairs said it was something like Humpty Dumpty. You are not going to put it back together again. After 9-11 it looked like we might have a consensus on foreign affairs, but that went away with Iraq. You have a proliferation of agencies that are brought into the interagency arena that you did not have back in the 1950s or even the early 1960s. Then the role of personalities, again, without that consensus, you see that emerging to some degree. It has always been there . . . NCS-68, one of the most significant documents of the Cold War, 1950, was an interagency blood-letting between the Secretary of Defense, Louis Johnson and the Secretary of State, Dean Acheson. They could not even talk to one another in putting this together. But you got it and you had a policy that most people could sign off on. You do not see that that often after Vietnam even though the Cold War goes on. Ed, in talking about Grenada, you made that point. Constantine Menges, in his memoirs, always talks about . . . he is an ideologue. He is a right-wing conservative, I do not mean a conservative, but a right-wing conservative on the Reagan National Security Staff. He has Ollie North along with him who is also a right-winger. Well, North is a reactionary. You have the pragmatists like Admiral Poindexter and Bud McFarland, a lieutenant colonel in the Marines, they are the pragmatists and these people are at each other's throats. Eagleburger was another pragmatist. You see this coming in once the consensus is gone, these issues can arise. And then finally, I agree with what everyone else has said, we study the case studies and then we forget. Who knows where this is going to end up? My guess is on a lot of library shelves collecting dust. It should not. It deals with crisis management, interagency, it deals with SOF, conventional, all sorts of things. Stability operations which is another thing that I felt, and John has as well, every time we go into it it is re-learning and taking the time to do it. We learned things quickly in Iraq. That is nice, but we should have known them going in. The idea of dropping Chalabi, and I am grossly simplifying, Colonel Benson would take me to task if I said there was no plan. Of course there is a plan, but we really have not learned the lessons of Panama for Iraq. Even though we talked about using the Panama model, very few knew what it was, and you can go back from Panama. Most people do not even realize there was a stability operation in Grenada, for example, and there was. And back to the Dominican Republic, my favorite, as Richard just said, we keep re-emptying the glass. We need to make a commitment to study it more. We pay lip service to it.

Audience Member

President Truman was the last president that publicly called his Secretary of State his senior secretary, promoting him above, if you will, the other secretaries in the cabinet and making him the de facto Chief of Staff, if you will, for his advice that he got from his cabinet members. That seemed to work well then and that was the last time, arguably, that interagency worked like it should I suppose, so why do you suppose ten presidents after President Truman chose not to do that and to level the playing field, if you will, amongst the cabinet members and sort of watered down in a sense the advice that he gets?

Dr. Yates

I would agree with the first part about Truman. First of all Truman had very little experience in foreign affairs. He has been in World War I as an Artillery captain and he cited that as his experience. He had a tremendous commitment to the United Nations. He had seen the failure of the League of Nations and was determined to have a United Nations that worked which is one of the reasons that we went into Korea because he felt the UN could not be allowed to fail. But he realized his shortcomings there and so did his staff, Clark Clifford, among others. Some would argue that he should have used the Embassy more than he did. He did not, but to the extent that it was used, he put the Secretary of State pretty much in charge of it. But he had two strong Secretaries of State, one of whom he idealized, George Marshall. This is the man of the century; that is a Truman-esque marginal note in his handwriting, the man of the century. Dean Acheson was equally up to the task. So he had two very strong individuals who Truman had no problem in delegating certain foreign policy issues to. Marshall threatened to resign over, he told Truman he would not vote for him over the issue of Israel, recognizing Israel. So Truman turned over the NSC (National Security Council) essentially to the State Department, its interagency, but there is the first among equals. Eisenhower changed the system, but he still had a strong Secretary of State in whom he had complete confidence that always had the first and the last word before Eisenhower made a decision. And because the two were in such synch, Eisenhower talked about that in his memoirs about how they sat around at the end of the day, he and Foster Dulles just in the Oval Office talking. Kennedy comes in, Eisenhower says keep the NSC pretty much the way it is. By that time you had the Jackson Committee which said all that the NSC is doing is turning out a lot of paper. We need a more dynamic National Security Council, one that is run from the White House, not by the State Department, and Eisenhower put the Vice President in charge of the meetings, not the Secretary of State, but one that is more run from the White House and you get the rise of the National Security advisors so that is where I disagree. It does not level the playing field. You elevate someone in the White House over the Secretary of State. Dean Rusk is considered a weak Secretary of State compared to McGeorge Bundy who was the first National Security Advisor, followed by Walt Whitman Rostow under Johnson. And then under Reagan, Brzezinski was very strong. Reagan tried to diminish the role, but it keeps popping up. Initially Truman and Eisenhower relied on very strong Secretaries of State.

Dr. Fishel

And part of that is that the Secretary of State is the senior cabinet official, regardless. Another part of what is going on is the constant Congressional whittling down of State's budget and the building up, over time, of the defense budget and other agencies so that the resource space available so far outweighs it on the defense side, the military side of things, that you get this very high-powered juggernaut on one side that has been doing its best in this administration, including time with Secretary Rumsfeld, to build back up some of the other agencies, and it does not work because again, it is underfunded. In the Truman era the policy planning staff of the State department did the things the NSC staff does now, and was created by Secretary Marshall largely to do what a Joint Plans and Policy element of a Joint Staff does. It worked kind of that way under Secretary Marshall, and then it started slipping because you did not have people with that kind of experience. *[Inaudible]* was the originator of it. So you have this larger scale thing that is creating diversifying power and concentrating power in the defense.

Dr. Yates

In 2005, my last year on the job, I was part of a task force that started at the Chief of Staff, went through TRADOC (United States Army Training and Doctrine Command), down to the CAC (Combined Arms Center) commander, and then to the college and it was on elevating stability operations to an equivalency with combat operations. I think that was what it was about. We had a big conference to start it off with and we all gave presentations. Someone was talking about how the State department needed to kick in more and do more and that the military was taking over and the State department representative says, "We cannot send you anybody. We do not have the money. We do not have the personnel. If you are going to get the job done, you are going to do it, the military. You have the people; you have the money."

Audience Member

Just a comment if I could because I am going to present this afternoon, but in fact, it is my understanding since I have dug into the history of this, the State department willingly gave the policy planning functions over to the NSC and it is in that 1949 to 1951 time frame which I think is incredibly important and then also to agree with the presenters on this, but to go back further, that shift in resources really began after the 1947 National Security Act and then in the 1949 reforms. And what happened was a lot of the key committees wound up having seconded military officers even then coming over and filling the functions and the third piece, and I think it will be interesting when Mr. Kimmitt is here on Thursday to raise this question because we have discussed this before too, is how much the State Department as an organization, whether for organizational culture, bureaucratic momentum, and so on has also continued that process of not stepping up to the plate when the opportunities were there. Again, it is my understanding in the reading of the history of this that the State department was offered coordination of intelligence role back when it was still seen as the entity that would be the primary coordinator in that 1947 debate and they said, "Thanks, but no thanks because it will make our State department functions more difficult." So there are some interesting pieces there in terms of the State department also doing things along the way that helped put it on that path.

Dr. Yates

The politics of the National Security Act of 1947 are incredibly intricate and fascinating and you wonder how Forrestal got to be the first Secretary of defense. It was because he was one of them that complicated things and Truman said, "Okay, you made this mess, you can deal with it." But you are absolutely right.

Audience Member

Gentlemen, just a comment. Not speaking for the Department of State or their ability to take on that increased role for stability and reconstruction, but it was kind of formalized within the last year or so with National Security Presidential Directive #44 in which he designated the Department of State as the lead for stability and reconstruction operations taking that whole of government approach to the interagency, kind of like what Goldwater-Nichols did for the whole of DOD approach. There are some growing pains with that. It is not fully fleshed out, but there is an Office of Stability and Reconstruction Coordinator at the Department of State and working closely with a lot of DOD agencies.

Dr. Stewart

And of course the first year after they created that, Congress zeroed out the budget to get back at them. It has gotten better, yes, but Congress is not thrilled with the whole idea.

Audience Member

Would you care to comment on one, Panama being a very strange case because you have the CINC (Commander in Chief) fighting for its own headquarters company. You do not normally have the CINC in the middle of the war zone. And the other problem is that there is no State department operational-level. It is either strategic-level or the Embassy which, in that country, is the tactical-level for us. So the missing operational-level, which in Panama you really have because it is right there, but in the other cases you do not. Do you care to comment on those thoughts?

Dr. Yates

One thing in Panama, given the CINC's regional responsibilities, with some reluctance they set up the Joint Task Force *Panama* (JTF *Panama*) to run the crisis on a day-to-day basis and to some degree, I would see that as an operational-level headquarters because under it you have various task forces that you would look for at the tactical-level, but it was doing . . . the problem with that was you had SOUTHCOM ten minutes away micromanaging it and there was a lot of friction there, especially between the J3 at SOUTHCOM and the J3 in JTF *Panama*. But that was the solution there. In terms of Washington, John has already mentioned that General Woerner dealing with Elliot Abrams who was the Assistant Secretary of State for Inter-American Affairs, not the Secretary of State.

Dr. Fishel

And the other thing, Jeff, is as Larry certainly knows, the operational-level headquarters JTF *Panama* became largely overwhelmed by the crisis and even before Fred Woerner was relieved they had gone to the 18th Airborne Corps and asked for the creation of a JTF based around the Corps. Little known, but JTF *Charley* was in fact activated prior to General Woerner's relief and then General Thurman re-activated his JTF side so that the operational-level headquarters actually was sitting at Fort Bragg for most of the time. I am not sure I would agree that an Embassy is really a tactical headquarters. I think it probably is, in and of itself, typically operational to regional strategic headquarters. That said, there is certainly something missing. The assistant secretaries for the regions in Washington simply cannot perform the function and part of that has to do with the fact that the Ambassadors do not work for the State department. They are personal representatives of the President and if they happen to be Foreign Service Officers, they technically resign from the State department for their period as an Ambassador. So essentially it is like dealing with the Chairman. You say you are in the chain of communication when you are talking to State, but not quite in the chain of command.

Counterinsurgency, the Interagency Process, and Malaya: The British Experience

by

Benjamin Grob-Fitzgibbon, Ph.D.
University of Arkansas

In his seminal work *Defeating Communist Insurgency: Experiences from Malaya and Vietnam*, Sir Robert Thompson laid down what he called the five basic principles of counterinsurgency. Rather than offering operational advice, each of Thompson's five principles concerned governance: first, the government must have a "clear political aim" in defeating insurgencies; second, the government must "function in accordance with the law;" third, the government must have an overall plan; fourth, the government must "give priority to defeating the political subversion, not the guerillas;" and finally, during the guerilla phase of an insurgency a government must "secure its base areas first."[1] In the attainment of each of these principles, it was essential that there should be a "proper balance between the military and the civil effort, with complete coordination in all fields."[2] At the heart of any counterinsurgency strategy, the interagency process must function flawlessly. This was Thompson's central message.

If there was an individual in the western world qualified to determine such, it was Sir Robert Thompson. When he published *Defeating Communist Insurgency* in 1966, Thompson had only returned from South Vietnam a year earlier, in March 1965, where he had operated since September 1961 as head of the British Advisory Mission to Vietnam (BRIAM).[3] Prior to that, he had worked in various guises in British counterinsurgency in Malaya from 1948 to 1960, eventually reaching Permanent Secretary of Defence with the acting rank of Lieutenant Colonel. He had begun his time in the region as one of the legendary Chindits of Britain's 3d Indian Infantry Division in Burma throughout the Second World War.[4] It was through his experiences in the Malayan Emergency that Thompson first developed his principles of counterinsurgency. It was a conflict which, unlike Vietnam, Thompson believed he had achieved considerable success in. Indeed, of all Britain's small wars at the end of empire, from Palestine, Malaya, and Kenya to Cyprus, Aden, Dhofar, and Oman, it was in Malaya that the interagency process was most advanced, and in Malaya that the British were most successful at countering an insurgency.

Malaya had been of interest to the British government since the earliest days of empire, when in the first years of the seventeenth century the newly formed English

East India Company opened a trading post at the mouth of the Kedah River. Such trade was increased a century later, when between 1765 and 1800 the British government signed a number of treaties with the Sultan of Kedah that provided the British Penang Island and a strip of land opposite in exchange for an annual income payable to the Sultan. Following an invasion from Siam in 1821, the British were temporarily expelled from the island, but in 1894 a British consul returned with permission of the Siamese government and in 1909 Kedah was once again brought under direct British control with the signing of the Anglo-Kedah Treaty. Meanwhile, the British had captured Malacca from the Dutch on Malaya's western coast in 1795. Expansion on the peninsula continued and in 1819 the British government signed a treaty with the ruler of Johore that gave the British the right to settle on Singapore Island. In 1867, Singapore was united with the west coast of Malaya to form the Crown Colony of the Straits Settlement.[5]

From the Straits Settlement Colony, the British government was able to exercise considerable informal control over the ostensibly independent Malay States, signing treaties with Perak, Selangor, and Sungei Ujong in 1874 and Pahang in 1888, treaties through which a British Resident was appointed in each state whose mission was to replace the traditional feudal structure of Malay society with western law and political and economic norms. In 1895, these four states merged to form the First Federation of Malaya, with a central government in Kuala Lumpur. Further treaties establishing British Residents were signed with Kedah, Kelantan, Trengganu, and Perlis in 1910, followed by Johore in 1914. In each of these states, British engineers, doctors, and civil servants were seconded alongside the British Residents, each responsible to the British High Commissioner in Singapore.[6]

Although unlike the Straits Settlement, the Malay States were not technically British colonial possessions, they were nonetheless an integral part of the British Empire, with the British Residents opening a Malayan railway system in 1884, clearing the swamps to prevent the spread of malaria in the late nineteenth century, introducing domestic and international flights with Empire Airways in the early twentieth century, and replacing coffee with rubber as the staple crop of the Malay economy. By the eve of the Second World War, Malaya (including the Straits Settlement and the Malay States) was exporting a quarter of a million tons of rubber, two and a half million gallons of latex, and 80,000 tons of tin and tin ore each year.[7] The British and Malays were not the only peoples to inhabit the peninsula, however. To find workers for the rubber plantations and tin mines, the British had embarked upon a large-scale immigration scheme from China and India, the result being that by 1945, Malaya's population of 5.3 million people included 49 percent Malay persons, 38 percent Chinese, and 11 percent Indians, together with 12,000 Europeans, most of whom were British. While the Chinese were willing to work for the British, they were indignant about being employed by Malays. Ethnic conflict thus became an ingrained aspect of Malayan society in the twentieth century.[8]

This Chinese separation from Malay culture and economics was further increased by the formation of the Malayan Communist Party (MCP) in 1930. Led from 1939 by Lai Tek, the MCP played a decisive role against the Japanese army following the latter's invasion and occupation of Malaya in December 1941. By 1945, the Malayan People's Anti-Japanese Army (MPAJA), as the communist guerilla force had become known, had grown to a strength of 7,000 men and was being directly supplied by the British from December 1943 onwards. Throughout these final two years of the Second World War, the British held numerous meetings with Lai Tek and provided the MCP with £3,000 a month to aid their anti-Japanese insurgency. Following the Japanese surrender, however, Lai Tek and the British war leaders had very different conceptions of what Malaya's future would look like.

Lai Tek had pledged to establish a communist republic, free of both Japanese and British control. The British, in contrast, were determined to regain control of their colony. Before they could do so, the MPAJA launched a final push against the Japanese. On September 8, 1945, they reoccupied Singapore and on September 28 they took from the Japanese the rest of the Malayan Peninsula. When the British government cut off funding for the MPAJA, staged a disbandment ceremony, and installed a British High Commissioner in December 1945, the MCP refused to acknowledge its authority to do so. The grounds for a communist insurgency were therefore laid. It was just three short years after this ceremony that the British government was once again embroiled in conflict in Malaya.[9]

The Malayan Emergency, officially declared in June 1948, was at first waged by the British as a conventional war, with soldiers seeking to hunt down and contain the Chinese Malayan guerillas, now reconstituted as the Malayan People's Anti-British Army (MPABA). The Malayan police force was not considered to have an operational role and the civil administration was left ignorant of military affairs. This was a campaign led by and carried out by the British Army, with limited support from the Royal Air Force and no input whatsoever from the various layers of civilian governance. By 1950, however, these methods seemed to be failing, with the MPABA holding its strength, its base of operations steadily increasing, and its having killed 850 Malay and European civilians, 325 Malay policemen, and 150 British soldiers. The British were not winning the war, they were losing it.[10] It was time for a radical shake up of the way Britain's counterinsurgency campaign had been run so far.

The idea of a coordinated interagency command in Malaya was first proposed on February 23, 1950, when Sir Henry Gurney, the British high commissioner in Malaya, sent a telegram to Arthur Creech Jones, the colonial secretary. He suggested that he had "for some time" been considering appointing a single officer to "plan, co-ordinate and generally direct the anti-bandit operations of the police and fighting services."[11] The reason, Gurney argued, was now that the conflict in Malaya had reached the stage of "protracted guerilla warfare" the civilian police commissioner was ill-equipped to direct operations, yet there was no other civil officer other than Gurney himself who had the authority to give directions to the army's General Officer Commanding (GOC) and the air force's Air Officer Commanding (AOC). Gurney therefore recommended that the government second to Malaya a high ranking military officer who could be appointed in a civil post with the following duties:

> He would be responsible for the preparation of [a] general plan for offensive action and the allocation of tasks to the various components of the security forces. In consultation with heads of the police and fighting services he would decide priorities between these tasks and general timing and sequence of their execution. He would exercise control through heads of police and fighting services and aim at achieving co-ordination and decentralisation by this means. . . . He would work directly under myself [British High Commissioner] and within the framework of the policy laid down by this Government. He would be in close touch with civil authorities responsible for essential features of the campaign, such as settlement and control of squatters, propaganda, immigration control and settlement of labour disputes, and would have [the] right to make represen-tation to me in such matters affecting the conduct of [the] anti-communists campaign as a whole.[12]

It was a position that was without precedent in British imperial experience, yet one that had been carefully thought through and had been designed by a man with the appropriate authority and experience; prior to taking the post of British high commis-sioner in Malaya in September 1948, Gurney had served as chief secretary in Palestine from 1946 to 1948, as colonial secretary in the Gold Coast from 1944 to 1946, and as chief secretary to the Conference of East African Governors from 1938 to 1944. There were few men in the British Empire with his experience of colonial governance, par-ticularly during periods of insurgency such as had dogged his years in both Palestine and Malaya.

Nonetheless, it was an inopportune time for Gurney to send his telegram. That day, a General Election had been held in the United Kingdom and Creech Jones had lost his parliamentary seat, at the same time automatically ceasing to hold his cabinet posi-tion. While the telegram had been copied to the prime minister, the war secretary, and the first lord of the admiralty, in the scramble of a tightly fought election in which the governing Labour Party performed poorly, the message of Gurney's telegram was lost. Despite this, Gurney persevered, and on March 9, less than three weeks after the elec-tion, he telegrammed to the new colonial secretary, James Griffiths, outlining his plan, suggesting that a lieutenant-general (serving or retired) be appointed for a minimum of one year, and recommending that this position be titled the Director of Operations with the same civilian rank as the chief secretary. Gurney was even so bold as to suggest to Griffiths the wordings of the press release that could be issued to announce the post. Significantly, the final line of the proposed announcement read: "His primary function will be to secure full and effective co-ordination."[13]

Field Marshal Sir William Slim, the chief of the Imperial General Staff, suggested to the cabinet that they approach his good friend Lieutenant-General Sir Harold Briggs, who had been in retirement in Cyprus since 1948. Briggs seemed the perfect candidate for the position. Only fifty-five years of age, he had followed a distinguished military career spanning thirty-four years and culminating in command of the Indian Army's 5th Infantry Division in Burma from early 1944 to June 1945, where he had gained familiarity with jungle warfare.[14] After initial hesitation, Slim managed to persuade Briggs to take up the post and on March 21, the government announced that the first Director of Operations for Malaya had been selected. Following a whirlwind series of meetings in London to explain and discuss the position, Briggs arrived in Kuala Lumpur on April 3, 1950.[15]

The retired general wasted no time in coming to grips with the situation in Malaya. After a two-week tour of the colony, meeting with military, police, and civilian authorities, on April 16 Briggs issued Directive No.1, which laid out the future direction of his proposed policy, eventually becoming the cornerstone of British strategy in Malaya for the next decade. Effective June 1, a Federal War Council would be formed, chaired by the director of operations and including in its membership the chief secretary, the GOC, the AOC, the commissioner of police, and the secretary of defence. The role of the Federal War Council was to produce policy. Each state was then required to form a State War Executive Committee (SWEC), chaired by the resident commissioner of that state and with a membership of the British advisor in state, the state's chief police officer, and the state's senior army commander. The role of the SWEC was to implement the policy laid out by the Federal War Council. In each district within the state, a District War Executive Committee (DWEC) would be formed, mirroring the composition of the SWEC only with lower-level officials. In addition to the SWECs and DWECs, a Federal Joint Intelligence Advisory Committee was set up. Its purpose was to examine "ways and means of strengthening the intelligence and Police Special Branch organization to ensure that the mass of information which exists in the country becomes available and is sifted and disseminated quickly and at the right levels."[16]

The idea of the administrative hierarchy, as articulated by Briggs in Directive No.1, was that this "joint conception" would be "followed at all levels, with the Civil Administration, Police and Army working in the closest collaboration and using combined joint operations and intelligence rooms wherever practicable."[17] A smooth functioning of the interagency process lay at the heart of Briggs' scheme for Malaya and was, in his mind, crucial for the successful operation of a counterinsurgency campaign. As such, within the SWECs and DWECs there could be no ranking of army and police personnel, with one claiming superiority over the other. Briggs made this very clear in his second directive, issued on May 12, where he explicitly stated: "It is immaterial whether the local military commander is a lieutenant-colonel and the local police officer is a sergeant or whether they are respectively a major and a superintendent; in each case they will establish a joint headquarters and will work in the closest cooperation also with the local administrative officer."[18]

Having established the administrative framework of the counterinsurgency campaign, Briggs—in coordination with the Federal War Council—turned to strategy and tactics, issuing on May 24 the "Federation Plan for the Elimination of the Communist Organisation and Armed Forces in Malaya," referred to at the time and by historians since as the Briggs Plan. At the core of this plan was the idea that the British government needed to demonstrate "effective administration and control of all populated areas" and wrestle the initiative away from the communist guerillas. Essentially, the British had to demonstrate that the western way of life was more appealing and could offer a higher quality existence than the communist way of life. This would be done through a six-step process. First, security would be maintained on the ground and the British government would demonstrate through the use of the police and armed forces that it was firmly committed to protecting Malaya against both external and internal attack and disorder. Secondly, Malayan squatters would be resettled into compact groups, where they could more easily be protected by the British security forces and given social welfare. Thirdly, local administration would be strengthened, so that it would become more effective and efficient than anything the communists could offer. Fourthly, road communication would be provided in isolated areas to link all Malayan subjects to the British administrative structure. Fifthly, police posts would be set up in these isolated areas, both to protect the population and to show the flag. And finally, a concerted propaganda campaign would be launched to highlight the negatives of the communist insurgency and the positives of British governance.[19]

Within this general framework, Briggs laid down four further objectives for his strategy: first, within the populated areas a "feeling of complete security" had to be built up, which would in turn lead to a "steady and increasing flow of information from all sources;" secondly, the MCP had to be broken up within the populated areas and denied access to greater Malayan society; thirdly, having been ejected from the populated areas, the insurgents would be isolated from their food and information supply; and finally, the insurgents would therefore have to attack the British security forces on their own ground, where they could be defeated without inflicting pain or inconvenience on the general population. With regards to the tactics used to implement this strategy, the police and army were tasked with working in coordination under the leadership of the SWECs and DWECs to create the feeling of security within populated areas, the army was tasked with establishing strike forces that could dominate the jungle within five hours journey either side of the populated areas, and the civil administration, supported by the police and army, would regroup or resettle a large number of Malayan squatters, with the resettlement program anticipated to be completed by the beginning of 1952. Briggs closed his plan with a warning that even when each of these successes was outwardly achieved, and when the population was securely resettled and under effective British administration and control, there could be a "rapid recrudescence of terrorist activity," and therefore "the danger of relaxing security precautions and of prematurely withdrawing troops must be realized."[20]

Briggs stressed throughout that any operations undertaken by the army or police had to be under civil control, had to be within the law, and the purpose for which they were being conducted had to be clearly articulated to the local population. For that reason, on July 4, 1950—just six weeks after the Briggs Plan had been formulated—he instructed the acting chief secretary in Malaya, M.V. Del Tufo, to articulate whom the enemy was in the *Federation of Malaya Government Gazette* with an explicit definition of terrorism. Following its publication, there could be no doubt of the British government's target. A terrorist was any person who:

> (a) by the use of firearms, explosives or ammunition acts in a manner prejudicial to the public safety or to the maintenance of public order;
>
> (b) incites to violence or counsels disobedience to the law or to any lawful order by the use of firearms, explosives or ammunition;
>
> (c) carries or has in his possession or under his control any firearm, not being a firearm which he is duly licensed to possess under any written law for the time being in force;
>
> (d) carries or has in his possession or under his control any ammunition or explosives without lawful authority thereof;
>
> (e) demands, collects or receives any supplies for the use of any person who intends to or is about to act, or has recently acted, in a manner prejudicial to public safety for the maintenance of public order; and 'terrorism' shall have a corresponding meaning.[21]

Briggs made it absolutely clear that the enemy was not only he who pulled the trigger, but also he or she who supplied the bullet for the gun, the food within the belly, or the bed at night. In so doing, he put the Malayan civilian population on notice that there could be no cooperation with the communist insurgents.

The Briggs Plan was set into action immediately, with the senior police, army, and civil representatives of the SWECs and DWECs meeting daily and the whole membership of the committees meeting weekly.[22] Their first task was to resettle the Malayan squatters, of whom there were upwards of 300,000. This was begun without delay and at the British cabinet's Malaya committee meeting of July 14, the colonial secretary Arthur Griffiths was able to report that already 20,000 squatters had been resettled, and the SWECs, working in "complete coordination" with the police, military and civil authorities, had laid the groundwork for many more resettlements.[23] By September 22, half of all squatters in the populated South Johore province had been resettled,[24] and by February 15, 1951, the job of resettlement in the priority areas was more than half done, with 67,000 squatters resettled and 52,500 remaining.[25] Once resettled, the squatters were incorporated into so-called "new villages," where they were brought under British administration and given social welfare, such as housing, health care, and education. These new villages, run in the short-term by resettlement supervisors and in the long-term by assistant district officers, were protected by the Malayan police force so that their residents could be shielded "against Communist physical and intellectual attack and helped to become contented communities." Once protected in this way, the superiority of British mores and administration would be self-evident and the communist menace in the colony would be defeated.[26]

These new villages were the epitome of the Briggs Plan and demonstrated clearly the importance of the interagency process in countering insurgencies. The Federal War Council determined which areas of squatter populations were high priority for resettlement, based on their perceived vulnerability to communist ideology. Within these areas, the SWECs and DWECs selected the exact site for the new village. The army, supported by the police, cleared the site and ensured that there was no communist guerilla influence within five hours of the area. Once cleared, the army took a lesser role and the police became the lead agency in keeping the perimeter secure. The squatters themselves were moved by the police, with army protection and directed by civil administration, into the villages, to which new and modern roads were built. Within the villages, the civil administration worked with the villagers to establish schools, hospitals, cooperatives, and businesses. As the villagers began to trust British administration, and were in turn trusted by the British administrators, they were incorporated into an unarmed Home Guard that served alongside the Malay police force. Following an unarmed period of time in the Home Guard, the villagers were allocated shot guns for defense of their settlement and were considered loyal subjects of the British crown. At this point, the insurgency—in every sense of the word—had been broken.[27]

On June 4, 1951, at the one year anniversary of the implementation of the Briggs Plan, Briggs, together with the high commissioner Sir Henry Gurney, composed for the cabinet's Malaya committee a combined appreciation of the progress of the counterinsurgency campaign. They wrote that after a long, hard year, British success in Malaya had finally reached a "turning point," with 240,000 squatters now resettled into new villages and generally doing well. In particular, the British policy of "committing them to our side by getting them to join our Home Guard organization, which they are doing in very great number[s], is bearing good fruit." Furthermore, in the first six months of 1951, there had been an 180 percent increase of communist guerilla surrenders and a 42 percent increase of guerilla casualties, compared to only an 11 percent increase in security force casualties. Civilian deaths had been reduced by 3.5 percent and civilian injuries by 33 percent. The strategy of drawing the insurgents away from the civilian population was working and, as a consequence, those resettled were becoming more cooperative with British rule each day. The Briggs Plan was succeeding and the British were winning the war. Although the Emergency would continue for another nine years, interagency cooperation had been shown to be the only path forward for a successful counterinsurgency campaign.[28]

The true test of this new approach came later that year, when on October 6, 1951, the communist guerillas killed Gurney in an ambush.[29] Less than three weeks later, a new Conservative government came to power in Great Britain under Winston Churchill, following the defeat of Clement Attlee's Labour government in a General Election held on October 25. With Gurney dead and a government run not by Attlee but by the more belligerent Churchill, the question was whether the interagency process would still be front and center of British strategy, or whether the army would once again be given primacy, with tougher measures called for against the communist insurgency and a corresponding drop in the input from the civil administration and police.

Shortly after Gurney's murder, representatives from the Federal War Council, the SWECs, and the DWECs met on October 17 to discuss future British policy in Malaya. Briggs suggested that full executive authority in Malaya be delegated from the high commissioner to the director of operations, negating the position of chief secretary and placing the director of operations in complete charge of civil administration, the military, and the police force.[30] To all intents and purposes, if Briggs' advice was followed, Gurney's successor as British high commissioner would be a mere figure head with no governmental responsibility, and the position of chief secretary would cease to exist. The new conservative colonial secretary, Oliver Lyttelton, approved Briggs' suggestion on November 1, just four days after taking over at the colonial office.[31]

The director of operations now had full and undivided control of all emergency policy, strategy, and tactics in Malaya. It would not be the architect of this development, however, that was its main beneficiary. Briggs grew ill in November and retired on the twenty-seventh of that month, overcome by exhaustion. Within a few weeks, he was dead.[32] His deputy, Sir Robert Lockhart, temporarily took on the new powers. The colonial secretary, arriving in Malaya on November 29 for his first visit, was dissatisfied with what he found. He believed that the figure-head high commissioner, by his very presence, undermined the authority that was vested in the director of operations. The only solution to this problem, he concluded, was to merge the two positions into a single all-powerful high commissioner, who would take on all the responsibilities and the position of the director of operations.[33]

Lyttelton returned to the United Kingdom on December 21 after a month in Malaya and immediately met with Winston Churchill to persuade him of this viewpoint. The two men then discussed the new position with Field Marshal Bernard Montgomery, currently acting as the inspector-general of NATO's European forces, who suggested General Sir Gerard Templer would be ideal. Templer, like Briggs, had a notable military career before him, being first commissioned in the Royal Irish Fusiliers in 1916, becoming the army's youngest lieutenant-general in 1942 in command of II Corps and eventually rising to become the director of military government in the British zone of occupied Germany. Churchill, leaving for his first official visit to the United States as the new prime minister, requested that Templer fly out to meet him, which he did on January 11, 1952. After a short and informal interview, Churchill invited Templer to take on the task in Malaya, which he immediately agreed to. Following a few weeks of preparation, Templer arrived in Kuala Lumpur on February 7 as the new British High Commissioner of Malaya and Director of Operations.[34]

Templer had dictatorial powers in Malaya but he used them with caution. Following Briggs' lead, he kept in place the administrative hierarchy of the Federal War Council, the SWECs, and the DWECs, and continued to hold the resettlement of squatters into new villages as the government's highest priority in the fight against the insurgency. He went further than Briggs and Gurney, however, quickly determining that the key to an effective counterinsurgency strategy was sound intelligence and this could only be obtained through a coordinated approach. On February 13, therefore, after less than a week in Malaya, Templer wrote to Lyttelton, informing him that a new position, the director of intelligence, would be appointed, who would have "the right of direct access at any moment to [the] High Commissioner." He would be given no executive control of any one intelligence agency, but would instead be responsible for coordinating the activities of police intelligence, naval intelligence, army intelligence, air force intelligence, and political intelligence. His role would include giving advice to each of these organizations, as well as being "completely responsible for collation and evaluation of all the intelligence available and for its presentation to those concerned in the proper form."[35]

By March 1952, at all levels of command in Malaya, in civil, police, military, and intelligence matters, the British High Commissioner had achieved absolute interagency coordination. Gurney, Briggs, and Templer had each realized that the key to defeating the communist insurgency was not an aggressive military offensive but rather a careful campaign to win over the civilian population. As Templer famously observed in 1952, "The answer lies not in pouring more troops into the jungle, but in the hearts and minds of the people."[36] In Malaya, the British government took these words to heart and made them a creed. The consequence was that it was able to erase all communist influence and declare the Emergency over on July 31, 1960. Three years earlier, on August 31, 1957, the British had granted independence to the Federation of Malaya in an orderly transition of power which kept Malaya in the British Commonwealth. Six years after independence, and three years after the close of the Emergency, the Federation of Malaya joined with the British colonies of North Borneo, Sarawak, and Singapore to form Malaysia, which like Malaya before it was an independent state within the British Commonwealth. Although Singapore was to secede in 1965, Malaysia has remained a stable state to the present day. For the remainder of the Cold War, it never again turned hot and British troops never again returned after their final departure in 1960.[37] In Malaya, the British government had combated an insurgency and had won. It had done so, in large part, through an effective use of the interagency process.

Notes

1. Robert Thompson, *Defeating Communist Insurgency: Experiences from Malaya and Vietnam* (London: Chatto & Windus, 1966), 50-58.

2. Ibid, 55.

3. For more on Thompson's time leading the British Advisory Mission to Vietnam, see Ian F.W. Beckett, "Robert Thompson and the British Advisory Mission to South Vietnam, 1961-1965," *Small Wars & Insurgencies*, Volume 8, Issue 3 (Winter 1997), 41-63; and P. Busch, "Killing the 'Vietcong': The British Advisory Mission and the Strategic Hamlet Programme," *Journal of Strategic Studies*, Volume 25, Number 1 (March 2002), 135-162.

4. For a firsthand account of Thompson's life and career, see Sir Robert Thompson, *Make for the Hills: Memories of Far Eastern Wars* (London: Leo Cooper, 1989).

5. Robert Jackson, *The Malayan Emergency: The Commonwealth's Wars, 1948-1966* (London and New York: Routledge, 1991), 3-4.

6. Ibid, 4-6.

7. Ibid, 6.

8. John A. Nagl, *Learning to Eat Soup with a Knife: Counterinsurgency Lessons from Malaya and Vietnam* (Chicago: The University of Chicago Press, 2002), 60.

9. For the origins of the communist insurgency in Malaya, see Richard Clutterbuck, *Conflict and Violence in Singapore and Malaysia, 1945-1983* (Boulder, CO: Westview Press, 1985), 37-41.

10. Michael Dewar, *Brush Fire Wars: Minor Campaigns of the British Army since 1945* (New York: St. Martin's Press, 1984), 33.

11. The National Archives [TNA], Foreign Office [FO] 371 / 84477, Telegram from Sir Henry Gurney, British High Commissioner to Malaya, to Arthur Creech Jones, Secretary of State for the Colonies, February 23, 1950.

12. Ibid.

13. TNA, FO 371 / 84477, Telegram from Sir Henry Gurney, British High Commissioner to Malaya, to James Griffiths, Secretary of State for the Colonies, March 9, 1950. The full text of the proposed announcement read as follows: "The Government of the Federation of Malaya has appointed (blank) as Director of Operations to plan, co-ordinate and generally direct anti-bandit operations of Police and fighting forces. The post is a civil one. The Director of Operations will be responsible for the allocation of tasks to various components of the Security Forces available for operations and for deciding, in consultation with heads of the Police Force and fighting services, the priorities between these tasks and general timing and sequence of their execution. His primary function will be to secure full and effective co-ordination."

14. Edgar O'Ballance, *Malaya: The Communist Insurgent War, 1948-1960* (Hamden, CT: Archon Books, 1966), 106.

15. Anthony Short, *The Communist Insurrection in Malaya, 1948-1960* (New York: Crane, Russak & Company, Inc, 1975), 235.

16. TNA, Cabinet Office [CAB] 21 / 1681, Directive No. 1, Director of Operations, Malaya, April 16, 1950.

17. Ibid.

18. TNA, CAB 21 / 1681, Directive No. 2, Director of Operations, Malaya, May 12, 1950.

19. TNA, CAB 21 / 1861, Federation Plan for the Elimination of the Communist Organisation and Armed Forces in Malaya, May 24, 1950.

20. Ibid.

21. Colonial Office [CO] 537 / 5984, Definition of Terrorism, laid down in the *Federation of Malaya Government Gazette*, July 13, 1950, No. 32, Vol. III, L.N. 302 in the Emergency (Amendment No 12) Regs, 1950. By command of M.V. Del Tufo, Acting Chief Secretary, July 4, 1950.

22. O'Ballance, *Malaya: The Communist Insurgent War*, 107.

23. TNA, CAB 21 / 1681, "Various Matters Discussed with the Authorities in Malaya," memorandum by Arthur Griffiths, Secretary of State for the Colonies, circulated to the Malaya Committee of the Cabinet, July 15, 1950.

24. TNA, CAB 21/ 1681, "The Present Situation in Malaya," memorandum by Arthur Griffiths, Secretary of State for the Colonies, circulated to the Malaya Committee of the Cabinet, September 22, 1950.

25. Progress was significantly slower in the nonpriority areas, where 50,000 squatters had been resettled but 280,000 remained. CAB 21 / 2884, Sir Harold Briggs, Director of Operations, "Progress Report on Situation in Malaya," February 15, 1951.

26. TNA, CO 1022 / 32, Sir Harold Briggs, Director of Operations, Malaya Directive No. 13: Appendix A: "Administration of Chinese Settlements," February 26, 1951.

27. Nagl, *Learning to Eat Soup with a Knife*, 75.

28. TNA, CAB 21 / 2884, Sir Harold Briggs, Director of Operations, and Sir Henry Gurney, British High Commissioner in Malaya, "Combined Appreciation of the Emergency Situation," circulated to the Malaya Committee of the Cabinet, June 4, 1951.

29. Noel Barber, *The War of the Running Dogs: The Malayan Emergency, 1948-1960* (New York: Weybright and Talley, 1971), 130-132.

30. TNA, CO 1022 / 7, Telegram from the Federation of Malaya (unsigned) to Oliver Lyttelton, Secretary of State for the Colonies, November 1, 1951.

31. Ibid. Lyttelton was named as Colonial Secretary on October 27, two days after the General Election had taken place.

32. Barber, *The War of the Running Dogs*, 141.

33. Ibid, 141-143.

34. John Cloake, *Templer: Tiger of Malaya: The Life of Field Marshal Sir Gerald Templer* (London: Harrap, 1985), 201-209.

35. TNA, CO 1022 / 51, Telegram from General Sir Gerald Templer, High Commissioner, to Oliver Lyttelton, Secretary of State for the Colonies, February 13, 1952.

36. Quoted in the page of Barber, *The War of the Running Dogs*.

37. For more on the aftermath of the Malayan Emergency and the formation of Malaysia, see Clutterback, *Conflict and Violence in Singapore and Malaysia*, chapters 15-20.

Panel 2—The Interagency Process: Southeast Asia

(Submitted Paper)

Counterinsurgency, the Interagency Process, and Vietnam:
The American Experience

by

Jeffrey Woods, Ph.D.
Arkansas Tech University

The American military and civilian officers involved in the counterinsurgency effort in South Vietnam, represented a myriad of bureaucratic cultures and interests. Some were conventional soldiers, trained to fight the North Vietnamese regular army. Others were special forces, proficient in counterguerilla tactics. Some were diplomats who had language skills and a deep understanding of the Vietnamese history and culture. Others were spies, skilled in intelligence trade craft. Some were agricultural, medical, or propaganda experts. Many were a mix of all of the above. Some served huge bureaucracies, others small. Some had their funding subject to strict public oversight; others spent money with little supervision. Some worked in contested areas, others in regions almost completely pacified.

Communication, cooperation, and coordination of these men and their agencies proved difficult from the beginning of the American advisory effort in Vietnam. Though they shared the common goal of defeating the Communist insurgents and building a viable non-Communist South Vietnamese government, their means often ran at such cross purposes that they were conveyed to their Vietnamese counterparts as fundamental, even irreconcilable political differences. A formal structure to better coordinate the activities of the various agencies implemented in 1967, helped to unify the message, but by that point the rifts were already entrenched and the opportunity for a war changing, unified effort had past.[1]

Mao Zedong famously tutored his guerillas that they were fish that depended on the water of the population for their survival. The Viet Cong (VC) adopted Mao's wisdom. Their primary goal throughout the war was to win the people, especially the majority rural peasantry, to the Communist cause. From the beginning, American counterinsurgency experts recognized this "people's war" as the conflict's center of gravity, and stressed the need to establish the authority of the South Vietnamese government in the countryside. They would deprive the VC of the water in which they swam. Establishing government authority, however, proved no easy task. Though recruited, trained, and supplied by the North, the VC were Southerners with kinship and friendship ties within the communities in which they operated. The South Vietnamese government, on the other hand, routinely deployed forces from the outside. Government officials, in addition, were often corrupt and unpopular. The exposing, capturing, and killing of VC by the government or its allied forces was, therefore, a complicated and potentially counterproductive activity. Security measures had to be offset with economic and social programs and carefully balanced for each of the various local, regional, and national South Vietnamese political environments.[2]

From the first days of the American involvement in the Vietnam conflict, counterinsurgency and pacification went hand in hand. Civic and political action had to be coordinated with military operations. The group originally deployed to oversee that coordination was the Saigon Military Mission (SMM). SMM was led by pacification and counterinsurgency guru Ed Lansdale who was fresh from helping Ramon Magsaysay build an army, suppress the Huk insurgency, and establish a government in the Philippines. Lansdale's SMM was CIA funded and administered, but it was independent of the Saigon Station. The members of the small twelve man SMM team, including Lansdale, were seconded to the CIA from the military or were given positions in the military as cover.

From 1954 to 1956, the overall American advisory mission was small enough that interagency coordination of counterinsurgency and pacification operations could be handled by the SMM; nevertheless, even then it was clear that serving the missions of the CIA and United States Army in support of a new leadership under Ngo Dinh Diem was politically complicated. Diem was Catholic in a country that was 90 percent Buddhist. He was virtually unknown as a public figure, especially when compared to Ho Chi Minh. He led a government that favored the urban affluent classes over the rural peasantry. He was an anti-French nationalist, but his army's officers were recruited from the ranks of soldiers who had served under the hated French. And he faced a possible North Vietnamese invasion as well as the stay-behind Viet Minh insurgency. SMM drew on Lansdale's experience in the Philippines, established PSYOP and intelligence commands, set up a civic action campaign, and began training programs to make the Vietnamese army and civil administrators more responsive to the peasants' needs. Lansdale also confronted or co-opted Diem's rivals in the Hoa Hao, Cao Dai, and Binh Xuyen sects. In the process, however, Lansdale alienated the French, who, though defeated by the Viet Minh and in the process of withdrawing from the country, hoped to retain some of its influence in Vietnam. France used its leverage as a key American ally to convince American embassy officials that Diem was too weak a reed to rely on. Lansdale ultimately had to exploit his friendship and direct communication channel with CIA head Allen Dulles and his brother, Secretary of State John Foster Dulles, to convince the Eisenhower administration not to abandon Diem as he consolidated power through 1955.

With Diem firmly established as the American backed leader of South Vietnam, the SMM was disbanded in 1956. The CIA continued to run counterinsurgency and pacification operations but mainly in small scale experiments like the Civilian Irregular Defense Group (CIDG) in the central highlands. This experiment and others like it were ad hoc and relied on local forces rather than the South Vietnamese Army. The army had been trained by American military advisors as regular conventional forces capable of withstanding an invasion from the North. With CIA help, Diem launched his version of counterinsurgency in the strategic hamlet program, where communities were moved to barbed wire enclosed compounds that could be cut off from the VC, defended, and controlled. Strategic hamlets, needless to say, made as many enemies as friends for the government. CIDG and the other local projects initiated by the CIA, on the other hand, were generally better received. They relied on what Saigon Station Chief Bill Colby called the "three selfs:" self-defense, self-development, and self-government. In the case of CIDG, specifically, the montagnard tribesmen of the central highlands, who resented the South Vietnamese government intrusions almost as much as the VC, were allowed to keep arms to defend themselves and maintain their relative political autonomy in exchange for their loyalty to Diem in fighting the VC. CIDG fought well against the VC, but as their military prowess grew, the regular army became suspicious of their motives. Diem eventually downsized the program. Like the SMM, the CIA backed CIDG program revealed underlying competing loyalties among the American advisory agencies that translated into political rivalries for the South Vietnamese. Where the CIA invested in irregular force programs like CIDG that bolstered local autonomy and decentralized power, the Army invested in regular military programs that bolstered a national sovereignty and centralized government.

The American presence in Vietnam increased from a couple of thousand advisors in the late Eisenhower era to over 20,000 in the Kennedy era. In those years, counterinsurgency and pacification efforts were divided among a number of different agencies. They were an alphabet soup of abbreviations and acronyms. The MAAG (the US Military Assistance and Advisory Group) and then MACV (the US Military Assistance Command, Vietnam) continued to provide the regular force training of the Vietnamese army and also from time to time supplemented their firepower needs. Political and economic components, irregular force training and the intelligence programs that provided the crucial links between combat and civil action planning, were continued by the CIA but were also taken up by several other agencies. USAID (the United States Agency for International Development), and its various iterations from IVS (International Volunteer Services) to USOM (United States Operation Mission), provided money, training, and logistical support for building economic infrastructure, roads, bridges, schools, hospitals, agricultural training centers, and the like. USAID also provided civil administration training for police forces and other civil administrators. USIS (United States Information Service) which became USIA (United States Information Agency) and eventually, in Vietnam, JUSPAO (the Joint US Public Affairs Office) provided communications as well as information, propaganda, and education programs. Counterinsurgency and pacification workers increasingly came from USAID and USIA after 1963 when the CIA was forced to withdraw from many of its covert operations and refocus on intelligence collection under Operation SWITCHBACK. The State Department contributed its share of key embassy personnel as well.

Each of the agencies provided much needed expertise, but cooperation could be sporadic and again the different organizational cultures, the different ways the agencies defined means and ends, often resulted in complications for the Vietnamese, sometimes even disaster. Mike Benge, a former Marine and agricultural expert who worked for IVS and then USAID in the central highlands, remembered in 1964 working with some nuns building a church near Loc Tien. He received word that some nearby villages had been bombed, so with the nuns' help, he picked up some food and medical supplies and went to the villages to distribute it. Visiting the province chief, Benge found out that the local USAID representative had had some success providing the district with money for rice, salt, fish and other supplies as part of the pacification effort, but that the villages had been bombed not by the VC but by allied aircraft acting on three month old reports of a VC PSYWAR team operating in the area. The lack of coordination between the intelligence, military, and civilian pacification components was tragic, but worse yet was USAID's reaction when Benge reported the incident. Instead of rewarding Benge for his identification of a clear communication breakdown, USAID banned him from the region. His complaints had made too many political enemies in the GVN province hierarchy.[3]

The struggle for a coordinated counterinsurgency and pacification was at its most difficult as the American military presence grew exponentially in the mid-1960s. The American military's advisory roll had been from the beginning larger than the civilian, but with the rapid military escalation from just over 20,000 in 1963 to some half a million in 1968, the American armed forces further dwarfed all the other agencies. MACV gained substantial control over the transportation systems, the buildings, the money, the guns, and the influence within the government in Saigon. Furthermore General William Westmoreland's attrition strategy relegated the counterinsurgency and pacification effort to a secondary role. The number and kinds of irregular forces grew overall from 1964 to 1966 but did not gain in proportion to the regular military. Neglect and mismanagement threatened to take the "other war" completely out of the strategic equation.

It was at this low point, however, that the push for coordinated national political, economic, intelligence, and political campaign was renewed. While agency cooperation under a unified pacification and counterinsurgency strategy had been hard to come at the national level, there were some important examples of success at the province, district, and village level. Led primarily by the civilian agencies and the local Vietnamese they supported, experiments in Long An, Kien Hoa, Hau Ngia, and Quang Ngai became models for new national programs. In Long An a USIS official named Frank Scotton organized survey teams that went from village to village learning about the community's particular needs and mapping its unique political environment. Team members lived and slept in the villages and questioned every family so as not to make any one a target for VC or GVN reprisals. Teams then coordinated CIA, USAID or MACV resources to provide the village with resources and security in the most politically sensitive way. Scotton had learned on a previous tour in Binh Dinh, while working with a Vietnamese army officer Nguyen Thuy, that it was most effective if his men helped the peasants plant crops, treat the sick, and build new schools, leaving it to the VC to condemn those projects and initiate hostilities. Communities that came under attack invariably rallied to denounce the VC and in many instances even helped repel the enemy. On some occasions VC even defected. In 1966 Scotton's methods became one of the cornerstones of the training curriculum for the Rural Development Cadre, a new national South Vietnamese force used for civic action and political indoctrination in the countryside.[4]

Boasting its own successful programs, certain members of the American military joined the civilian irregulars in calling for new national pacification and counterinsurgency efforts. Perhaps the most famous of the military efforts were the Marine Combined Action Platoons (CAPs) that helped pacify several areas of I Corps near the border between North and South Vietnam. CAPs lived and worked in the communities they were helping to pacify and, as a result, like Scotton's teams, came to know the unique problems of the village and were better able to coordinate efforts to provide for specific needs. The success of CAPs and other pacification programs eventually led Army Chief of Staff Harold K. Johnson and then General Creighton Abrams, Westmoreland's deputy, to advocate a major reassessment of America's Vietnam strategy. Johnson and Abram's study, "A Program for the Pacification and Long-Term Development of Vietnam," known as PROVN, was published in March 1966. PROVN declared that there was "no unified effective pattern" to American actions and called for a greater emphasis on pacification in the allied war effort. Under the advice of returned counterinsurgency and pacification guru, Edward Lansdale, Ambassador Henry Cabot Lodge quickly added his support and helped persuade President Lyndon Johnson to make the PROVN arguments to the South Vietnamese leadership at the 1966 Honolulu Conference. The most important force behind the new initiative, however, was Robert J. Komer a smart, intense, ambitious man who had been running the Middle East shop in the NSC. Komer and CIA Far East Division chief, William Colby, made sure that field reports from counterinsurgency and pacification officers like Scotton made it to Washington around the MACV command who typically weeded such reports out.[5]

Shortly after the conference in Hawaii, President Johnson named William J. Porter to be deputy ambassador to Vietnam and head of the revitalized pacification program, but in the latter half of 1966, Komer persuaded the White House to turn the operation over to him. Komer gave the program a new title, CORDS, short for Civil Operations and Revolutionary Development Support, and began building the first combined civilian-military command in US history. CORDS encompassed all of the typical pacification activities: economic improvement, security, and political development, and its officers held both military and civilian ranks. Province and district level advisors were recruited from MACV, USAID, USIA, CIA, and the State Department, and CORDS acted as a liason between those agencies. Special forces were also heavily represented, especially in the Provincial Reconaissance Units and Phoenix programs that were designed specifically to root out the VC infrastructure. By early 1967, the coordination of the American counterinsurgency and pacification effort had reached its apex.

CORDS offered many solutions, but it was never the panacea its supporters had hoped it would be. First, the organizational cultures of the military and civilian agencies continued to cut divergent paths in the Vietnamese political landscape. Coordinating pacification forces and regular military forces remained particularly difficult. Second, the 1968 Tet offensive changed everything. Tet left the VC decimated, but it also sapped the political will of the United States to continue the war. With tragically bad timing, Washington made its first steps toward Vietnamization at the same time Hanoi chose to deemphasize the guerilla insurgency in favor of a purely political campaign in the South that would be backed by conventional North Vietnamese regular army operations. CORDS officials recognized the opportunity to fill the political vacuum while the Viet Cong were weak, but Saigon refused to act quickly enough. Thus CORDS was never really tested in the context in which it promised the most success.

As much as some would have liked it to, CORDS could not replace the existing advisory agencies in Vietnam. The best it could offer was a bureaucratic overlay that facilitated better communication. Personnel from USAID, USIA, CIA, State, and MACV seconded to CORDS experienced generally improved coordination, but the different agency cultures, their varied means of fighting the Communists, still remained a barrier to unified action. The rift between regular military forces and the civilians and irregulars focusing on counterinsurgency and pacification, in particular, remained a problem. Ken Quinn, a foreign service officer with no military training who served four tours with CORDS in the Mekong Delta, posed the dilemma this way: "There was always a little cultural difference between civilian and military [agencies], but one of the great lessons of CORDS was that it was not just the different colors of your clothes but where you sat that made a difference in your attitudes. The army guys who were in the MACV team and I generally saw eye to eye, but it was a different view than army guys in same town who were advisors to the ARVN 9th division. The two groups saw different wars from different perspectives with different counterparts."[6]

Bruce Kinsey, a foreign service officer working for CORDS in Long An, recalled similar problems. Despite generally good relations between personnel working for CORDS, he remembered miscommunications with the regular military that had very real and tragic consequences:

> There was a village that I worked with like crazy. I had a cadre team in there. We strung up barbed wire and threw down tin cans on the perimeter so if the guerrillas came in you could hear them more easily. And we set up a school and the VC blew it up. It was fighting tooth and nail. The third brigade of the ninth division came in and set up at the end of the road that went through this hamlet. They ran these huge deuce-and-a-half trucks through there full of garbage, and ammunition, and god knows what else. They were scared to death, so they ran them at fifty miles an hour. They killed like eleven Vietnamese kids. I talked to those people until I was blue in the face. And I put up signs saying, "US Drivers—Friendly Hamlet—Slow Down" and they wouldn't.[7]

Despite these ongoing issues, CORDS successfully unified the various agencies' efforts in many key regions, especially in the months following Tet. In Tay Ninh Province, Terry Lambacher, an ex-Green Beret who served a combat tour in Vietnam before joining USAID, was among the many CORDS district representatives who took advantage of the post-Tet political vacuum. When the local VC leadership exposed itself in the Tet attacks, Lambacher's teams, including CIA, Special Forces, and Phoenix personnel, were able to pinpoint the top ten VC leaders in the district and target them individually. In each case they used the technique that effectively neutralized the enemy while still maintaining community support. In some cases they embarrassed family members into turning in VC relatives. In other cases they deployed national police to make arrests. In others still, they lured the Communists into firefights. In all cases they made sure the South Vietnamese government took the credit for the operations and government officials justified their actions to the community. Lambacher made sure the district chief, in particular, received most of the accolades. That favor could then be used as leverage in convincing the chief to limit corruption. As Lambacher remembered it, CORDS went beyond "coordination," and even beyond "joint" action. It was a unified counterinsurgency effort that was winning the war in his district.[8]

Larger political forces, however, intervened. Nguyen Van Thieu, who had joined with his political rival Nguyen Cao Ky to wrest control of the government in 1965, had been chosen chief of state in September 1967 having gained only 35 percent of the vote in an election that was only marginally free and fair. Just months into his presidency, Tet constituted an early referendum on Thieu both in South Vietnam and in the United States. Thieu was slow to act, showing more concern for his urban constituency and his personal power than for the "people's war." It took American leaders one hundred days to convince Thieu just to move his forces back into the countryside after Tet, and when he finally did, the soldiers arrived as occupiers rather than pacifiers. A full six months after Tet, Thieu finally accepted a CORDS proposal to initiate an Accelerated Pacification Campaign (APC), a plan designed to pacify 1000 new hamlets and meet new Phoenix quotas for eliminating VC infrastructure. Once implemented, though, the Accelerated Pacification Campaign bypassed real rural political development for short term security and control measures. The government's return to the countryside became just a land grab to be used as leverage in negotiating a settlement with Hanoi and creating a decent interval under which American withdrawal and Vietnamization could take place. Thieu effectively abandoned the people's war just as the Communists, so weakened militarily by Tet, could only mount political operations in the South. The South Vietnamese government had belatedly gained a greater physical presence in the countryside, but the VC were filling the political vacuum.[9]

CORDS continued through 1972, but Thieu's continuing reluctance to launch political operations devastated morale. A small group of American pacification and counterinsurgency experts that included Ed Lansdale, John Paul Vann, and Daniel Ellsberg advocated finding a leadership alternative to Thieu, one that could negotiate a compromise that might save South Vietnam. They found a champion in Vietnamese national assemblyman Tran Ngoc Chau. Chau had long experience studying guerilla tactics. He had been a guerilla himself with the Viet Minh, studied American tactics based on Ed Lansdale's experience in the Philippines at Ft. Benning, and saw firsthand British counterinsurgency on two trips to Malaya during the Communist crisis of the 1950s. As a province chief in Kien Hoa, he had collaborated with Rufus Phillips, a Lansdale disciple and USOM advisor, in systematizing a pacification formula that became among the most effective in Vietnam. Chau's Census-Grievance program was at the heart of the formula. It was a formal census consisting of interviews with the heads of households about local social, economic, and security conditions. Census-Grievance teams, recruited from the local population, elicited complaints about both the VC and the GVN, which allowed Chau to custom design civic action, encourage citizen loyalty through responsive government, and leverage intelligence for surgical strikes against the VC. Chau brought his Census Grievance program to the national Rural Development Cadre training center at Vung Tau where he briefly served as head of instruction. Returning to Kien Hoa to begin a career in politics, Chau was elected a national assemblyman and mounted a significant opposition to Nguyen Van Thieu. In the words of Historian John Prados, Chau would be "among the first to understand the impact of Tet would be to move the United States to search for a way out of the war." Chau would push for "reasonable negotiations and a settlement while Saigon still retained bargaining power." Thieu, of course, worked to prevent any such settlement. The president eventually had Chau arrested for treason on trumped up charges that Chau was conspiring with his brother, a Communist agent, to overthrow the government. Ironically it was CORDS head Bill Colby, CIA Saigon Station Chief Ted Shackley, and Ambassador Ellsworth Bunker who signed off on Chau's arrest. For Chau's supporters in CORDS, it was the last and most disastrous of the American political choices. As Doug Ramsey, a USOM worker who would be captured in 1966 and spend seven years in a VC prison camp, put it, the United States ultimately supported "traditional bribe-soliciting and patronage dispensing military politicians like Nguyen van Thieu . . . instead of socialist-leaning counterinsurgency and village administration experts like Tran Ngoc Chau." They had lost the key ingredient in winning over the rural peasantry, a "solid, honest, civilian democratic modernizer" in the presidency.[10]

One of the crucial things we can learn from the Americans who were on the front lines coordinating the counterinsurgency war in Vietnam is that success and failure is sensitively dependent on political conditions. That is no revelation in and of itself. The inherently political nature of warfare has been recognized in military doctrine from Sun Tsu to Clausewitz to Mao. But the front line American counterinsurgency warriors in Vietnam offer us a glimpse at the mind boggling complexity of the political contexts in which they had to operate. Though they all shared the common goal of defeating the Communist insurgents and building a viable non-Communist South Vietnamese government, the bureaucratic cultures of the agencies they served and the varied tactics they employed translated into fundamental, even irreconcilable political differences when applied in country. Where some encouraged bottom up, rice roots, local self defense, and decentralized control, others supported top down, national army, and centralized authority from Saigon. Both approaches were necessary in fighting the insurgency in the South and preventing an invasion from the North. But the separate paths fostered factionalism that in part prevented South Vietnam from achieving a decisive political victory in the countryside. When the American advisors achieved a new level of pacification and counterinsurgency coordination with CORDS, it was too little, too late. For all its potential, CORDS could not completely erase pre-existing agency rivalries, could not overcome the new strategic reality caused by Tet, and could not find the crucial balance of local and national government.

Notes

1. The secondary literature on the counterinsurgency and pacification campaigns in Vietnam is extensive. Most important for this study were Thomas L. Ahern, Jr., *CIA and Rural Pacification in South Vietnam* (Washington, DC: Center for the Study of Intelligence, 2006); Richard Hunt, *Pacification: The Struggle for Vietnam's Hearts and Minds* (Boulder: Westview Press, 1995); Neil Sheehan, *A Bright Shining Lie* (New York: Random House, 1988); Zalin B. Grant, *Facing the Phoenix: The CIA and the Political Defeat of the United States in Vietnam* (New York: W. W. Norton, 1991); John A. Nagl, *Learning to Eat Soup with a Knife: Counterinsurgency Lessons from Malaya to Vietnam* (Chicago: University of Chicago Press, 2002); William Colby and James McCargar, *Lost Victory: A Firsthand Account of America's Sixteen-Year Involvement in Vietnam* (Chicago: Contemporary Books, 1989); John Prados, *Lost Crusader: The Secret Wars of CIA Director William Colby* (New York: Oxford University Press, 2003); Lewis Sorley, *Honorable Warrior: General Harold K. Johnson and the Ethics of Command* (Lawrence: University Press of Kansas, 1999); Cecil B. Currey, *Edward Lansdale: The Unquiet American* (Boston: Houghton Mifflin, 1988).

2. "Insurgency" is most often defined as a revolt against an established authority or government. Definitions of established authority and government could vary wildly among different people at different times and in different regions in South Vietnam. In some areas of the South, it could be argued that the Viet Minh and then the Viet Cong held more authority for a longer time than the GVN.

3. Author Interview with Mike Benge, Telephone, 10/30/06.

4. Author Interview with Frank Scotton, October 12-14, 2007, Deland Florida; Author Interview with Frank Scotton, April 8-9, 2008, Fayetteville Arkansas.

5. Author Interview with Jean Sauvageot, January 5, 2008, Washington, DC; Lewis Sorley, *Honorable Warrior: General Harold K. Johnson and the Ethics of Command* (Lawrence: University Press of Kansas, 1999), 238.

6. Author Interview with Kennth Quinn, November 7, 2007, Telephone.

7. Author Interview with Bruce Kinsey, September 20, 2007, Telephone.

8. Author Interview with Terry Lambacher, October 16, 2007, Telephone.

9. Merle Pribbenow, *Victory in Vietnam: The Official History of the People's Army of Vietnam, 1954-1975* (Lawrence: University of Kansas Press, 2002).

10. Tran Ngoc Chau, Unpublished Memoir Provided to the Author; Interview with Jean Sauvageot, 1/5/08, Washington, DC; Douglas K. Ramsey, *Moth to the Flame* (unpublished memoir, 2006), IID, 3, IVC 11-15, 31-35, 45-49, IVD 22.

Panel 2—The Interagency Process: Southeast Asia
Question and Answers
(Transcript of Presentation)

Benjamin Grob-Fitzgibbon, Ph.D.
Jeffrey Woods, Ph.D.

Moderated by Donald P. Wright, Ph.D.

Dr. Wright

I will start off with the first question. What came to mind, as I was listening to both of your papers, has something to do with the work that we do at CSI and all the attention paid to both Afghanistan and Iraq. I am curious to know, when you looked at the interagency processes, both Malaya and Vietnam, to what degree did these processes feed into a better intelligence effort? There is a lot of discussion right now about the creation of fusion cells in Iraq and the ability to pull in the CIA, coalition intelligence agencies, you name it, all kinds of acronyms together, and unless you do this you are not going to really be able to create a coordinated effort against insurgents as well as a larger, something they call a "pacification effort," if you will. So any comments on that as you looked at your particular cases?

Dr. Grob-Fitzgibbon

One of my favorites, which I had to cut. The time restraints led to that. Briggs actually became ill in November 1950 and had to leave his position as Director of Operations. He died shortly thereafter and Gerald Templer came in. Templer looked at the situation in Malaya and said what Briggs has started is fantastic as far as the new villages are concerned and Briggs is spot on as far as drawing the commies and insurgents away from the villages to take on the British Army and isolate the position so that the civilian inhabitants would be spared the warfare, but he said there was no consideration given in Briggs' plan to intelligence coordination. So shortly after Templer took over in 1952 he established intelligence centers throughout Malaya with the intention being that they had army representatives, RAF (Royal Air Force) representatives, navy representatives in coastal states, police representatives, civil administration representatives, and those intelligence networks would have liaisons throughout the Malaysian community where information would be brought into the same building with police officers sitting next to an army officer sitting next to a civil servant sitting next to an RAF officer where they would all be reading the same intelligence and sharing the same information, and it could then be taken out to the plans on the ground.

Dr. Woods

I hate to be flippant and too general about this, but Vietnam was about families. Families are so important so this local-level approach, interagency approach, you do something good for the community, build them a road, build them a school. You make the Viet Cong into the enemy by having to attack that. It makes you friends. It makes you people that you would talk to and chances are they are going to be related to somebody who is a communist, like Chau himself. This guy who was championed by some of these CORDS officials; his two brothers were communist intelligence agents. He spoke with them a couple of times, in 1965 and 1966, and Thieu used that as evidence that he was conspiring to overthrow the government when really it was just the opposite. He was gaining information from those guys. He was using it against them so it was getting into that family structure and I think you cannot do that just with what he can do with guns. You can provide security for people which they can appreciate, but it is almost like you have to become part of the family. You have to be there with them. You have to know enough about them to have a relationship that allows that kind of intelligence to happen. So you need CIA, you need USAID, and you need those other agencies that have expertise in doing that, and plus who have the language skills and know the historical culture well enough to be able to do that.

Audience Member

I have a question for Dr. Woods. I understand your approach to Thieu. He is not exactly a lovable character, but at the same time, after Tet he did, with some prodding, move out into the countryside, and I am not sure, you may be overstressing the fact that he abandoned the people's war. This was after all the man who, on our advice, set up the People's Self-Defense Force Militia going beyond the regional forces, popular forces sending, literally, hundreds of thousands of small arms out to the countryside. Again, not entirely trusting them, but willing to take the chance with them and by the end of about two years the People's Self-Defense Forces were fairly strong, fairly trained, and fairly well armed. This is not the act of somebody that fears that all those people are going to come up and rally against him. So I think he was willing to take a few more risks than you have given him credit for, a combination of that program and the Revolutionary Development Program which continued to expand dramatically in 1969 and 1970, I think, indicates that he was continuing to see the possibilities in mobilizing the people. Any thoughts?

Dr. Woods

I think that is fair. I am generalizing a bit from having to do the paper in this context, but it did take him a hundred days just to get out of the cities after Tet. And it took him a full six months to approve something like the APC (Accelerated Pacification Campaign) which could have been done immediately afterwards. Well, maybe not immediately, but faster than six months for sure. What I think is that that time was crucial, particularly if you are going to have an influence both on Vietnamese public opinion and American public opinion about those things. If you can fill that gap quickly you might have an effect on the political ramifications of Tet. Now, maybe not; we do not know. We did not test that. It did not happen.

Audience Member

Well, attacks continued into May. So he was still on the reaction side of the house, but I understand your point.

Dr. Woods

Not quite as severe attacks.

Audience Member

Thank you.

Audience Member

If I got it right, this guy Briggs shows up, a retired military guy. Things are not going well. He runs around for two weeks, comes up with a structure which seems to work until the end. Then they do some studies and then within a month or two he comes up with a plan which pretty much works until the end.

Dr. Grob-Fitzgibbon

Yes, that is correct.

Audience Member

I think that is pretty amazing that that happened.

Dr. Grob-Fitzgibbon

We have to be careful because, of course, the war did not end immediately after Briggs came in.

Audience Member

That is where I was going actually. You said by 1952 the coordination was absolute and very good, and I do not doubt that, but even given all that, the right structure and the right approach, it still took another seven or eight years to make it happen.

Dr. Grob-Fitzgibbon

Yes, absolutely, and when Briggs actually became sick and then died, he was extremely pessimistic upon his death bed because out of the 300,000 Malayan squatters that were present when Briggs' plan was implemented only 25,000 had been resettled and he had originally laid down that the Briggs Plan would be completely implemented by the end of 1952. As it was, the settlement did not take place until the end of 1954 so it was almost double the time frame that Briggs had expected because of the usual sort of delays every general imagines with implementing that sort of plan. But what we did have, and I can give you the exact information if you need it. Again, I skipped over this part for the sake of time. Although the emergency did not end until 1960, we do have a gradual betterment of the situation in certain provinces. So by 1957, the Federation of Malaya was granted independence so we still have a counterinsurgency coming and going for another three years after that. From a British perspective there was an emergency going on, but from which the command of the forces and interagency processes were transferred over to the Malayan government. That was six years after the Briggs Plan was implemented. But again, you are absolutely right. The British forces and the British administration found a hierarchical structure that worked and found a way to coordinate them. Even within that coordination, as Briggs himself said, this is going to take time, even when it appears that the Malayan populations are settled, are peaceable, or are in favor of some sort of British administrative structures, we cannot at that moment let up our guard because we have to ensure that the ideology on the line of insurgency has been defeated before we can give way on some security measures.

Audience Member

Dr. Woods, I will use a Thompson sort of quote, a version of it, but basically, looking at Vietnam a little later, he said that in Vietnam the counterinsurgency had three components—nation building, pacification, and military affairs, and that nation building was the most important because it built the capability of the nation to do what it needed to do. Pacification provided the linkage between the central government and folks in the countryside, and of course, the military component for security was always essential. And then he added that the Americans, when they did those, if they did them all, tended to do them in the reverse order. And while you are talking about CORDS and pacification there, I understand that, but do you have any insights about the interagency on the nation building side of that? Not just the pacification program that came in 1967 through the CORDS program.

Dr. Woods

I wish I did. That is actually the side that I am working on right now. Vietnamese politics is very complicated and the literature on it is not real sophisticated right now. So that is the area that I am going into. We do run into problems. I focused a little bit in my research on Bill Colby and trying to get some insight and what is he doing in Saigon and the current influence he is building, particularly in the CORDS years and his relationship with the Interior Ministry and that sort of thing. Unfortunately, I do not have access to documentary evidence and things like that so I have to do it through an interview kind of process. All I can tell you is that I am picking away at it. I do not know at this point, but you are right. That is the fundamental question.

Audience Member

You can get some hints of that, at least from the American perspective, by going to the foreign relations documents, but again, that is just from the US side.

Audience Member

I have a question for both of you gentlemen. In the interagency process, given that both nations had a consensus as we heard earlier from Dr. Yates, maybe one that has not returned, but a consensus nationally that these were emergencies and needed to be addressed. Both nations have their focus on NATO and the central front and both of these operations were supporting efforts. Did either nation have an advantage over the other in the process? What is the difference?

Dr. Grob-Fitzgibbon

That is an excellent question. Let me think on that for just a minute.

Dr. Wright

At least your paper, Benjamin, argued that one of the advantages that the British commissioners brought was their times in other colonies. I think Palestine and some of the African nations.

Dr. Grob-Fitzgibbon

Yes, I think there are a couple of advantages. I think that is one, that the British had been running an empire whereas the Americans had not. So, when somebody like Gurney comes in he is not approaching how to deal with colonial subjects fresh; how to deal with some type of occupation policy. He is somebody who has been in Palestine for two or three years and dealt with the Palestine insurgency or Jewish insurgency against the British. He has been in Africa. He has been in Burma. So you have this tradition, if you will, of colonial administration which I think is a sizable advantage for the British. I also think in the situation in Malaya, what you have is that the British have been in Malaya in some aspect since 1608 and certainly controlling certain Malayan states and colonies since the 1800s. So what that does is the majority of the Malayan population, when the emergency begins in 1948, are familiar with British rule, are familiar with certain British officials. They already have British administrative structures established there. So rather than seeming as occupiers in that land, it is a certain minority of Chinese who seem to be rebelling against the system that was already in place rather than a system being imposed on the population which was never there previously; so a combination of what you brought up and that experience of the British in Malaya for a long period of time. I also think the British in the 1950s were not any way shy or embarrassed about being an empire, actually having long-term control over Malaya so I think for the British in 1951, 1952 to say that they were bringing civilization to Malaya and that the western way of life was self-evidently better than the colony's way of life. It was perfectly fine in the moral aspect of foreign policy to compel Malayan entities into these villages to give them British administrative structures, to give them British social health, education, and that sort of thing. It was fine. That is what the Brits had been doing for 250 years. I think to the Americans the whole idea of, I know we like to spread our values, but the idea of actually obtaining a colonial situation was almost squeamish in this country.

Dr. Woods

Is that answering your question?

Audience Member

It does. I understand that the British do have a colonial tradition. Obviously, we are both speaking the same language, sort of. The point is we had been, the United States, had been in Vietnam really since 1944 and so by 1964 this was not a new process. We had administered the Philippines and parts of China and Cuba. This is not a new process, but yet it seems continually just in the short period say from 1962 until 1968 to be an incredible thrash, although we had as you said, political consensus earlier and I do not understand why that process was not there.

Dr. Woods

Part of the problem is that the specifics of all the situations are so different. You can help control what happens in say Japan and Germany in the late war because you have a big occupying force and they have admitted to defeat and that sort of thing. Vietnam is a limited kind of conflict by nature. I think that is part of it. The Philippines, it depends on what you think of the interpretations of this kind of stuff, but Lansdale thought it was all about *[inaudible]*. You had to find the right person who could be both in touch with the people and provide a coordinated national leadership and that was the success for *[inaudible]*. In terms of the larger issues that Ben is talking about, yes, we have this kind of . . . our anti-imperial culture is as prominent as anything in all of this. You do not want to go in and tell people exactly what you want them to do. Lansdale's approach to it is that you are bringing people to realize what is good for them already. There is sort of this anticommunist assumption there, but that is kind of what he is arguing now and he is an advertiser. What you do in advertising is not sell people stuff; you give them access to the things they already want. So it is a different kind of process, and there is a real reluctance to be that heavy-handed. And I think that in some ways constricts the interagency process. We want it both ways. We want to be able to go in and help and control the thing, and provide control, but at the same time we do not want to be imperialistic about it. That may be the fundamental conflict there.

Audience Member

I have a comment and then a question. Again, one of the things that I came across that is interesting in the National Security Act of 1947 and during the war looking to the British model for some of the interagency coordination, military and civilian-military, but what emerged, and came out afterwards, was also the distinction between the parliamentary or the cabinet system and the presidential US system where you have one executive and a cabinet that has a built-in political coordination capacity on the American side and I just throw that out. That is probably a challenge and I will hit on it a little bit for us now too, not contradicting anything either one of you have said, but puts it kind of in that bigger shell that we have this problem in terms of the political organization and the apparatus.

Dr. Woods

I keep going back to the *Federalist Papers* for some reason. I do not know why.

Audience Member

So the comment and question, if I may, I think I understand from both of you, but I just wanted to put this question out, how much of it, if this is a fair question, how much of it then on the back of the envelope sketch could you say is the result of individuals that were at the right place at the right time or the wrong place at the wrong time as opposed to structures and processes that were, in fact, survivable even in both cases, as ongoing institutions and institutional processes? Thank you.

Dr. Grob-Fitzgibbon

I am a big believer in personalities in history so I would certainly subscribe to that interpretation. In the case of Malaya the choice of Briggs was very fortuitous in that he came along and he had the idea and the forceful personality to push that through. And of course, when Templer comes in he is even more of a forceful character and I think throughout the British imperialist experience you see the impact of personalities and recognition of that impact. My larger work expands beyond Malaya to other counterinsurgency operations in the empire, and you see time and time again whether it is Malaya, Cypress, Kenya, Aden, Dhofar, memorandum cautioning against the lessons learned in an organization or an apparatus fashion because what they say is that in each of these campaigns, whether it be Cypress, Kenya, Malaya, what have you, the situation on the ground was so completely different, the context so completely different that they need to be managed in a different manner and the type of government, whether you use strict interagency processes in Malaya where we used more dictatorial power, in Cypress where we used the local sector population, and in Kenya that was determined by the situation on the ground so you could not take the DWEC/SWEC model and put that on Kenya or put that on to Cypress. You also could not take somebody like Templer and put them using the same methods of management in Cypress or in Kenya because that would not work. But again, throughout all these memorandum, back to what Jeff was saying, you see a patronizing attitude, if I may call it that, that I think brought American history to be somewhat uncomfortable with what the British have always, right through to the 1970s have been quite convinced that their way of government and life were superior to all others, including the American system, and that it was their right and duty to compel others to take them.

Dr. Woods

I think the heart of your question is the really important thing and that is, what of these things can you control? When you start to bring up personality, does that become more a sense of chance and luck, like you said, right person, right time? You cannot make that happen most of the time. The people I interviewed were mostly civilians from the various civilian agencies. They have the hardest time, I think, talking about this kind of thing. They want to make a champion of somebody like Lansdale because he was in there early and seemed to have a good idea. It seemed to have worked in the Philippines. Why could it not work in Vietnam? But then he is pulled out of there. Then you wait for someone like Komer and I hear people talk about how crucial Komer was to the CORDS project. Yes, that is true, but there are a lot of other people who are pushing for it as well. He just happened to be, how should we put it, very vocal about his . . . "Blowtorch Bob." So yes, it was sort of a good circumstance that he was in there, but could Bill Colby have done the same thing? He has as much influence, but he is not the spokesman that Komer was. So there is a degree of chance, but the problem with that is that you lose control. In a way, I am more interested in things that you can control, I guess. I think there is a lot you cannot, and a realization of what you cannot is vitally important, but that said, I think control is what we are all searching for, right?

Audience Member

I want to press this issue of causation because I think there is a danger here in concluding that interagency cooperation worked in Malaya because it worked, and in Vietnam was, as you argue, it was too little too late because it obviously did not work. In one mission it succeeds; in one mission it fails. And I wonder if I could press both panelists to talk about how we can look at these events historically and determine not whether or not these succeeded or failed, but whether they were decisive or how they were decisive or whether there are alternative explanations, things that have to do for instance with the internal coherence of the insurgency or take your pick of any number of other factors. So it is obvious, I think, that cooperation is, in and of itself, a good thing. My question is, how can we test historically to see if it was a decisive thing? Thanks.

Dr. Grob-Fitzgibbon

I think in the case of Malaya we have a nice breakdown in emergency and ask historians to study it before the Briggs Plan was implemented and after the Briggs Plan was implemented. We have the years from 1948 until 1950 where a particular military approach was taken where the British essentially considered this to be a continuation of the campaign that we fought during the Second World War, just delayed by about three years and with a slightly different enemy, but essentially the tactics to use against Malayan communists were safe to use against the Japanese, and that was actually a misunderstanding of the situation because the Japanese had been viewed by the Malayan people as occupiers and thus the British fighting the Japanese occupiers were not really the enemy, but ethically the Chinese guerillas were somewhat less occupiers. Then Briggs comes in and implements his plan and again, I think he as a person is as important as his ideas and plans, but he is actually only on the ground managing that for seven or eight months before he moves, yet it is his idea which continues. Prior to interagency coordination the British were losing ground in Malaya, the communist organization was growing which attracted more members, numerous states were falling into a status—security-wise—which would be considered close to anarchy, the British not really having control, and then once the Federal War Council was established according to policy, and once the SWECs are established, once the DWECs are established, although it is going to take another nine years, we do see a gradual regaining of ground by the British and gradual ostracization of the Malayan Communist Party within Malayan society. So whether that is answering the first question which you cautioned against as to whether it was success or failure, I think it does answer the question as to how decisive was the British plan. Certainly the situation on the ground changed greatly after 1950, yet the Malayan Communist institution itself did not have a dramatic restructure and they did not get any leadership in 1950, none of the other variables were changed.

Dr. Woods

Cause and effect in history is incredibly complicated and I tell my students the one thing you can learn from history is that things cannot be simplified. You cannot break it down, and unfortunately, when we come in here to conferences and things that is kind of what we do. We break down elements and things like that, but there are always other causations. There are always other things going on. In this particular case, my instinct is to look at implementation on the ground, at the local level and how it affects the people that you are dealing with. So when I juxtapose somebody like Bruce Kinsey's experience and his problem in trying to conduct pacification efforts at the same time the military seems to be killing civilians in his region versus somebody like Terry Lambacher who is taking advantage of Tet and getting great cooperation. Both of those are under CORDS. Both of those are supposed to be under a kind of cooperative environment. They are different places with different cultures and influences, but that is the juxtaposition that I am setting out. Like I said, CORDS is not a panacea. I cannot make the statement that it is going to save everything. I cannot. But it was one of those untested areas, one of those areas that was tried, but I think under different circumstances might have had a better opportunity. Again, that is a wishy-washy answer, but that is history.

Audience Member (From the Blogosphere)

Given the mixture of views on how effective interagency has been throughout the historical vignettes discussed during this symposium and our trend to backslide on war-time lessons learned in times of peace and tighter budgets, i.e., in World War II we had a Military School of Government in Charlottesville, Virginia collocated with the University of Virginia to train civil administrators. Is it finally time for legislation, a Goldwater-Nichols for the interagency?

Dr. Grob-Fitzgibbon

I know from the British perspective, they have always been hesitant to make any sort of formal lessons learned procedure, sometimes to their detriment. I know in Cypress in 1956, you still have the Malayan emergency going on, you have had the Kenyan emergency going on for five years, the NLF (National Liberation Front) in Aden, four or five insurgency campaigns being waged simultaneously and the governor of Cypress, who had himself been in Malaya just the year previously in an official capacity, wrote back to the Chief of Defense and said, we are doing this in Malaya and it worked quite well, should we try to attempt an interagency system in Cypress? And he was shot down very quickly. He was told in no uncertain terms that Malaya is Malaya, and Cypress is Cypress. We get into situations and different wars and whenever we get involved in insurgency situations we have to look at the situation on the ground. We have to look at local contact and we have to develop a new form of plan. Now in saying that, does that mean that we should say that interagency works in some situations and not in others? I do not think so. I think we can look at situations of Malaya and central Kenya and other areas and say for each of them that interagency coordination of a certain type was attempted and a certain type worked. And in any situation where an interagency cannot promote the objective of the military only approach the British tended to do worse, but I think as far as any systematic we should always have a SWEC, we should always have a DWEC, we should always have a Federal War Council. I do not think we can bring that sort of certainty to the process. So any sort of legislation in recommending abstractly how insurgency campaigns should be run or how interagency processes should work, I would shy against them.

Dr. Woods

I certainly do not want to be a political advocate here. I do not want to advocate for any kind of legislation or anything, but I do understand the question though. We were talking about this actually before we came in here. This curious notion of lessons learned. I had a professor in graduate school who was very against that whole idea that you could get lessons learned from history. He was sort of willing to leave that to a political scientist and others, but he said what you could learn from history is instinct. And instinct was built on very particular knowledge and a very particular place to a very particular time. So you could use the methodology to gain that kind of instinct on the ground and it allows for adaptation, right? But if you do a strictly lessons learned kind of thing, you will run the mistake of repeating past failures in the wrong context.

Dr. Wright

Does anyone want to be a political advocate for an interagency Goldwater-Nichols Act? We have one.

Audience Member

I want to touch on some of that. It is not particularly an advocate for Goldwater-Nichols to the interagency. There may be other things much broader than that, but it comes back, I think, to the point that we need to look at structures and processes that we created most of what we have in 1947 and the years after. It is at least worth looking at whether or not what we created and that which we can legislate and change is worth looking at again. So whether it is Goldwater-Nichols, I will argue that taking all of that off the table is very dangerous because eventually it says that we have to deal with what we created ourselves from now into perpetuity whether or not the problems we are facing are at all relevant to the same structures and processes.

Panel 3—Interagency Efforts at the National Level

(Transcript of Presentation)

21st Century Security Challenges and the Interagency Process: Historical Lessons About Integrating Instruments of National Power

by

Robert H. Dorff, Ph.D.
Strategic Studies Institute

Thank you very much. It is a pleasure to be here. Indeed, I am going to take up somewhat indirectly at first, but I think then more directly toward the end of my presentation, the very question that came up with the person coming in from the streamed video world out there. I want to take a look, as the title suggests, to the 21st century challenges in the interagency process, and to do so in something of a historical approach which I will jump right in and say as a matter of disclaimer that first of all my views are not the official policy or position of the Department of the Army, Department of Defense (DOD), US government, or probably any other thinking human being. I am also a political scientist; therefore, I can violate all those good principles our fellow historians were telling us about—lessons that we should not try to learn from history.

I can just jump right in there and make all kinds of mistakes and draw all kinds of wrong lessons and so on and that is just part and parcel to what we as political scientists do anyway. I will say also that I had hoped with this paper, and I will still have because hope is not a method as we all know, at least three historical cases—the National Security Act of 1947, the Goldwater-Nichols Reforms of the 1980s, and then the contemporary case. My idea was to begin with setting a framework for looking at the historical attempts to redesign, or to design and redesign and reorganize and see if there is anything we can learn there that would help inform this very relevant and current debate that we are having right now about integrating civilian military capabilities in a whole range of operations, but they change the names so fast. It was only recently SSTR (Stability, Security, Transition, and Reconstruction). Stability and Reconstruction came first, and then SSTR. I think we are back to Complex Operations now. You know we are really in trouble when we come up with these large umbrella terms that can mean almost anything, but at least what we have now with the attempt of DOD and the State Department and USAID (United States Agency for International Development) to work together and create this consortium called Complex Operations suggests maybe that is the new term. But the idea was to how, if at all, looking at some of these cases might inform that. I will stick to at least part of that. I will introduce a very brief framework that will not be filled with social science jargon. I am sure most of you will be pleased to know that, but I want to use it to look especially at the National Security

Act of 1947, and I will gloss over Goldwater-Nichols very briefly in hopes that I can spend a little bit more time on the contemporary, and end with that as something to draw our discussions. I will take a broad view of the interagency process, which is not only what I call at the strategic level, which would be in Washington, DC, but also on the ground and argue perhaps that in the end what is happening on the ground is often more important than what is happening in Washington, DC. I will make the argument that if what you are doing on the ground is not also supported and sustained at the strategic level, the likelihood that it is going to be sustainable over time and therefore successful in these complex operations is not very good.

The paper is largely complete. The historical cases took up much more space than I had planned, probably stemming from practicing without a license as a political scientist in the history realm, but I am still trying to struggle to find the right balance between those. And let me say this up front, my interest here is not in describing the interagency process. So this is not a historical study of what it is and how it was created. I am really interested in looking at the role of the interagency process in integrating. This will become an important point later because I am going to make an argument that the difference between coordinating and voluntary coordination and the goal of integrated use of national instruments of power is one of the fundamental challenges that we face. And I will change a little bit from my planned presentation if time allows. I have some slides, especially on the contemporary challenges, which I will try to get to. So you are going to be staring at that one for a little while and perhaps I am going to conclude before you stare at any other slide. And you can ask yourself, "I wonder what, if anything, came behind that title slide?" All right, all of that aside.

The impetus for this paper, some work I have been doing for a while now is, as I said, really in the recent calls for reform, reorganization, what have you of the interagency process. Certainly, a lot of us know that the attacks of September 11, 2001, caused a lot of rethinking in terms of who the enemy was, how we were going to fight, what we needed to do across the board in a number of different ways other than just simply military operations. And that certainly accelerated a debate about whether or not we are organized properly to conduct national security strategy and policy in the 21st century.

But my first point, as I have argued before that really fits with the historical framework, is that there were many arguments that long pre-dated 9-11 about the changes we needed to be thinking about in terms of how we are organized to conduct national security and defense policy. They go back at least to the end of the Cold War. Some go back even further than that, but they have their origins in some things that were personal interests of mine, notably arguing, I guess 15 years ago now, that one of the new strategic imperatives that we were really going to face is what I will come back to later and call the good governance deficit. That is the whole panoply of things such as failed, failing, fragile, and weak states, ungovernability, and a host of problems that grow out of them. We will come back to that, as I said, if time allows. I am going to do my best to make sure it does.

The point made earlier and I just want to mention this again. I am one of those that adheres to the view that there is no such thing as the interagency as a noun. It is not a person, place, or thing, but that is an important thing to say. It is not amorphous, often shifting and changing set of representatives from US government agencies, and increasingly, from nongovernmental agencies and nonagency organizations. I am sure my fellow panelists here will help us understand very importantly how international nongovernmental organizations play in this realm as well. Sometimes they meet formally. Sometimes they meet informally. Sometimes they meet in places as obvious as Washington, DC, or sometimes as obscure as a remote PRT (Provincial Reconstruction Team) setting in rural Afghanistan. I hope that, again, my remarks will address and cover those variations.

Let me start then with the basic framework for looking at institutional change. There are really five components to this and I have stolen heavily and shamelessly from a friend and colleague, Doug Stuart of Dickinson College, who if you have not read, has just recently completed a book which is excellent. He has an article out ahead of it, but in the framework that I am going to be drawing on from him is not in that book *Creating a National Security State*, but it is an excellent historical study of the National Security Act of 1947. The article is "Constructing the Iron Cage," *which* should be published by SSI (Strategic Studies Institute) by the end of this year or early 2009. What he argues is that there is really a five-phase model of institutional design. It begins with an initial goal or problem. It is then followed by an impetus or a trigger event. There follows a period of tests and models, in which in response to that crisis or trigger event, you began to tinker around with things and try to find stuff that will work. I generally think of that as adapting and flexing on the fly. You then, on the basis of that and what things get through that testing and modeling stage, start to construct institutions and then you formalize those and move into the fifth phase which is the initial operation and adjustment phase. That is when you go back and try to fix and, again, tinker further with those and ideally fine tune them. Stuart goes on to explain that that initial goal or problem combines with a trigger event to create a public theory. And that public theory for our purposes is little more than an agreement, both in the public's and in the policy-makers' and leaders' minds that this is a general explanation for what is going on. That then provides the framework for guiding the search for the tests and models piece, what you try, develop and use, and that then butts up against the reality of institutional creation, and you go into the final phase, as we said, also scientifically known as the shake-out period in which you basically find what works and what does not work and adjust accordingly. I would like then to use that as the very simple framework for this historical piece on the National Security Act of 1947.

Stuart and others make the argument that if you really want to look at the thought processes that exist and the definition of the problem that occurred it certainly did not occur in 1945 and the immediate post-Cold War era. It also did not occur in December of 1941 with the Pearl Harbor attack. A lot of the thinking had already been expressed, written, spoken, and otherwise circulated in the 1930s. So a lot of that had to do with the rise of totalitarian states, obviously the growing power that Nazi Germany and Japan had, and the threats that those countries and their ways of life posed to the United States. The way they were organized, the powers that they had, and so on, seemed to be threats, and the concern was that the old concept of national interests needed to be rethought and there was a new concept introduced into the, at least policy, lexicon. The concept, of course, was national security, something broader than national interests. That then was the problem. The goal was to find out how to make the United States, in essence, more competitive with the totalitarian threats and the new technologies in their hands that we saw rising before us at the time. So we needed to be more competitive in foreign and defense policy realms. The trigger event, as we all know, was Pearl Harbor on December 7, 1941. More than just an impetus for change, as we will see we have had later, this one was a trigger event of grand proportions that would become both a symbol of what was wrong with the existing security architecture and what was needed to address it. So what we find in this period is that there was a confluence of the early thinking that had already gone on, a trigger event that really drove interest in doing something about it, and that forced thought into action. And the confluence created the public theory that I referred to, namely, that we needed to do some of these things, perhaps moving away from national interests and national security as a concept to drive it and to take a look at the four lessons that came out of it. The institutional design effort, this testing by the way phase, was one that we could then conduct during the actual war itself. So you could see the institutional design effort beginning in 1941 in how the US organized to fight the war, how it developed and coordinated the immediate post-war policies, and then in the attempts to construct the actual post-war foreign and defense policy system.

The four lessons, we will just go through these quickly because we all know them. Obviously, one was we needed new and better institutions for collecting and analyzing information so there could be no more Pearl Harbors. We also needed a permanent and more influential military voice in the peacetime making of foreign and security policy, something that was very new to US culture historically and would probably have been almost impossible were it not for the Pearl Harbor event. Third, the US needed to ensure the interservice as well as civilian-military cooperation would be seamless and ongoing in peace and in war time. And fourth, the US needed new procedures for integrating the capabilities of the domestic economy, especially industry and science in the new national security state. Now, as I said, the first elements as we got into that, the public theory was formed, the testing stage is really what we can see in how the United States organizes to fight the war. We created what one author has referred to as an accidental Joint Chiefs of Staff as we needed to coordinate with the Brits. As it turned out there was not a similar system so that we could communicate with their equivalent staff. Shortly thereafter an ad hoc Chairman of the Joint Chiefs of Staff emerged and as the issues of postwar reconstruction and occupation began to require attention, the need for improved civilian-military cooperation was addressed through the creation of the State-War-Navy Coordinating Committee, and I cannot even pronounce this acronym since there is no vowel in there, SWNCC, which I am sure many of you are familiar with. It is interesting and I alluded to that in a comment earlier today, that committee, again, according to at least a couple of historians was one that presaged a trend, and that trend was State was outnumbered two to one within that committee by the Armed Services and the institution itself became gradually to be dominated by the military. It was an institution that also foreshadowed this lesson learned and that is the application of a continued and strengthened military voice in US foreign and defense policy. Now, there is a lot more detail in all of this. I want to just fast forward a little bit to simply point out that with all of those changes as we fought the war, we came out of the war with what seemed to be a consensus, a very important term alluded to earlier, that we really had to institutionalize this for the new world. The experiments that were relatively easy to implement in war time, however, that test and model stage, now moved into the construction stage where concrete institutions and processes had to be proposed, debated, constituted, and eventually put into action. And as Stuart himself notes, "The fact that policy makers agreed on the need for institutional reform did not make it any easier for them to agree upon the details." This could be words of wisdom for today's ongoing debates.

What ensued, and let me go through this very quickly. Many of you in here are familiar with it. From 1945 roughly until the 1947 Act was passed in late July of that year and signed into law, the effort to achieve military unification, a key point of that reform, clearly dominated in terms of time, energy and attention all of the other elements of that National Security Act of 1947. We know by now that Army leaders abdicated a full merger of the Army and the Navy, a problem that many of us in this room can probably see no problem with today, right? The creation of a Joint Chiefs of Staff with a Chief of Staff at the helm that would serve as the principal military advisor to the President. This idea which seemed to work fairly smoothly, probably because it was modeled a little bit after our Combatant Command model of today, was not going to survive very long after the conclusion of World War II. The Navy switched fairly quickly over to a system that emphasized coordination over merger, and I ask you to bear with me on that because it is a key word for all of what I am going to argue here. It emphasized coordination over merger. President Truman and his supporters of full unification and integration eventually got far from all of what they wanted. Forrestal's efforts that I alluded to earlier today to stave off unification and the Secretary of the Navy was very successful and it points to the difficulty of forcing comprehensive change on a system designed to innovate at the margins, if at all. Forrestal's victory is the result, and I quote, "of his ability to garner Congressional support and to channel it using tactics that frequently verged on insubordination." And yet, according to Stuart the very institution that Forrestal had succeeded in creating in opposition to full and complete unification, the national military establishment which with amendments and new legislation and so on would later become the Department of Defense wound up with Forrestal himself serving as the first Secretary of National Defense, and it could not even stand the test of the first six months of operation. The initial operation and adjustment phase began almost immediately and it did so in earnest. What Forrestal had done using a friend of his and a well-known advisor at the time, Eberstadt, wrote a report that shifted the focus of the debate from the military per se to civilian-military coordination at the top of the Washington policy community. The central issue of all the organizational and reform efforts at that time and since then have been clear and consistent, improving civilian-military coordination.

I will fast forward again, August 5, 1949, major amendments to the National Security Act were passed trying to fix the very first problem, I think, August 11th, 16 days after the National Security Act was passed. I believe the Army circulated a memo having to do with significant military roles and missions that did not mention any offensive operations on the part of the Navy. That launched the first of the modern interservice rivalry battles. In August of 1949, the Department of Defense emerged from NME (National Military Establishment), the Secretary of Defense was created having more authority, direction, and control and fewer statements about how he would coordinate, and the service secretaries were eliminated from the cabinet and so on. Now, other entities created in that National Security Act, the National Security Council, the Central Intelligence Agency, and the National Security Resources Board are three. I will not dwell on them. The National Security Act has also been referred to in this because, and the CIA got so little attention compared to military service unification that the NSC almost emerged as something of an afterthought. I will make the point that in my readings of the history on this, that the original proponents of this system and the National Security Council envisioned it as the coordinator of the overall system. And what is interesting is that over time the NSC became essentially a presidential or executive tool for managing foreign and security policy. It never became, I would argue, the coordinator let alone the integrator of all these various national instruments as even some people today think that it is. The original language that had it specified as that never made it through the debate in the markup stage, and as I said, over time it became a presidential tool for executive management and the same legislation gave the CIA relatively clear guidance in terms of what it was supposed to do, here is the term again, coordination information, never gave it the authority over into other intelligence services to do so. So for the CIA, it never really got its intelligence coordination function off the ground, and the debates we had after 9-11 were nothing new if you look back on this. In fact, what the CIA did under very deft leadership and opportunistic, I think, use of resources was to go heavily into the covert operations role which was not anything that was mentioned in its original mandate and downplayed the intelligence coordination role which it simply could not do because it all was hinged on voluntary coordination among institutions that probably did a lot of that during the war but were less inclined to do so once the war ended. I mentioned the State Department in a comment I made this morning. It is interesting; again, there are several people that point out that the State Department seemingly made every choice it could to erode its own role in the still emerging and evolving Pearl Harbor system. Part of that was its own problem still seeing peace and war dichotomous and that the new postwar system would be very much the same. Quickly, let me mention that is the system we got. There were other changes; some of you have alluded to them earlier today. When I look to Goldwater-Nichols, and I was going to apply the same framework, I could not see the trigger event and the same notion that we obviously had some failures or viewed failures in Vietnam and others that were impetus to change. But the main point that I came away from here is first of all that most of what Goldwater-Nichols is not about integrating national instruments of power; it is really about unifying the Pentagon. It is almost exclusively operational and it is almost exclusively military operational. That does not mean it is not important, but it does mean that it is going back to one piece of the problem and not the problem that I said I am most interested in here. And even with that the cases that were mentioned this morning are very interesting. Panama, again, from my reading demonstrated how much we had improved in joint operations and unity of command, but equally, and I think John Fishel and others were alluding to this, there was little evidence of real effective postconflict planning and execution. So again, we fixed some things, but the bigger problem I am interested in was not. And Afghanistan, I think we could look at Haiti in much the same way. Afghanistan in the initial phases was a real joint warfighting success, especially at the small unit level, but again a postconflict, if not failure, at least something much less than a great success. Let me say that the problem to that then is, before we are casting blame everywhere, a lot of it is Goldwater-Nichols. It was designed to improve civilian-military control. It was designed to give better professional advice to civilian leadership, and in many ways it did that and by empowering the Chairman of the Joint Chiefs by strengthening the Joint Staff. Many critics then argued that the civilian control was actually weakened. Whether that is true or not, some of those same people say that the failure to act just as Elliot Cohen says, it is the civilians not the soldiers that have abdicated their responsibilities. So again, it is not so much the pointing of fingers; it is how do we get this integration. After Goldwater-Nichols I think if you look at this five-step framework I think it is surprising how with the end of the Cold War, maybe not a trigger event like Pearl Harbor but certainly a significant shift in the international system, and even with the dissolution of the Soviet Union no serious attempt emerged to really revisit the institutions and organizations and processes that we had.

Let me fast forward now. I am probably already over my time; it is very close to the end. There is our institutional design model (Slide 1). If you come from the Army War College you have to have a Clausewitz quote in every presentation so here is mine (Slide 2). Actually, this one helps a lot because not only in applying to war, the fundamental premise that we begin with in getting institutions and structures and processes and so on right is understanding the nature of the security system that we are confronting. And that is how I would paraphrase Clausewitz for what we are doing here today. To get strategy right you have to understand the nature of the security environments correctly and then it is lining up your resources and using them, the right resource in the most effective and efficient ways to accomplish your objectives. I will skip those last two bullets.

These are gaps and challenges that I want to cover very quickly, and let me do that right now (Slide 3). This is the contemporary setting that I am looking at. Many of you may disagree with this (Slide 4). I believe terrorism is certainly important, vitally important. I believe that insurgency is vitally important, but I am not convinced that terrorism in and of itself and insurgency in and of itself are really the kind of strategic kind of imperatives that we are dealing with. I believe that this is the new normal and that is a world in which it has all been spelled out. We have non-state actors empowered with all kinds of weapons. We have serious issues of, I still argue that a lot of the problems of terrorism and insurgency grow out of the various forms and fashions that failed in fragile and illegitimate states that we have, and we can talk about that in discussion if time allows.

If you have not seen "wicked problems" I encourage you to go to Wikipedia and look this up (Slide 5). I heard Larry Sampler talk about this when he was speaking to the consortium on Complex Operations, but there are some interesting things in here. I love the second bullet there—every problem interacts with other problems and is therefore a part of a set of interrelated problems, a system of problems. I choose to call such a system a mess. The point here is that these are not Cold War problems with a single definable enemy. They are not just about winning the war. They are about winning the peace, but it is about winning the peace in ways that are very much more complex and complicated than I think we have dealt with in the past.

We will come back to this. I think that therefore, and this is truly Robin Dorff's view on this. I believe that our grand strategic objective should be promoting effective legitimate governance (Slide 6). To grow that community where you have effective legitimate governance and therefore the gaps that I want to just quickly toggle through. Gap one is a gap between the reality and the understanding of the strategic environment. If we are at war with terrorism, a war and certainly a concept in general I do not have a problem with except I do not think you go to war against a tactic, and I think that is what terrorism is. I want to know who the enemy is that wants to use terrorism against us. But if you define it only that way, my problem is that in the "wicked problems" world that we are dealing with in these complex operations we are going to miss a lot of the other important things that we need to be able to do.

Here it was alluded to this morning, yes, we have NSPD-44 (National Security Presidential Directive) (Slide 7). The President does designate the State Department as the lead agency in coordinating these operations. Yes, we even have DoD 3000.05 that equates SSTR in reconstruction operations, the equivalent of combat, but where is the strategy? What we have, and I can go back to Clausewitz on this, yes, policy must drive strategy, but right now we have a lot of policy, but we do not have a lot of strategy. Now, we have generated some doctrine at the counterinsurgency level, but a lot of that has really been about how we are going to operationally conduct things on the ground to deal with the problem at the moment. We have yet to see resources flow significantly from either NSPD 44 or 3005.

Here is something I borrowed, I spent some time working as a senior advisor for a company that has done education and community development work internationally for 35 years, long before any of this stuff came out (Slide 8). They have now been on the ground in Iraq, in Afghanistan, and many other places around the world and their strengths are working exactly on the community development, education, and local governance kinds of things and if time allows I will speak to that too. I think these are organizations that need to very much be brought into this integrating argument that I am making. The point here is that you need organizations that can operate effectively in both of those spaces—battle space and humanitarian space. You especially need organizations that can work in those areas that overlap the two. The implications then of those two operational spaces, here is the gap I am pointing to, the gap between policy and guidance and operations in the field.

I will let you read these (Slide 9). The joint civilian military concepts—doctrine, training, and execution. I am a huge fan and this relates to the last slide and to this slide. We have been doing some great things in the field. I have friends that are very much involved in some training of soldiers about to deploy in PRTs and I think it is making great progress. The problem is, again, it is largely military only. And why? Because of some of the things we talked about earlier. The civilian organizations, the governmental organizations do not have the training float to pull people off and go through programs like some of you in this room I am sure have gone through or are going through. Somebody talked about the people problem that State has. I think the best statistic I heard about this, I assume it is true, I have not been able to go out and count it up, but I am told that there are more musicians that the Pentagon has access to than the State Department has deployable State Department people. That is kind of a worthless comment, but it does make a point. That whole aspect of I have been a big beneficiary of watching military officers be able to take 10 months off to do a senior service college rotation at the O-6, soon to be O-7 level. There is very little float in civilian agencies to even come close to that. A two-week course at the Foreign Service Institute is maybe all they can cram in. Who can work that stabilization piece? The whole of government is required, but we do not do it very well. When I say making it up in the field, we are learning great things about how to do things on the ground, but my earlier point, if we are making it up on the ground, but it is not getting back to the top, then it is not going to be resourced over time and it is not going to be sustainable. And if it is not coming from the top down and so on, then I am coming back to the cases that I have seen today and the question that I ask because the making it up on the ground then depends almost completely on who is on the ground at the moment. And it is not really in the kind of training and I make a vague reference here to training jointly. I mean jointly civilian military predeployment as well.

The fourth gap here then is this gap in useable military civilian instruments (Slide 10). That gap exists and it is expanding. I just alluded to that. I mentioned the policy driving strategy that is not fully resourced. Executive and legislative branches, this is what the history piece learned. If you figure out where your allies are, you can prevent just about anything from getting through any kind of a redesign or design phase, and once again, making it up in the field, but is it sustainable?

I will not talk to this one, but what we have also done to fill some of these gaps then is to go to contracting (Slide 11). And contracting raises at least these questions and a number more, so finally, the functions that we need in US government versus contractor functions.

I have already noted that where this story began with the National Security Act of 1947. (Slide 12)

My last slide really on this piece and here are the things I think we need (Slide 13). And unfortunately, it is leadership, leadership, and leadership. And I am fairly confident of leadership in a couple of these areas. I am less confident in a couple of other areas. And that is as specific as I will get.

This I will leave up there if we want to post these later (Slide 14). These are some of the major reform efforts that are being addressed out there. Some of them are Goldwater-Nichols II which is kind of Goldwater-Nichols for the civilian agencies. PNSR, the Project for National Security Reform. It talks about starting all over again and creating a whole new National Security Act of 2009 or 2010. Others talk about just a little better coordination and so on within the organizations that we already have.

I think that is it (Slide 15). Yes. Let me wrap up then. I apologize. I believe that if we are not able to come to grips more fully with the requirements to integrate rather than stovepipe, to develop operators with multidimensional skills sense that cut across these traditional stovepipes, and by the way, we are now calling those, is anybody familiar with this one, cylinders of excellence, and engage the broader national community in these efforts the project is likely to resemble the sausage-making exercise that has all to often characterized that interagency process. I will leave you with this in terms of coming back to the framework for analyzing institutional change suggests that the ability of the US to adapt to these 21st century security environment challenges will be huge and perhaps insurmountable. While no structures and processes alone can ensure good decisions and outcomes really bad ones can make them almost impossible or very difficult to achieve. If we look back on the National Security Act as it was first proposed, as it was passed into law, as it has been modified and amended since then what we see is this principle of voluntary coordination dominates all of our thinking and we try to find ways to make us voluntarily cooperate better. What that fundamentally comes down to is we give agencies, people, czars in the White House, and so on responsibility without authority, or responsibility, a little authority, and no resources as SCRS (Office of Reconstruction and Stabilization) is in the State Department is today. My argument is not that we necessarily need to throw out everything we have and start over again. My argument is that we need to go back in terms of what the 1947 Act did, and that was to have a very serious debate and discussion about what the challenges, threats, and opportunities are, what we need in the way of capabilities to address them, and how we can best organize in order to use the capabilities we have, generate the ones that we need, and take our strategy and policy into the 21st century.

The last point, I promise. I will get that Bill Clinton "thank you" in conclusion. Wild applause. If we do not get this integration through some sustainable structures and processes, still relying on right people being in the right places at the right time, that is an interagency process for the 21st century, I believe that you in the military, and especially the Army, will continue to very, very busy and I also believe you will continue to be very, very under-resourced. And there is a whole debate to have in there about whether or not the civil affairs, especially civil military pieces, especially of the Army, if this is going to be a lot of the work of the future if 90-plus percent of it should reside in the non-Active Component of the military. I strongly hope that military leadership, and especially Army leadership, will play constructively in this debate, both on the need for taking this serious look at the problem and in terms of what the content of that should be. Thank you very much.

INSTITUTIONAL DESIGN

Initial Problem or Goal=>Trigger Event=>
PUBLIC THEORY=>
Tests and Models=>Construction=>
INSTITUTIONS=>
Initial Operation & Adjustment

Slide 1

Context

- "The first, the supreme, the most far-reaching act of judgment that the statesman and commander have to make is to establish . . . the kind of war on which they are embarking." **Karl von Clausewitz**
- Strategy is the calculated relationship among: Ends (Objectives), Means (Resources), and Ways (Concepts)
- Interagency in Broadest Sense
- My Personal Interests

Slide 2

The Challenges and the Gaps

- The Strategic Environment
- Policy and Operations
- Interagency: National and Field
- Capabilities and Resources: Military and Civilian
- US Government and Contractor Roles
- National Security Organization
- Leadership

Slide 3

The Strategic Environment

- Terrorism is important but….
- Insurgency is important but….
- The "New Normal" and Good Governance Deficit
- Wicked Problems
- A New Grand Strategy: Promoting Effective, Legitimate Governance

GAP I: Reality and Understanding of Strategic Environment

Slide 4

Wicked Problems

- Wicked problems have incomplete, contradictory, and changing requirements; and solutions to them are often difficult to recognize as such because of complex interdependencies. Rittel and Webber stated that while attempting to solve a wicked problem, the solution of one of its aspects may reveal or create other, even more complex problems.
- "Every problem interacts with other problems and is therefore part of a set of interrelated problems, a system of problems.... I choose to call such a system a mess."
- "a Social Mess is a set of interrelated problems **and other messes**. Complexity—systems of systems—is among the factors that makes Social Messes so resistant to analysis and, more importantly, to resolution." Source: Wickipedia

Slide 5

The Strategic Environment

- Terrorism is important but....
- Insurgency is important but....
- The "New Normal" and Good Governance Deficit
- Wicked Problems
- A New Grand Strategy: Promoting Effective, Legitimate Governance

GAP I: Reality and Understanding of Strategic Environment

Slide 6

Policy and Operations

- NSPD-44
- DOD 3000.05
- Where is "The Strategy"?
- Implications: Two Operational Spaces

GAP II: Policy Guidance and Operations in the Field

Slide 7

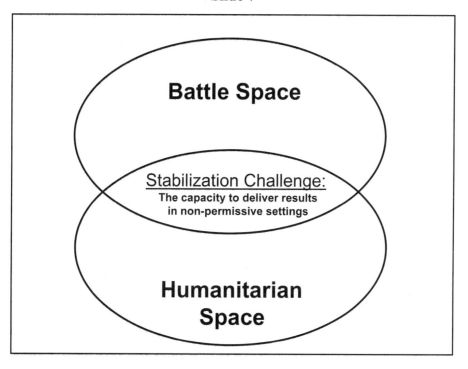

Battle Space

Stabilization Challenge:
The capacity to deliver results
in non-permissive settings

Humanitarian
Space

Slide 8

141

Interagency: National and Field

- Need "joint" civ-mil concepts, doctrine, training and execution (e.g. TRADOC and PRT training)
- Who can work the "Stabilization Space"?
- Whole of Government required—but we don't do it very well.
- "Making it up" in the field but sustainable?

GAP III: IA Org & Process (DC) and On-the-Ground Capabilities Integration

Slide 9

Capabilities and Resources: Military and Civilian

- Gap Exists and is Expanding
- Policy drives a strategy that is not fully resourced (not just $)
- Executive and Legislative Branches
- Reluctance of Civilian Agencies (historical, cultural, bureaucratic, etc.)
- "Making it up" in the field but sustainable?

GAP IV: Useable Mil-Civ Instruments

Slide 10

US Government and Contractor Roles

- Sourcing vs. outsourcing
- If outsourced, how coordinated?
- Can you "contract" ***functions*** without "contracting" ***strategy***?
- Sustainable?

GAP V: USG and Contractor Functions

Slide 11

National Security Organization

- The legacy of NSA 1947
- The lessons of NSA 1947
- Affects not only how we coordinate:
  - Define problem
  - Develop strategy
  - Implement strategy
  - Adapt and adjust

GAP VI: USG National Security Architecture

Slide 12

Leadership

- Military
- Civilian
- Executive and Legislative
- American People

GAP VII: Leadership Across the Board

Slide 13

Some Current Reform Efforts

- Executive Branch Key Players: Adjust the Balance of Roles and Responsibilities
- Interagency Coordination and Integration: Foster Horizontal Integration
- Interagency Coordination and Integration: Create a New Coordination Body
- National Security Decision-Making: Require Greater Rigor
- National Security Strategy: Strengthen the Guidance
- Resources for National Security: Create a National Security Budget
- Congressional Oversight: Reorganize
- Start all Over Again: National Security Act of 2009

Source: CRS Report for Congress RL34455

Slide 14

Observations

- Assumes: 1) We really believe these are the strategic threats, challenges and opportunities we face, and 2) we really want to do something about them. Neither is a given.

- If so, where is political will to lead, and convince the American public – and with them Congress—that this is what we face and we must resource the strategy.

- Absent some significant changes in ability to integrate civ-mil capabilities, the mil is likely to continue having to do many of them. Implications?

- If mil continues in that vein, what about the respective roles and missions? And what about skill sets that may reside in the Reserve and Guard?

- Clausewitz Revisited: "…extension of politics by other means." Neither "Deliberate planning" **alone** nor "flexibility above all" **alone** will resolve these tensions.

Slide 15

145

The Independence of the International Red Cross: The Value of Neutral and Impartial Action Concurrent to the Interagency Process

by

Mr. Geoff Loane
Head of Regional Delegation to the United States and Canada
International Committee of the Red Cross

Abstract

The International Committee of the Red Cross (ICRC) is an exclusively humanitarian organization. Its mandate to help victims of armed conflict derives from international humanitarian law, of which the four Geneva Conventions are the foundation.

In this paper, author Geoff Loane acknowledges the ever-expanding role of integrated approaches to furthering recovery and reconstruction efforts by combining military, political, and economic activities. Despite this trend, he argues that the delivery of humanitarian assistance to persons affected by armed conflict must be carried out in a strictly neutral and impartial way. He explains that this is the only way to provide help in a timely effective way and not place persons most in need at greater risk.

The ICRC is able to work in such a way due to its strict adherence to the fundamental principles of the Red Cross Movement, of which humanity, neutrality, impartiality, and independence are paramount. In addition, the author explains ICRC's operational approaches, including proximity, universal vocation, dialogue, and agreement among all parties to a conflict to ICRC's role. In this way, the ICRC can secure the support and agreement of all the parties to a conflict and the acceptance of the ICRC as a neutral intermediary.

Introduction

Over the past two decades, the field of humanitarianism has experienced a seismic shift in terms of size, composition, and diversity. The hundreds of organizations now assisting persons in need, whether due to a natural disaster or situation of armed conflict, represent a disparate mix of actors. They include locally affected communities and individuals as well as governmental and nongovernmental, civilian and military, religious and philanthropic, and charitable and corporate agencies. Today, they collectively harness an unprecedented level of resources. Moreover, each actor brings its own particular vision, mandate, and working modalities into the common humanitarian space in which all operate. While the sheer growth of humanitarianism has enabled more people to access higher quality assistance more predictably than ever before, threats to aid effectiveness abound. Actors struggle with delineating their roles and boundaries, sharing information, coordinating activities, involving and empowering affected communities, and linking relief aid to development programs. This diversity of operating bodies with widely differing mandates poses a significant challenge for the community of actors that work based on a strictly humanitarian impulse, particularly that aim may conflict with other agencies' goals linked to foreign policy or political objectives.

As an agency whose *raison d'etre* remains exclusively humanitarian, the ICRC bases its action exclusively on the fundamental principles of humanity, neutrality, impartiality, and independence. The ICRC ensures that the delivery of assistance sends an absolutely neutral message. This both means and allows it to work on all sides of a conflict simultaneously, identifying needs first hand, and having the support of all arms carriers, on behalf of the largely civilian victims of conflict.

Other agencies do not follow these precepts nor could they, and some might argue should they. Governmental agencies, for example, often deliver aid in accordance with the national priorities as determined by their political authorities. Within the United States Government (USG), the dialogue has increasingly focused on what is often referred to as the "interagency process." This approach attempts to capture the combination of all the activities and resources of civilian entities like the US Department of State and US Agency for International Development (USAID) and military components of the US Department of Defense (DOD), including the regional combatant commands.

The challenge for the ICRC is finding a way to relate to and coordinate with different interagency actors providing assistance, especially military ones, while safeguarding its fundamental principles, which ensure its acceptance by all arms carriers and access to all victims, on every side of a conflict. These principles are more than words; they are practical tools that enable the organization to reach communities in need. In any situation of armed conflict or political tension, actions by outsiders as well as by local agents, are systematically analyzed for any political content, and judged accordingly. The very act of providing emergency assistance to civilian communities in armed conflict will be carefully monitored and analyzed by political leaders and armed groups for any potential impact that act may have politically. The bombing of both the United Nations (UN) and the ICRC delegation in Baghdad in 2004 was also an attempt to intimidate those providing assistance. In general though, because it remains apart from political influences, can gain the trust of all sides, and serves as a neutral intermediary in the midst of hostilities, the ICRC is more often accepted in providing protection and assistance, on the exclusive basis of need.

The following exploration will argue for the need to understand and respect the essential value of this neutral and independent form of humanitarian assistance, in order to ensure effective access to and support of all victims, regardless of their location, political persuasions and beliefs.

Law & Principles: The Basis for the ICRC's Work

One finds the legal basis for the ICRC's work in international humanitarian law, of which the four Geneva Conventions are the foundation. These universally ratified treaties provide the ICRC[1] with a humanitarian mandate from the community of states[2] to help victims of armed conflict. The States also gave the ICRC the responsibility of monitoring the faithful application of international humanitarian law in armed conflict. As a guardian of humanitarian law, the ICRC takes measures to ensure respect for, to promote, and to reaffirm this body of law.

This universal ratified legal basis thus provides ICRC with the authority and legitimacy to negotiate with state and non-state actors about its humanitarian role. The two primary objectives of this negotiation is firstly, the promotion of the faithful implementation of humanitarian law. The second is to ensure acceptance of the humanitarian role of the ICRC in providing assistance to conflict victims across all front lines and ceasefire lines to all communities.

Central to this are the principles of neutrality and independence, which allow the ICRC to make its exclusive humanitarian decisions without the influence of political, financial, or other interests. These core humanitarian principles mean that for the objective of acting as a neutral intermediary or provider of assistance the ICRC requires acceptance from all arms carriers, irrespective of their ideologies or political persuasion, the freedom from external influence to choose where to go, who to assist, and how to operate. These operating principles are at once the strength of the ICRC, and the single hardest aspect to negotiate. States and non-state armed actors believe in the justness of their cause and are regularly distrustful of the provision of assistance to the enemy, ostensibly as it may be construed to support the war effort. The ICRC aims not only to reach needy communities, but also to distribute assistance in ways that are transparent and controlled by its own, often-expatriate staff, thus ensuring no misuse or diversion of aid commodities to the wrong persons.

On this basis, the ICRC works actively to earn and retain the trust of all States, parties,[3] and people involved in a conflict or other situation of violence.[4] This trust is based on an awareness of the principles and practices of the ICRC, and through an observation of how the ICRC acts in the field over long periods. The ICRC gains trust through continuity and predictability. Combining effectiveness and credibility irrespective of time, place, or range of needs is a continuous challenge for the organization, because it must be able to prove it can always meet the standard of assisting all in need, regardless of the context or circumstances.

Experience has taught the ICRC that this approach offers the best, if not the only, chance of acceptance by all sides during an armed conflict or other situation of violence. Arms carriers and their political leaders would simply not allow an agency to travel freely across front lines distributing assistance with motives other than humanitarian ones.

The Operational Aspects of ICRC Work

In order to implement this policy of independence, neutrality and impartiality, the ICRC must build confidence with arms carriers and their leaders through its operational approaches. Primarily, the ICRC favors field-based engagement. It meets, works with, and lives in the midst of communities whose lives have been affected by conflict. The *proximity* of ICRC to those affected is of central significance for it helps to analyze and understand the particular needs of communities. Proximity also ensures access to local leadership, often the building block of political decision-making. In addition, the ICRC systematically consults local community members to establish their priorities, and to ensure that they are associated in the action taken.[5]

The ICRC's work has a *universal* vocation. It is not limited to certain places, or to certain types of people (such as children or refugees). It is present in all conflict zones, often for many years through successive changes of regime. This long-standing presence means that the ICRC interacts with armed opposition who may become political leaders, as well as with political leaders who may end up as armed opposition or security detainees. Such relationships, managed in nonjudgmental and humanitarian contexts often serve considerable value as roles change. Responding to the specificities of each context and working tightly in a needs-based approach are key aspects of analysis and strategy.

The ICRC engages in *dialogue* with all those involved in an armed conflict (or other situations of violence) who may have influence on its course, whether those persons are recognized and/or acknowledged by the community of States or not. All those who may have an influence are consulted because multiple and diverse contacts are essential for assessing a situation and for guaranteeing the safety of ICRC activities and personnel. Such dialogue does not confer any formal status on anyone, as the discussions focus exclusively on the humanitarian situation.

The ICRC requires the support and *agreement* of all the parties to a conflict, as well as access to all communities affected by the conflict before engaging in humanitarian operations. Any humanitarian operation can only begin if both (or all) parties to the conflict understand, accept, and support operations on both (all) sides of the conflict.

One benefit of these operational approaches is ICRC's ability to serve as a neutral intermediary. Because of its acceptance by all arms carriers, the ICRC has recently been able to negotiate a halting of armed activities by the armed opposition in Afghanistan to facilitate UN-led vaccination campaigns and secure the release of dozens of hostages in Afghanistan, Ethiopia, and Colombia. The ICRC has also acted as an intermediary in the repatriation of mortal remains and detainees between sovereign powers. A current example is the ICRC-facilitated phone calls between detainees in Guantánamo and their families, often located in remote and inaccessible areas.

While doing what it can to help people in need, the ICRC also considers the efforts of other agencies working in the humanitarian world. The main objective of *interacting* with other providers of aid is to make the best use of complementary efforts in order to meet needs. Interaction based on transparency, equality, effective operational capacities, and complementary, reality-based, and field-centered relationships is essential.

The ICRC's Understanding of the USG Interagency Approach

The current tendency among many other actors, most notably in the governmental sector is to "integrate" or combine the military, political, and economic instruments of power to achieve a set of predefined political objectives, including conflict prevention, conflict resolution, and postconflict nation building. In the United States, this process is often referred to as the "interagency process." Indeed, Provincial Reconstruction Teams (PRTs) in Afghanistan and Iraq are the most visible representation in the field of this integrated approach. Such an approach is based on multiple objectives, many of which reach well beyond the humanitarian imperative into realms of good governance, strengthening civil institutions and reinforcing compliance with "good behavior." Such an approach is not confined to western democratic states. Other actors such as Hezbollah and Hamas have incorporated civil society interventions, including the provision of social services, into their efforts to reinforce their relationships and influence over their constituencies. To the extent that such mechanisms exist in places of conflict, they will suggest a direct link between delivery of humanitarian assistance and political objectives. This linking of emergency assistance (development aid has long been associated with political conditionality) with political condition is a new departure for a system of assistance that long held the value that suffering in disasters merited absolute and unconditional assistance as soon as possible.

The current developments stem from the US Government Interagency Initiative[6] in 1997 when President Bill Clinton approved Presidential Decision Directive (PDD) 56. This directive laid the foundation for "interagency planning of future complex contingency operations." Focusing curricula, budgets, and exercises on interagency practices and procedures, PDD 56 called upon government agencies to institutionalize lessons learned from complex contingency operations in Haiti and Bosnia. The current impetus, it is understood, stemmed from National Security Presidential Directive 44 (NSPD 44), signed by President George Bush, which detailed responsibilities for governmental departments in supporting stability and reconstruction. NSPD 44 also charged the State Department with developing "detailed contingency plans for integrated US government reconstruction and stabilization efforts" and to "lead US government development of a strong civilian response capability including necessary surge capabilities." This has led to the establishment of the multidisciplinary and interagency structure housed in the State Department under the Coordinator for Conflict, Reconstruction, and Stabilization (State/CRS), whose primary objective is the harmonization and implementation of postconflict reconstruction efforts.

DOD has also moved increasingly toward an interagency approach to deal with postconflict reconstruction, and in doctrine has elevated stabilization operations to the same as warfighting. By way of illustration, in 2001, the term "interagency" was mentioned 13 times in the QDR, whereas "interagency" is mentioned 47 times in QDR 2006. Extrapolating lessons learned from Iraq and Afghanistan, QDR 2006 places specific attention on stabilization and reconstruction efforts. This is a fundamental and categorical shift in the alignment of priorities if viewed over time and is likely to have a major impact on how the US military interprets and implements its role in the generation to come. The recent response of the US military to the conflict in Georgia and South Ossetia to distribute relief (on one side of the conflict) reflects this role DOD has identified for itself.

The years to come will be instrumental in clarifying the architecture of the interagency process and its relationship to the humanitarian community. The results will have a significant impact on humanitarian operations in the coming generations.

Problems of Interagency, Utility of Independence

The overarching question that now persists is how to respond to the humanitarian needs of the civilian populations in situations of armed conflict in ways that are effective in terms of providing succor to affected populations, while guaranteeing the acceptance of humanitarian assistance as a neutral activity, aimed at supporting victims of conflict and without any political value. To what extent do different models of intervention, be they rooted in political interagency interests or in notions of independence and full access, prove mutually reinforcing and to what extent do they undermine each other? For clarity's sake, the following line of questioning focuses exclusively on humanitarian assistance, as longer-term development assistance by its very nature is conditional to the political interests of States that fund and of States and communities that implement.

Evidently, no one organization or agency can respond to the entirety of humanitarian needs. In the face of unmet needs resulting in the suffering of populations, who then is in a position to say how and by whom should humanitarian assistance be delivered? What are intervention criteria, given this reality?

Second, how are humanitarian needs determined? And by whom are criteria designed? Are they based on what agencies consider to be needs, or what agencies have in warehouses and/or consider appropriate responses? Do staff actually listen to those in need or do agencies actually think on their behalf?

Third, what happens when agencies move on? What are residual obligations to the recipients of aid and how are they determined? Availability of resources? Mission mandates? Political interest? In every case, agencies must consider the expectations created when operations started. The aid community is accountable to its constituents for the aid it delivers and must ensure that accountability is managed with the rigor and control as in any professional field.

Moreover, how does humanitarianism assistance relate to local government structures or foreign occupiers, who after all are responsible for the delivery of health, security, and welfare to all persons falling within their territory? Is that leadership role (as opposed to those who grant permission to foreigners to distribute aid) recognized and respected? For that matter, what are the obligations on agencies to provide sustainable resources to manage and run the programs they set up in an emergency?

Lastly, in this debate is the question, who should provide assistance in situations of armed conflict, the most extreme manifestation of political disagreement between and within States? Is it possible to eliminate a conflict of interest between a military fighting a war and providing assistance at the same time?

The answers to these questions are as complex, yet clarity around these core issues is vital to make humanitarian assistance effective and thus respected by governments and recipients alike. A one-stop shop for addressing humanitarian needs in times of crisis is unthinkable. Crises will generate extra needs for which the voluntary and external humanitarian community and its supporters are required. Responses to humanitarian emergencies therefore will be a function of complementary activities by agencies and the factors that will determine those activities and relationships will include the nature, scope, and size of the emergency and the agency capacity and motivation to address the problem.

That being said, humanitarian agencies—and this includes the ICRC—should accept the idea that the USG interagency, with its politically motivated approach as epitomized by the PRTs, is not going to fade away, even if PRTs are unpopular in some quarters because they combine political and military objectives with the delivery of assistance. Many Western countries are adopting and developing this model. As with all systems, however, some risks are inherent in the linking of humanitarian assistance with this approach.

Humanitarian action should not be used as a substitute for sustainable political action. This lesson has been learned through all major evaluations of large-scale emergency operations. The international community has tended in the past to overcompensate for its political shortcomings through an increased humanitarian response. This does not help address the underlying political issues which the international community has an interest in, if not a responsibility for, and also suggests that some humanitarian needs are more important than others. International political failures in Rwanda in

1994, the Balkans in the 1990s, and in Somalia in 1992 all saw a surge in the provision of humanitarian assistance. At no time could such assistance address the root causes of the tragedies that encompassed those countries and failure to do so resulted in unacceptable levels of human suffering.

Related to this is the acknowledgement that humanitarian and other problems in a country at war are primarily the responsibility of the parties to the conflict. Humanitarian action simply substitutes itself temporarily for the limits of governments to take care of its own population. This does not change the fact that governments are responsible for their populations and their decisions in all spheres must reflect their responsibilities towards their citizens. Such awareness, in contrast to that of responding to needs as a way of satisfying agency mandates or interests, would generate a very different assistance community. In the long term, only sustainable and proactive local government action will address the problems generated by conflict, and assistance resources need to consider this when looking at choices for intervention. While not all agencies should or indeed could adopt an independent and apolitical perspective like the ICRC, all agencies should acknowledge the primacy of responding to human suffering, irrespective of political interests.

A further danger in adopting a political approach is that the armed opposition or parts of the population will perceive humanitarian agencies as instruments of a foreign and aggressive agenda. They will likewise consider them as legitimate targets in their war. Tragically, increasing numbers of humanitarian workers are killed in circumstances that suggest their role is perceived as promoting a conflicting political approach.

Conclusion

Humanitarian assistance addresses the basic needs of individuals and communities to cope in situations of extreme stress. Activities target those who are most in need and provide a range of life-saving, life-protecting, and recovery support to immediate victims of crisis. Actors who take on such work to benefit communities living in situations of armed conflict or political tension must provide such support unconditionally, whether or not a natural disaster exacerbates the level of suffering. Such contexts include the obvious countries like Sudan, Somalia, Iraq, Afghanistan, Georgia, but also more subtle contexts such as Abkhazia, Zimbabwe, Pakistan, and Nepal.

In such situations, limiting delivery of humanitarian assistance to agencies that can demonstrate an absolute neutrality and independence is in the interest of all involved, as securing the support of all parties is the only way to ensure impartiality. Only agencies with strictly humanitarian objectives can assure access to all people in need regardless of differing political interest. This necessarily excludes representatives of governmental interests, who will have political and partisan interests.

Security of agency staff requires an acceptance that all staff involved in humanitarian assistance operate with that agenda. If governmental officials represent other interests in similar activities, then the security of humanitarians is compromised, as humanitarians may be perceived of pursuing interests that exceed humanitarian concerns. In a charged political environment, such concerns become security threats.

The protection of the recipients of assistance depends on the perception that they are selected based on need alone. Should armed groups consider that certain communities receive assistance because of their collaboration with political or military entities, a real risk exists that they too will be charged as collaborators and their physical integrity can be compromised. Should this occur, then the ones suffering only suffer more, the least desirable outcome of any humanitarian effort.

Placing conditions on the delivery of humanitarian assistance, as is likely with government-supported assistance, introduces a dangerous precedent when lives are at risk. In the post-9/11 world, a number of countries have attempted to limit distribution of assistance in case aid might reach communities who may support antigovernment or even terrorist-type activities. Great care must be taken, however, while respecting political realities, to ensure that urgent humanitarian needs during times of armed conflict can be met in a neutral and impartial as well as a timely and effective way.

Notes

1. The ICRC is often considered to be *sui generis*: legally, it is neither an intergovernmental nor a non-governmental organization. It is a private association under Swiss law with international mandates under public international law.

2. International humanitarian law expressly confers certain rights on the ICRC, such as that of visiting prisoners of war or civilian internees and providing them with relief supplies, and that of operating the Central Tracing Agency (see Arts 73, 122, 123 and 126, GC III, and Arts 76, 109, 137, 140 and 143, GC IV). In addition, international humanitarian law recognizes the ICRC's right of initiative in the event of armed conflict, whether international or non-international (Art. 3 and Arts 9/9/9/10 common to the four Geneva Conventions). The ICRC's role is confirmed in Art. 5 of the Statutes of the Movement. In situations falling below the threshold of international humanitarian law, this article of the Statutes alone recognizes that the ICRC has a mandate to take action.

3. In this document, "parties" or "authorities" should be understood to mean all entities (*de jure* or *de facto*) having obligations.

4. See Art. 5.3 of the Statutes of the Movement. In its capacity as a specifically neutral and independent humanitarian organization, the ICRC examines whether it is better placed than other organizations to respond to the needs arising from these situations, such as visiting security detainees in cases where information or rumour indicates there may be poor detention conditions or ill-treatment.

5. They should contribute for example to decisions regarding priorities and regarding the implementation, management and assessment of programmes.

6. Source for background information is taken from paper on "Interagency Leadership – The Case for Strengthening the Department of State by Lieutenant Colonel Shannon Caudill, USAF, Major Andrew Leonard, USA, and Sergeant Major Richard Thresher, USMC, 15 April 2008.

Panel 3—Interagency Efforts at the National Level
Question and Answers
(Transcript of Presentation)

Robert H. Dorff, Ph.D.
Mr. Geoff Loane

Moderated by Mr. Kelvin Crow

Mr. Crow

Mr. Loane, I would like to completely abuse my position as moderator and ask you about your recent trip to the North Pole, but instead I am going to ask Dr. Dorff if we have reached the trigger point that you were talking about with a change in our thinking to the new reality, the new stresses of the new normal?

Dr. Dorff

That is one of the things that concerns me a lot. I do not think we have, at least in terms of the trigger event. I certainly would have thought that 9-11 was that and I had every reason to believe at the time that it was. But I think if we look around right now, I am not sure that we still have that momentum to generate the kind of consensus that I think Dr. Yates referred to earlier. And part of that is because, I think, there is still some confusion to what 9-11 meant. I know that may sound strange, but if you think back, for some of us, including myself, cannot think back to Pearl Harbor, but through studies of Pearl Harbor you know that the American public really wanted just to go beat up on the Japanese right away. And it took some very astute political leadership, I think, to translate the trigger event into what led to the kind of action, both in the fighting of the war and how it was prosecuted, to then try to build the structures and so on afterwards. I think right now the challenge is that some of what got lost was in, I think, the focus on 9-11. In the short term that was fine, but in the longer term I think we had opportunities to broaden that into a trigger event that really could have generated that broader discussion about how we really need to be organized. We will see, but right now if I can make the observation looking around the campaign trail today at the Presidential level I do not hear a lot of discussion about any of this. There are some very good people out there working, I just flipped up that one quick slide on it, about the reforms and the reorganizations that are needed, but I am just not sure that the oomph is still there to translate that thought into the huge effort it is going to take.

Audience Member

I am observing that you suggest a trigger event. Your discussion of the National Security Act of 1947 really shows how marginally it changed the structure. Post 9-11 we get two major pieces of legislation that changed structure. One was the intelligence community. We used that to bring a number of ideas together, but more importantly was the one that created DHS (Department of Homeland Security), and that frankly, is an even greater structural change than the National Security Act. The National Security Act brought together essentially three institutions, two institutions, split off one and made a third, and brought them together under one roof. The Homeland Security Act takes 22 agencies of completely different cultures—some police cultures, one military culture, some intelligence cultures, not as many people, but certainly a whole bunch and 22 organizations and says, "Work together." That strikes me as a really tough job and in some ways actually more revolutionary than anything the National Security Act did.

Dr. Dorff

We should talk about this because I think in some respects you are right in terms of the 22 organizations that existed that you put together and cats and dogs and so on fights that you might have with this, but I take a strongly different view. Taking 22 existing organizations and throwing them into one pot and declaring them the Department of Homeland Security was the easiest way out. Instead of really taking a look at what the true functions and capabilities that were needed were, we essentially rearranged some organizational charts and I happen to work with then special advisor to the President, later Secretary Ridge, as he was going through that process, and that was one of the things that really troubled him greatly because there was a rush to take all of these existing organizations and throw them together instead of maybe, again, pausing for a moment and asking ourselves what is it we really need to be able to do? And I would point to that, and my feeling on that, in precisely the same way, not just because Hurricane Katrina was certainly the next disaster that brought that home, but the rush to figure out who is really in charge and what reporting line even. And after Katrina we have FEMA (Federal Emergency Management Agency) having this direct line back to the White House that is supposed to fix it comes back to this problem, I think, of voluntary cooperation, and in many respects I think Secretary Ridge felt like he never was able to get completely through that, that it was still almost like 22 organizations within one voluntarily cooperating with others. That is not to say it was not, it was hugely significant in that it shook up a lot of people in Washington, but I am not sure it was the kind of process of really taking the strategic imperative and translating that into the organizational need. And intelligence, I am not an intelligence expert, but people who are tell me that just creating yet another Director of National Intelligence, all that is doing, again, is layering another layer on top of it. So clearly we did something, but I think that was one where it may have been a push from Congress especially which politics drives all of this. Clausewitz tells us that too, but . . .

Audience Member

But how is this any different really from 1947?

Dr. Dorff

I think it is very different. Thanks for following up with that because in fact, the institutions that did wind up showing up in the National Security Act of 1947, there really were not any of them as such that did exist before the war. Now, I was making the point that yes, the War Department, but in effect there was no JCS (Joint Chiefs of Staff), there really was not a Central Intelligence Agency. I think the precursor of all of them were things that were created, some of them in collaboration with some of our British colleagues in the war effort itself, so I think that was fundamentally different. And probably coming back to this question too, the transformation of an American public to now be thinking about national security as a peace time, not just a war time. That to me was an even bigger transformation although that is more attitudinal than philosophical, but absolutely essential and why I am saying I do not think we have made the big transformation after 9-11 because other than the fear and everything that we all rightly had of being attacked again, has not really necessarily translated into dealing with some of the problems we both have been talking about which I think humanitarian and disaster relief in semi-, if not still non-permissive, environments, ones that are going to be caused by bad people who are every bit as much our enemies as some of the others are going to be the problems that almost all of us are really going to be dealing with in the future.

Day 2—Featured Speaker

Why Interagency Operations and Reform Are Hard To Do[1]
(Submitted Paper)

Richard W. Stewart, Ph.D.
Chief Historian, US Army Center of Military History

Yesterday we heard from a number of panels and speakers about how interagency coordination actually worked, or didn't work, in a number of instances. Certainly Washington these days is focused on this issue with Jim Locher of Goldwater-Nichols fame, among others, generating books and papers about the need to reform the entire National Security structure to improve the planning and conduct of interagency operations and more efficiently focus the elements of national power on today's crises. I've been tangentially involved in this Project for National Security Reform for several years, contributing to this collection of case studies on interagency operations in the past that came out from Praeger a few months ago entitled, alliteratively, *Managing Mayhem.* One fact seems to stand out from this study and the others presented so far in this conference is: coordinating actions through the interagency maze as it currently exists is HARD! And changing the rules through interagency reform to make coordinating easier is, perhaps, even HARDER! So I wanted to take a few minutes this morning to discuss with you some of the reasons why I believe it is so hard to get things done in the interagency or to reform the process effectively and, at the same time, hold out hope for some eventual improvement in the interagency process.

Two conferences ago here at CSI I presented my analysis of CORDS, an on the whole successful interagency experiment in Vietnam to coordinate all aspects of US support to pacification during the last five years of that war. I was asked in the question and answer session what I thought of the prospects of some kind of interagency unification act like the military's Goldwater-Nichols act. My answer then was fairly brief. I stated that based upon what I had seen in the documents about the two-year struggle between Washington and Saigon, and between DOD and the State Department, and all the squabbling that it took to create that single manager headquarters in a foreign country despite the fact that there was a clear and present danger that we were going to fail in pursuit of a major policy of our government, that such institutional interagency reform in Washington seemed highly unlikely. In fact, I believe that I said something along the lines of, "a 20-year struggle with a lot of very political, savvy people in Washington who will die in many ditches" before that happens. Perhaps I was a little too emphatic, since change in Washington does occur, albeit slowly, and today there are some signs that politicians and think tanks are beginning to recognize that something needs to be done along the lines of interagency reform. Although, I submit that 20 years for any major changes to occur is still not too far off as a guess. I wanted to take a moment in this setting today to step back from the fray and explain why I still think that finding a permanent interagency solution like a "Goldwater-Nichols Act for the Interagency" is, if not impossible, at least very, very hard.

Why Is It Hard?

First, we have to realize that our Federal government was established initially to perform only a few basic functions without being a threat to our liberty. One of the fundamental reasons why interagency cooperation and coordination is so hard is simply this: the founding fathers who established the basic structure of our government were more interested in *preventing* an efficient government than they were in creating something that would function smoothly, efficiently, and to their minds potentially tyrannically. The very concept of separation of powers ensures that the legislative branch checks the power of the executive branch and that the judicial branch watches both of them. And even within the executive branch the various cabinet officials do not report to any single powerful coordinating body but directly, as individuals, to the President. The National Security Council, an important organization that tries to coordinate policy and ideas through a series of policy committees, the deputies committee, and the principals committee, cannot direct a cabinet-level official—an official appointed by the President and confirmed by Congress—to do *anything* without the direct intervention of the President himself. That explains both why interagency coordination is so hard and why entire departments can so easily ignore policy initiatives with which they happen to disagree. And, of course, the art of "slow-rolling," of public agreement and private noncompliance, is as much a Washington art form as the careful calibrated "leak." Only the office of the President truly has the power to enforce policy across the executive branch. No cabinet officials, men and women of considerable personal power and experience, can be forced to fall in line and do what any other cabinet official might want them to do without knowing that they can use direct access to the President to slow a policy or initiative with which they disagree. I think that the founding fathers preferred gridlock to an overly powerful, efficient, and potentially oppressive governmental structure. (Remember, these are the same founding fathers that originally had the man who garnered the second highest number of electoral votes in a presidential election become the Vice President, thus ensuring some measure of conflict at the highest level of the executive branch.) This underlying dynamic ensures interagency friction and presents an institutional and cultural barrier to all well-intentioned attempts to force different parts of the interagency to work together.

Those in favor of a "Goldwater-Nichols Act for the Interagency" are also deluding themselves somewhat in believing that what worked fairly well for the military services—remarkably similar cultures when you get down to it—will work for the other parts of government. Let's take a quick look at the obvious success of the Goldwater Nichols Act. Looking back, it is easy to say what a great idea this was. Create more unity among the services, consolidate some functions, force them to pull together and fight joint, live joint, breathe joint, think joint: what a great idea. However, this Act was passed in 1987 against very strong opposition of almost all of the service chiefs.[2] And,

I submit, it is still a work in progress. Joint doctrine is still being written to implement the joint warfighting aspects of the Act. The services have begun, slowly, to fight a little more jointly, but even in IRAQI FREEDOM the Marine's I MEF and the Army's V Corps attacked toward Baghdad essentially on two separate avenues of approach. Cooperation between the forces is probably at an all time high (except at budget time in Washington) but even the two ground forces still basically fight their own campaigns with three separate air forces (the Air Force, Naval air, and Marine air—sometimes answering to a single-air asset manager; sometimes not) flying in support.

Two other key aspects of changing separate military cultures are joint schooling and joint assignments. After 20 years we are just now starting to make a dent in unifying higher level staff work, and joint assignments are only now seriously being viewed as essential for promotion. The Navy up until 2005 was still being granted waivers every year on joint education and joint assignments required by Goldwater-Nichols for promotion to flag rank. They saw no reason, in their heart of hearts, to value "jointness" nearly 20 years after Goldwater-Nichols became law. And as for joint education, I can tell you after my year at the National War College, that the Army, Air Force, and Marines take such education seriously as part of a progressive and sequential joint education system with education opportunities throughout an officer's career. The Navy still seems to send people to the premier strategic, joint, educational opportunities—the Senior Service Colleges—almost by default: they pick those officers who are between assignments and have nothing better to do. I submit that their culture is still a long way from truly valuing "joint" education.

So we can see that the road to full implementation of Goldwater-Nichols has been slow. And this was merely an attempt to force more cooperation and unity of action upon military cultures that, when you get right down to it, are more similar than different. They were all uniformed services; with a deep emphasis on leadership skills; extensive career-long education and training programs; focused on mission accomplishment, self-sacrifice, and "can do;" with progressively challenging leadership and management assignments and with nearly identical pay, promotion, pension, and performance rating systems. Perhaps most importantly, the services were all within the same cabinet department. Nothing had to be enforced or negotiated across cabinet lines. So Goldwater-Nichols, despite fierce service opposition, was an overall successful attempt to harness and focus remarkably similar entities that were already within the same command structure and had been since at least 1948. And even then it still took years to pass the necessary legislation mandating change and decades to implement that change with much still undone.

If accomplishing greater service unification was and is a challenge, how much greater is the challenge of coming up with a structure that can provide for unity of effort in national security from such wildly different cultures as that of the State Department, USAID, Justice, USDA, Treasury, and the hodge-podge that is Homeland Security, an organization still trying to straighten itself out?

Look at the size of the resources differences. The State Department consists of a total of about 57,000 employees, of whom 37,000 are foreign nationals, 11,400 are Foreign Service Officers and 8,700 are Civil Service employees. Subtracting the foreign nationals, this means that the State Department has about 20,000 US national employees at most, less than 1 percent of the size of DOD's 1.4 million Active Duty personnel and 700,000 civilians.[3] The total State Department budget in 2006 was only $10.4 billion and of that $2.3 billion went to pay for dues to the United Nations and other international organizations and support to Peacekeeping operations overseas, leaving about $8 billion to conduct America's foreign affairs and pay its employees. This is well under 2 percent of the $700 billion+ DOD budget.[4] And number of employees and budget does matter when called upon to support unplanned contingencies or additional missions such as Provincial Reconstruction Teams or ministerial assistance teams whether in Haiti, Iraq, or Afghanistan.

State has certainly never had the numbers of employees or the budget to be able to devote dozens, let alone hundreds of its personnel to support contingency operations or even planning cells. Nor has it had the culture, the organizational structure, or the mindset to plan extensively, organize large programs, manage extensive resources, or even to interact effectively with the well-staffed regionally oriented Combatant Commands.[5] At the risk of over generalization, the State Department culture is one of small meetings and discussions, extensive negotiations, and one-on-one understandings, of compromise and nuance, intellectualizing problems, and sending "cables" of insights and observations in hope of influencing larger events. They are at home in embassies and consulates, at receptions and seminars, at fora and colloquia. (These are all valuable skills, and should not be discounted, but they are very different skills and reflect a very different culture than that of the military.)

As Lieutenant General David Barno, former Commander of the Combined Forces Command–Afghanistan said in a recent interview:

> there's a vastly different culture there. In the military, you kind of grow up in the teamwork culture where accomplishing a mission, which is to get something done, is the overarching order of the day. In the State Department, to caricaturize it a bit, it's a culture of 'observe and report,' and the highest-value outcomes are the well-done reporting cables back to Washington. [6]

And those written cables are examples of highly individualized skills, not teamwork efforts.[7] Is it small wonder that numerous officers since 1998 have referred to this divide as: "Defense is from Mars, State is from Venus." (I guess that leaves USAID and their NGOs and PVOs as intergalactic denizens likened by one military observer as "those guys . . . from that bar in Star Wars."[8])

As a symptom of those vastly different cultures, I know I was extremely struck by the now famous remark made in November 2007 by a State Department employee at a public meeting with the Foreign Service Director General during which a number of employees were told that they might be directed to go to Iraq in conformance with their oath of office to serve anywhere in the world. The employee stated, in a horrified tone, that an assignment to Iraq was a "potential death sentence" and "Who will raise our children if we are dead or seriously wounded?"[9] The implication was strongly that he didn't sign up for this sort of thing! It doesn't take many instances of that to shake the confidence of military personnel—whose every posting these days is a "potential death sentence" since soldiers don't generally get sent in large numbers to serve in Luxembourg, Aruba, or Tahiti, but in some such "garden spots" as Somalia, Kosovo, Haiti, Iraq, and Afghanistan.

If the State Department culture is somewhat different from the military one, then let us compare the truly different culture, training, and orientation of the US Agency for International Development (USAID). USAID is even smaller than the State Department (and has, for all intents and purposes, been rolled up under State although many don't admit it) with an even smaller budget. USAID has only about 1,800 full time government employees, augmented with only 700 contractors and 7,500 foreign nationals working overseas in the 90 some odd missions. The budget for their various programs, separate from civilian salaries, is on the same scale as the State Department, a mere $10 billion, or less than 2 percent of the DOD budget.[10]

Unlike their predecessors during the Vietnam War, many of whom were assigned directly to various postings in villages and districts involved in government capacity building and direct contacts, USAID representatives in foreign cultures now spend most of their time on development and humanitarian aid projects as contract managers by facilitating the operations of nongovernmental and private volunteer organizations (NGOs-PVOs) and other contractors. If they can be said to have an institutional culture, it probably would be one more closely characterized as "Save the Children" than either State or military cultures of national security. The people who make up USAID are often suspicious of, if not downright hostile toward, that 800 lb. DOD gorilla that shows up briefly in distressed countries that may have had USAID missions for 20 or 30 years, throws around massive resources, establishes huge depots of supplies and armed camps, only to depart in a few months or years only having muddied up the waterhole, so to speak, leaving USAID and their associated NGOs and PVOs to pick up the pieces. The military's often good intentions do not enter into their equations since good intentions or not, our military presence inevitably distorts and complicates any situation from perspective of aid-givers and development personnel.

But there is little doubt that USAID people do live in a separate culture. To give just one example of, shall we say the lack of operational or planning "mindedness," we interviewed a member of the staff of Combined Forces Command–Afghanistan in March of 2007 and asked him about his working relationship with USAID. He related to us a story of USAID planning for the Phase I Ring-Road opening ceremony in Afghanistan for which USAID was the lead. President Karzai, Ambassador Khalilzad, and USAID Chief Natsios were all to attend the ceremony in December 2003. USAID put together an elaborate ceremony with parking, speaking agendas, food, etc., and when asked by CFC-A HQ what they needed, the USAID planning staff responded that everything was squared away—no problem. The military staff officer later told us: "Hey it was all squared away, [but the] bottom line was, there was no security [planned] for this event. I mean, clearly a strategic level type event, but USAID had done no coordination for security. We essentially went into a crisis action mode several days before the ceremony."[11] There was simply no culture of planning, and no real understanding of how to conduct operations in hostile or semihostile environments.

And as for other potential interagency partners—Justice, Education, USDA, Homeland Security or Treasury—one is hard-pressed to assign them "cultures" but—to perhaps again over general—each of them are very Washington-centric in their scope and programs, and focused more on policy generation, wrapped up with heaping doses of legalese, than on operational mission accomplishment in austere environments on extended deployments. (Pieces of Homeland Security are the exception, especially the Coast Guard, Customs, and Border control, but DSH, as a whole, is a confused mishmash of bureaucracy with a variety of mostly nonoperational cultures.) If one is looking for organizations filled with experienced planning staffs, deployable teams, and "can-do" action officers, one has trouble finding very many examples outside DOD.

In addition to cultures that do not stress planning, deployments, and mission accomplishment in hostile overseas environments, there is the additional complication in the varying levels of training and education available to the military as opposed to any other parts of the government. The military structure for training its officers is well known to most in this room. Over the course of a 20-plus year military career, an officer—starting with a variety of sources of commissioning each of which stresses teamwork and leadership—returns every few years to an advanced course, or Staff College, or War College, or Pre-Command Course to gain months of peer contact, progressive education and training, and inculcation with doctrine.

Compare the military training and education system—focused on career long training in leadership, staff development, doctrine, planning, etc.—with the civilian leadership and development structure. I will now discuss the leadership and education systems of the other players in the interagency . . . . (long pause) Well, that's about it. No, it's not quite that bad. State has the Foreign Service Institute although many of its courses focus on cultural studies and languages. But State and the other agencies have the twin problems of limited training dollars and, more importantly, the challenge of who will staff the empty slots in the agencies and embassies while the incumbents are going to school? Unlike the military with its hundreds of thousands of soldiers and civilians, neither the State Department nor any other agency has a large training budget or a flexible personnel account. For example, let's look at the Army's TTHS (Trainees, Transients, Holdees and Students) account where, on average, at least 50,000 soldiers who are assigned to schools, in training, or on medical hold can reside for months and even years at a time. Think of it: a "float" of 50,000 soldiers in transit, awaiting assignment, in schools, or on medical hold. You have fewer that 20,000 Foreign Service Officers *and* Civil Service employees to staff the Department of State headquarters in Washington and all of the embassies and consulates world-wide. It is very hard to siphon off even a few hundred a year to undergo a series of lengthy professional development and leadership programs that would in any way be comparable to the military system. In point of fact, the *entire State Department*, including Foreign Service Officers, Civil Servants, political appointees, contractors *and* foreign national employees overseas, is smaller than the 62,000 soldiers on average that were in the Army's TTHS account in 2007.[12] So, the Army has more soldiers in "limbo" than the State Department has employees. It is hard to overlook this dramatic brake on any attempt to expand the educational and training structure of the State Department that would help them be more effective interagency partners. And they probably have a better training system than many other Federal agencies outside DOD.

In short, because of the actual structure of our government, our vastly disparate cultures, resources, missions, goals, focus, training opportunities, and size of the different players in the interagency process, getting them to coordinate anything is a miracle. Forcing them together into anything like a unified structure akin to Goldwater-Nichols is nearly unconstitutional and will be very difficult, if not actually impossible. Even if legislation occurs that generates better methods to manage the flow of plans and concepts through the National Security Council system, this will do little to change the actual management of operations or the blending of widely disparate cultures.

Now, having pronounced interagency reform at the national level nearly impossible, let me take a moment and show some areas of light that may indicate that interagency operations can occasionally work. Perhaps the prime examples of this phenomenon are in instances of overseas operations where the various interagency partners—State, CIA, USDA, USAID, and DOD—are working together as a team, focused on an important mission, and are able to ignore or downplay many of the rigid lines of authority within the beltway. With the right personalities, in the right setting, focused on the right mission, interagency can work despite the lack of a true Washington consensus. Prime examples of this from history include CORDS in Vietnam and even, on a much smaller scale, the PRTs in Afghanistan and Iraq. These small interagency and international teams—providing little more than band-aids for the gaping wounds of these countries due to the lack of a robust organization and limited resources—are managing to work together on the mission of providing some reconstruction aid to war-torn areas. However, cultural differences even then often create tension between military commanders and their civilian interagency counterparts working on their teams.

Another example of where the interagency worked despite the lack of a formalized structure is that of Task Force *Bowie* in the early days of Operation ENDURING FREEDOM (OEF) in Afghanistan. Though less well-known than Task Force *Dagger* or Task Force *11*, the "White" and "Black" SOF task forces at that point in time in OEF, Task Force *Bowie*, commanded by experienced Special Forces officer Brigadier

General Gary Harrell, was an experimental Joint Interagency Task Force, or JIATF. This JIATF, created at the behest of the CENTCOM commander, brought together CIA, FBI, NSA, and Special Operations personnel to focus on the generation of actionable intelligence in a "fusion cell." This fusion cell, consisting of representatives of all the major intelligence and Special Operations players, sat in one room, and shared intelligence from a variety of sources with virtually no barriers in an attempt to put together timely, actionable intelligence in the search for high value targets (HVT) in Afghanistan from December 2001 through June 2002. This example shows that when the mission is critical enough, and the personalities mesh, interagency cooperation, indeed integration is possible. But successful cooperation and integration are still heavily reliant on just that: the right mission and the right personalities—personalities that won't play the "I'm going back to Washington on this issue and you can't stop me" game—and the right organization. The mission and personality issues, I submit, cannot be legislated, but the organization can, and in the future perhaps model structures can be developed where interagency individuals can actually be *assigned* to an organization, not just temporarily loaned, with all that means. The military officers would write the evaluations of the civilians assigned to them or, depending on the mission, the senior civilians would write the evaluations of the military personnel assigned to them. It must work both ways and the personnel assigned must have the right skills, the best training, and the right "bias" for action and mission accomplishment. Only then can an interagency task force be positioned for success on the battlefield or during crisis operations.

So, in conclusion, what can we do in the next two decades as we wait for potential institutional change to begin the long process of cultural change within the various agencies of the government? We can start by focusing on improving the quality of people within all the agencies of our government by expanding training and educational structures, opportunities, and funding. We can then begin throwing personnel from all agencies of the government together more often in training and planning exercises and assignments, building more interagency task forces focused on specific issues and plans, and creating the modalities of working together effectively on the ground overseas to make up for a basic and continuing lack of interagency coordination back in Washington. This will require more resources (money and people) in agencies throughout the government outside of DOD, more formalized educational requirements for DOS, DOJ, the Treasury, etc., so that their people can be trained in leadership, management, and planning procedures on a par with their military counterparts.

In short, we need to work on the practical workarounds that will force us to work more closely together and not wait until we have a "Goldwater-Nichols" for the interagency. Even if such legislation was passed today (highly unlikely) it would not have a serious impact on changing the cultures of government in any less than 20 years. The stakes of the current global struggle are too high to wait for legislative grand schemes, but we can focus on fusing the interagency cultures as we have done over the past 40 years in DOD: create more interagency assignments, more interagency educational opportunities, and recognize the need for more money and people for the other parts of the interagency outside DOD. Only then can we slowly create the functional structure that makes the interagency work during critical national security operations. The stakes are too high to do otherwise.

Notes

1. The views expressed in this paper are those of the author and do not reflect the official policy or position of the Department of the Army, the Department of Defense, or the US government.

2. See especially James R. Locher III, *Victory on the Potomac: The Goldwater-Nichols Act Unifies the Pentagon* (College Station, Texas: Texas A&M Press, 2002), pp. 3–12.

3. US Department of State Bureau of Human Resources Fact Sheet as of 09/30/2007, "Facts About Our Most Valuable Asset—Our People," at http://www.state.gov/m/dghr/c24894.htm.

4. US Department of State Budget Summary 2008, US Department of State Web Site, http://www.state.gov/s/d/rm/rls/bib/.

5. For an interesting proposal about how to organize the State Department into regional bureaus to match Combatant Command theaters with a projected improvement in planning and working together, see Commander John M. Myers, "Singular Vision: A Plan to Enable CENTCOM and State to Work Together," *Armed Forces Journal,* March 2008, 43–45.

6. *Enduring Voices: Oral Histories of the US Army Experience in Afghanistan, 2003–2005*, Interview with Lieutenant General David Barno, unpublished manuscript, Christopher Koontz, ed. (Washington, DC: US Army Center of Military History, 2008), p. 119.

7. Ibid.

8. Ibid. Interview with Colonel Tucker B. Mansager, p. 123. And see Rickey L. Rife, "Defense is From Mars State is From Venus," Army War College Research Paper, 1998, p. 16.

9. Jack Crotty, a senior Foreign Service Officer, as quoted in "Diplomats Decry Being Sent Against Will to Iraq: Hundreds Boo State Department, Liken Policy to 'Potential Death Sentence,'" MSNBC, 1 November 2007.

10. USAID Facts Briefing Charts provided to the author by Mr. Thomas Staal, Chief, USAID Iraq Office, 7 March 2008.

11. *Enduring Voices*, p. 181, Interview with former Chief of Staff, CFC–A, Colonel Thomas J. Snukis, 1 March 2007.

12. For more on the size and structure of the Army's TTHS "investment", see Scott T. Nestler, "TTHS is Not a Four–Letter Word," *AUSA Landpower Essay*, Institute of Land Warfare, No. 04–7W, November 2004.

Day 2—Featured Speaker
Question and Answers
(Transcript of Presentation)

Richard W. Stewart, Ph.D.
Chief Historian, US Army Center of Military History

Audience Member

I found your discussion of the President having to play fascinating. I just call the attention of everybody to a book that first came out in early 1960 or maybe 1959 by Richard Neustadt called *Presidential Power*. In that he pointed out that not even the President has the power to order things to happen. In fact, it begins with an anecdote about President Truman waiting for President Eisenhower to take office and President Truman says "Poor old Ike; he'll find it is nothing like the Army. He will say do this and do that and nothing will happen." I submit it has not changed all that much.

Dr. Stewart

I also submit though that even in the Army, and here I look for correction from the senior folks in this audience, even in the Army the number of times when a senior general can say, "Do this and do that," and instantly everything falls into place are few and far between. There is still a lot of consensus building of discussion of putting folks' minds in the right place to not only get them to do what you want but for them to believe in doing what you want, which is a very different process. But you are right about Presidential power and yet, if he cannot do it, no one else can. If he is not actively engaged . . . we saw that in some of the case studies the other day. In Grenada it was Reagan who was involved not once or twice . . . but was involved in several parts of the process and pointed them each time in the right direction. With CORDS it was the President who over 18 months could not direct. He knew he was not going to get away with directing. He slowly built the consensus and allowed other people to try out their experiments first. They failed, then he began pushing until he finally got what he wanted. He did not direct; he built the consensus. The President has power; the only person who is the decider in the interagency. There must be a better way to give him a better staff so he can make better decisions perhaps. That comes down to the structure of the National Security Council. That is probably the best area for structural reform in my opinion.

Audience Member

Your initial remarks about the philosophical underpinnings of American government, the emphasis on constraining effective government led me to pause and think for a second. It seems that many other regimes, some with authoritarian philosophical underpinnings, do not do any better in the interagency. To what extent is it just about bureaucratic dynamics as opposed to philosophy?

Dr. Stewart

Good question. You can never discount bureaucracy. A lot of my doctoral work was done on ways in which an arms bureaucracy was created and tried to do things and it fell into the same bureaucratic traps you run into today. The power of bureaucratic structural organization, turf fights, budget wars, all of those things are there whether it is autocratic or, in our case, a more democratic state. But I am indicating that the founding fathers perhaps pushed more in that direction than authoritarian regimes who want to get things done after all, who want to have more efficient governments, and I am not sure that a lot of the players of our government are still that convinced of that efficiency. Because again, they have their own missions, their own focus, they know what is important, but according to law they have to do "X" and suddenly the military comes along and says, "We would like you to help us with 'Y'." That is fine but I do not have a lot of resources. I will help you a little bit with "Y," but if it takes away from "X," you are not going to get my full attention and "X" is mandated by a series of, not just the structure of government, but the laws that have been created since that time which always restrained the active players.

Audience Member

It would seem then that the US founders were closer to getting what they actually wanted and prescribed than some of their authoritarian counterparts who wanted effective regimes and were able to pull it off.

Dr. Stewart

Amazingly so, but remember these are the same great people that when they set up the electoral college had the person getting the largest number of votes becoming President, but the person who ran second becoming Vice President, even trying to guarantee gridlock in the executive branch. It was only after trying that a couple of times that they realized that this is not going to work at all that they backed away from it. So the default position was definitely that.

Audience Member

AFRICOM (United States African Command)—I would like to hear your thoughts on that. Do you see this as a forcing function to create a force interagency effort in its study or experiment? As we look at it and see it from where we stand we certainly see a lot a challenges it has. I would be interested in your thoughts on AFRICOM and how they are trying to organize that command.

Dr. Stewart

I have only been watching AFRICOM from a distance. It seems like an interesting experiment with a lot of words that sound really good. Oh, we are going to have more interagency partners and all that. They have had trouble getting people to volunteer from the other interagency organizations, haven't they? I wonder why? It is because they know they are going to be outnumbered 500 to 1, that they are going to end up being rated by some people of a very different culture, and that their own organization will then look upon them with a certain measure of suspicion when they return. Again, it is all things that we saw in the military culture back in the early 1980s when you were assigned to work for an Air Force officer or a Navy guy was assigned to work for you. You were supposed to defend your service and not get too carried away with this other organization having a mission. I submit that with AFRICOM they are going to have to be very careful to both provide good advice but not blend too much with that culture. If they increase the numbers of interagency players within their organization, that might help, but where are these numbers are going to come from? You need 100 USAID (United States Agency for International Development) people in your headquarters to make a difference in Africa. Okay. That is 1/18th of the entire staff of USAID. Are you going to get it? No, all good intentions but it founders upon the problem of different cultures, different resources, different numbers, different mission. A number of folks within USAID in State have not bought into the idea that a military regional command for Africa, which up until now has been pretty much their turf, is going to help the situation. Some of the African nations who initially said this sounds like a good idea are beginning to back away from that as well. So I think there is a lot of work; like every good historian, ask me in about 10 years and I will tell you how things are going today.

Audience Member

My question is, what capabilities would you like to see in the interagency? Let's talk about State and USAID in particular. I am not sure DOJ (Department of Justice) is going to be sending a lot of folks out with muddy boots anytime soon. Do we just want State to be able to plan better and come to meetings on time or is there a specific set of capabilities that we are actually talking about there? I know my students very much enjoy complaining about the State Department. You can derail any seminar for a good hour with that topic. But the idea that there is something more important for them to do besides diplomacy, more important than maintaining relations with our NATO (North Atlantic Treaty Organization) allies or something, it strikes me as a bit of a false dichotomy. There is actually very important work that they are supposed to be doing, and the problem is that we do not have enough folks to dig wells which is, apparently, nobody's job. So my questions is, what exactly are the capabilities in the interagency that we are lacking and you would like to see?

Dr. Stewart

You are raising an excellent point because we cannot go around making fun of the State Department and trying to say that the mission they are doing is not important; it is. And that is what their resources are focused on and that is an important thing for them to do. They have to continue to do that and, in fact, I submit they need more resources to do that mission. But planning and schooling and assignments would be a good start. Certainly, the difference between journalists and historians—with journalists three days is deep research, and with historians, three years you are just beginning to define the problem. With State a couple of days at a seminar, that is planning for them. That is all it is. Whereas military staff in Washington and out here as well, can spend months and years updating plans, revising, coming up with new assumptions, new facts. So there is a difference in perception about what planning is. But more planning, more in-depth planning, having enough people trained and skilled in planning and the planning methodology that are able to set aside 10 of them with widely varying skills in a room in the State Department for six months and work on various local and regional plans. And remember, with every local plan you have, you are going to have a number of ambassadors who will deny all knowledge that this will at all be helpful or work. You need more people doing that, more training in how to do that, more appreciation within their own organization from their senior folks that this is an important thing to do which at the moment is not there. Even if you start the planning cells today it will take 10 or 20 years for a change of culture. During Vietnam, having State and USDA and USAID people ready to go out on various missions in the countryside, check on facts, deal with people, and attempt to build government capacity—who better to set up a village development program than maybe a State Department person who knows about how government is supposed to work? That would be useful, but at the moment they do not have the people, they do not have the time, and they do not have the training to teach those particular skills. We keep talking that building government capacity is a great thing; we ought to be doing more of that. Okay, who is going to do that? The captain and major had a civics class in high school, but he is not trained on how all elements of putting together a government work, or someone from State Department or USAID even who have at least a notion of participatory democracy and structures. You will not find too many of the old State Department types like an ambassador I ran into at Fort Bragg who will go out to the landing zones during Operation DRAGON ROUGE in the Congo with a .45 strapped to his hip, an operational State Department guy ready to take on all comers in pursuit of his diplomacy. That is not going to work probably. But a little more operational mindedness and mindset will help as well, but that too comes in time and through assignments and through exposure. You cannot expect someone to jump in on a hot landing zone right from day one from the State Department and expect them to function. Does that answer your question? It is sort of rambling a bit. Thank you.

Audience Member

We hear a lot on the subject from Thomas P. M. Barnett who talks about the SysAdmin

Force. Now the interagency is going to come in and do the stability and reconstruction and the State Department has a lead on that by NSPD-44 (National Security Presidential Directive 44) and there is an expectation among the students, majors, that somebody else is going to come in and take over this country after the invasion is done, after the occupation is done.

Dr. Stewart

I wonder where they get that idea.

Audience Member

Which to me seems to be an unrealistic expectation. Nobody is going to come in and take it over for the reasons that you mentioned; nobody else has the capacity to do it. During World War II, Roosevelt did not want to give the reconstruction mission to the Army. The Army stepped forward and seized it, basically building a military government force and making plans for the occupation of Germany against Roosevelt's wishes.

Dr. Stewart

With the MPs (Military Police) in the lead, but that is another story.

Audience Member

Right. I guess my question is, when you talk about culture, when is the culture in the Army going to change so that we realize that this is going to be our mission, we need to be the lead on it, and we need to step forward and take the lead and finish off the invasion through the occupations that we begin?

Dr. Stewart

Good question. It is not a mission that for many years we were at all comfortable with. I remember talking with the civil affairs planners with DESERT STORM when they said, "Well, after we occupy these large chunks of Iraq, what are we suppose to do next?" They were told, "Well, do not worry about it because there are not any civilians there and we do not have to plan for them," even though the Tigris and the Euphrates Valley had been occupied with civilians for some time. The ability of the Army and DOD to deny that there is going to be a problem or that in six months, to use as an example, we will be pulling out anyway so it will be somebody else's problem de facto is remarkable. I think recent experience has once again filled that particular glass up to be about half full with the fact that we need to be able to do this thing. The difficulty is when we have the capacity, when we have the doctrine which we are getting developed, when we have the people who are beginning to be trained to know how to do this, the fear by a number of folks and senior Army leadership is, since we know how to do it, we have the people, we have the capacity, we are there, we are always going to get it, no way around it. In other words, it will stick to us like glue when we, in our minds, should be training to do other things. I think that is a false argument. I do not think it matters whether we want to or not, we have it, we are going to get it the next time, we might as well work on a more robust civil affairs structure although I submit that most of them still need to remain in the Reserves rather than Active Duty. I would argue that with anyone who would like. More robust civil affairs structure, a clearer doctrine, which I believe is beginning to be developed now on our stabilization and reconstruction even within the military and we capitalize upon this wonderful amount of experience we have received from our captains, majors, and everybody in the Army at all levels on how it actually works on the ground in an unstable environment from Iraq and begin capturing that experience, instilling the practical tips of how it works in addition to the doctrine of how it should work and putting the two together and bouncing them off each other for awhile. So I think we are heading in the right direction. I think we have to head in that direction in order to actually make things work, not just for the rest of our time in Iraq and Afghanistan, but for the next one, and we do not know where the next one is going to be. But we need to be prepared for it even if it looks very different than Iraq or Kosovo or Bosnia.

Audience Member (From the Blogosphere)

He says that you had mentioned it would take 20 years for the "silver bullet" to become effective and his question is, the incentives are not in line be it for DOD military, tax free, etc, and for State Department and civilians, it is not tax free in the combat zones. So is there any short-term actions, either legislation or regulatory changes that could effect or entice incentives-wise civilians to go do this rather than the State Department saying it might be a death sentence?

Dr. Stewart

And there is a disparity, it goes back to the cultural question. The people in charge of the State Department, I do not think, have been pushing for this change in part because if you suddenly incentivize this and have a thousand extra people volunteer and you suddenly have to triple your budget for this particular thing, where is the money going to come from? From a Congress that keeps zeroing out your budget line? Congress will not turn down a military appropriation for any length of time. They will do it postural for awhile but not for any length of time because they know the American people will look at that and say, "This guy voted against protecting our troops." The legislature or the Congress can turn down any number of bills that expand or improve the State Department, one of which is funding, any time they like. So the incentives should be there. I would hope the State Department leadership and others within the interagency are working to incentivize that because there needs to be more money to ensure that their children are being taken care of and that they have better health care when they get back. A number of State Department folks have gone over there and gotten shot at and gotten wounded, come back and the State Department says "Well, I guess that is it for you. Sorry about that." They do not really give that same level of health care that in the military we are used to thinking of automatically. It is not there in their culture. That culture needs to change. That mindset needs to change from the highest levels of leadership down to their operational level, down to the troops in the field so they do fight to get those resources to take care of their people, to make them believe that this is a career-enhancing assignment, and that we will protect them financially and physically while they are there and when they come back and promote them and look upon this as a good thing. These are the people that are the rising stars within the State Department rather than with someone who moves on to the next diplomatic reception in Monaco. Does that answer the question?

Audience Member

Yes Sir.

Dr. Stewart

It is a big issue and it will be going on for a long time, even if the legislation was passed today. Especially if it is not passed, you need to start those things anyway and not wait.

Audience Member

I have a question relating to the 20 years that you think it might take to make the institutional adjustments.

Dr. Stewart

That is just a pull out of the hat, because that is how long it has taken for Goldwater-Nichols to begin to make substantive change but . . .

Audience Member

Right Sir, I understood long term. It made me think of the Army objective force which, I think, in its original conception was supposed to take about 30 years to transform from what we had in early 2000 until what we are moving toward in the objective force. Obviously, that has been greatly accelerated lately due to the external stimulus of the long war. The question I have is, are there any useful models we can use to possibly apply a way to make those institutional changes in a shorter amount of time from say the military form of movement from the late 1970s or early 1980s, or does it always take a large external stimulus to change the inertia from organizational behavior and governmental politics that preclude those changes in a shorter time period?

Dr. Stewart

One of the interesting things about the discussion yesterday was even when there is sometimes a tremendous stimulus which you would assume 9-11 was, sometimes it does not have the long-term effect that you would think it would have. So when you are dealing with cultures like with nation building, it takes generations. You have to sow the seeds; you have to be patient. It is a hard thing to shortcut. If you start coming up with shortcuts, you think you have achieved something but then when you move on or when your successor decides that it is not as important, that change begins to evaporate. So there needs to be a systematic, long-term, consistent cultural change beginning, probably, with mid-level officers here and mid-level officers at the State Department in order to get the right sustained change over time. I submit you cannot go a lot faster than that. The 30-year thing for the future combat systems is interesting although I think 15 years of that has to do with dealing with the contracting office and the acquisition community. Do not get me started on that. Compared to that community the interagency is an efficient team. Thank you.

Audience Member

Sir, my background has been a combat engineer for my first decade of service in the Army, but I had the opportunity to work for the National Geospatial-Intelligence Agency on my last assignment in the Washington, DC, area. This was my first exposure to the strategic intelligence world. I was wondering if you were aware of, or monitoring, the review of the IC (Intelligence Community) and how they are encouraging their employees to do two-year tours or joint jobs within the IC. You referenced Goldwater-Nichols but the IC is also moving in that direction as well. I just wanted to get your comments on that.

Dr. Stewart

Interesting, I did not know about that. That is an interesting initiative which, again, shows that instead of waiting for somebody in Washington to come up with the better structure they have recognized that there is a problem and are beginning to make some incremental changes. It does start with people; culture is just a collection of people trying to learn to do things differently. So they are beginning to change the culture one person at a time. That is a good initiative. Even then, of course, as you run into it anywhere in the intel world there are not just regulatory, but legislative barriers sometimes in the sharing of information between different communities. So even with people being trained up and disposed to share, they have to be very careful that they do not end up in trouble for it. But it is a good initiative and I am glad to hear about it. I will check into that. It sounds like something interesting. Thank you.

Audience Member

It seems to me that Army officers are very well adapted to dealing with the kind of situations you are talking about, coordinating disparate elements, non-Army. They grow up in a combined arms framework where they have to coordinate all kinds of things and as they grow and interact with battalion, brigade commander, division commander, they get better and better at this kind of coordination, and it certainly is showing up in Iraq where captains and lieutenants, taking on completely strange jobs working at a village or working on what have you, adapted very well. This should be built on and no doubt will be built on in the future, so I do not think that things are all that bad. It seems to me, however, that the superstructure of the Department of Defense gets in the way because Army is just one of the services and they have all the other services. Things have to be done equally and command jobs parceled out equally pretty much. The Army has done a lot of training of the other services. You will find years ago the Army's method of analyzing doctrinal problems and systems problems was picked up by other services. They all finally got a TRADOC (United States Army Training and Doctrine Command) kind of thing. I do not know what to make of this, but I think there is a lot to be hopeful for as we go along, and maybe incremental change and adaptation by people is going to solve some of these problems.

Dr. Stewart

I can only hope you are right, Sir. Certainly the amount of time it took to adapt to new environments and new situations in Iraq was surprisingly swift when you think about a huge institution based upon, if you ever knew the old combat-based requirements system and the old doctrinal developments model, it could sometimes take five years to turn out a new chunk of doctrine, whereas Iraq turned things around a lot quicker than that, a lot more agile than that. But again, they were building in some cases on a very experienced deployed officer core from Bosnia and Kosovo, and even though there is not a direct mixture, they were amorphous enough situations that a different set of skills were tapped to deal with those environments and then they became a little more flexible and in Iraq had to be even more flexible. Still I submit that when we go to Afghanistan there will not be another full culture shift, but there needs to be a certain amount of adaptation as well to deal with that situation.

Audience Member

Which I think they are doing.

Dr. Stewart

Yes Sir, I think so, but again, talk to me in about five years on that. We will see how well the institution captures that experience, puts it into something that is both reasonable and not locked into one time or place but is more generalizable, and then transmits that to the next generation of officers. That is always the key to a training institution such as this and such as the command schools—how well do they capitalize upon current experience and get it back into the system so that the next generation knows that as well as the new skills they have to learn?

Audience Member

Secretary Gates has used a number of vehicles and venues to try and emphasize increased resources for Department of State. I would like your comments on whether you think that will even have an impact on Congress or whether they will respond?

Dr. Stewart

You would think it would have an impact on Congress when the Secretary of Defense says, "Look, this is important enough for me to give a 100 million dollars out of my budget to the State Department today, saying please take this money, use it because I know you need this capacity; go for it." I am not even sure if Congress allowed him to reprogram that money but he probably got away with it. But it did not seem to shame Congress at all. They are almost unshameable in that regard. I do not see any huge expansion, some incremental expansion. They have authorized a couple hundred more Foreign Services Officers, for example, and that is a start. But a hundred million dollars, again, a good start, but up in Washington you cannot institutionalize anything for a hundred million dollars. You cannot set structures in place. You can use that to fill current shortcomings maybe, but you need a sustained funding flow of 100, 200, 500, a billion dollars a year in order to set the structure in place and expand it permanently. When talking about training soldiers you cannot just recruit another 10,000 soldiers. You need to figure out how much that is going to cost over the course of their career for schooling, housing, development, doctrine, and expansion of the training base to deal with them. The State Department has to do that and it must have dedicated funding. It is a good gesture; it is a good start. It needs to be followed up by Congress with something sustained.

Audience Member (Internal Circuit Television)

What are your thoughts on this historical application of the US Information Agency and whether it needs to be expanded or resurrected in today's information environment?

Dr. Stewart

Excellent question because the US Information Agency, the US Information Service, the name changed over the years, was now looked upon by historians as one of the great institutions of the Cold War that helped us not just build confidence in our system of government, but bucked up wavering allies, helped us win the information war, as it were, for the Cold War, the hearts and minds war for the Cold War. And if right now we are in a struggle for some hearts and minds around the world, getting our message out, why are we ignoring this public communications thing that we have assigned to the State Department, but again not giving them the resources to do it when you need something like the US Information Agency, run by people with the confidence of the administration, with a consistent message (once we find out what that is), and thus use over time to tell the world our side of the story? It will be seen by some as propaganda, fine, but by repeating our message in a number of venues I think that we are missing a bet by not resurrecting something like the US Information Agency. Yes, keeping it within the State Department, but giving them the resources to make it actually work. This is a valuable tool. It is not PSYOP (psychological operations), but information that in the long run will help, I believe.

Audience Member

We have talked about resources and the fact that they are not getting any and they do not have enough, I would like your opinion on the argument that the State Department and these other organizations are not effectively communicating their needs and that is why they are not being met. They are either not speaking the right language and we can provide a translator for them, or they are not making it a priority that what they need . . . is there a way to intertwine it with the 700-billion-dollar DOD budget that you are talking about and state it in terms of what they can do to support the overall military effort worldwide?

Dr. Stewart

You have made a good point because for any argument to succeed you both need an effective and skilled arguer on one side, but then a set of ears on the other side that is willing to listen and then turn that into some degree of action. At the War College, we had Carlos Pasqual drop by who was one of the heads of the initial stabilization organizations that they tried to set up at State. A more articulate, passionate, devoted individual for his organization and its goals and needs it would be hard for you to meet. For two years he met with Congress, which I submit probably wined and dined them, but also talked to them at committees, individually, in an attempt to get them to realize this was a long-term need of the nation not just of the State Department. Like two weeks after he talked to our War College class—I do not think we had anything to do with it—he resigned in disgust because his information was being thrown out there on the ground and none of it was sprouting. So if he is an example of some of the others at State, they are very skilled, they are very articulate, they know the levers of power that they can try to pull, but it has to fall upon ears that are willing to listen and Congressmen with bases in their districts; they will listen to a military spokesman. Where is the State Department base in Peoria, Ohio? It is not there. It is a very weak arguing position to begin with so State has a problem right from that; not their fault but the way that it is.

Audience Member

Real quick follow-up. Is there any benefit in tying it to an overarching goal? Something that gets more response that they can basically hitch their wagon to and try and receive some of the benefit of that?

Dr. Stewart

Yes, probably so. They have not been able to do so successfully so far because it seems that in the budget there is only a certain amount that people are willing to give out on the budget and I do not believe we should stint the DOD budget necessarily because that is critical to our national survival, but Congress needs to come up with more creative ways to put a little more in the other side of the house. So far they have not shown a huge inclination to do so.

Audience Member

At any given time there is some amount of DOD manpower detailed to the other agencies be it State or intel agencies. What is your observation or assessment of the extent to which that manpower facilitates interagency coordination or perhaps cooperation, or is it making a difference at all?

Dr. Stewart

I have nothing quantitative to deal with it. From the people I have talked to it seems to make a difference within the offices to which they are assigned, but like everyone else, everywhere else within the State Department, there are desks and subdesks, and subcommittees and divisions and all that. Within a small organization a handful of military folks will begin having an impact upon them. But remember it is military in their culture; we have to learn to adapt to their culture and that is good for us, but they are not really adapting much to the people that we have working for them. I submit the only way around that is for them to have enough people so that some of their people can be assigned more than just one POLAD (Political Advisor) at CENTCOM. Come on now, you need a hundred folks from State, CIA, a variety of agencies assigned to various headquarters, and then go back to their headquarters and spread the word about the new culture they have run into. So yes, military folks assigned in a wide variety of things do help us as a community, I think. I think it helps us more than it helps them, but it at least exposes them to another viewpoint. They need to come and become part of our culture as well for long-term change.

LTG Caldwell

As we watch what our Chief of Staff of the Army is doing, where his effort was to elevate the importance of stability operations to be equally as important as offense and defense and he publishes a doctrinal manual to do that. He approves it and publishes it and then he goes on and has us working on stability operations where, again, it will be codification of the requirement for the military to be able to, in fact, do many of these type activities that perhaps the interagency would want them to do, but with the inability of them to have the resourcing, recognize that we are going to pick up a lot more of this. As we look back over history, and again, we continually peak things, are we on the right track to try to learn these lessons we have observed over the last six years, five years in combat, or are we doing what has been done before and we are really not on the right path yet? I know you can look back in 5 or 10 years and give me a much better answer, but today, what would you say?

Dr. Stewart

It appears that we are on the right track now in a sense of trying to enshrine some doctrine, do some training, try to capture the experience of what is going on now, but then we did the same thing in the 1960s with Vietnam. Down at Special Operations Command I had file cabinets filled with all the new initiatives, counterinsurgencies, special warfare, PSYOP, the blending of this and that, tremendous training effort. In 1975, apparently someone came and said, "This is no good and we are just going to throw it away. We are never going to do that mission again." If this mindset occurs again after we have pulled most of our forces out of Iraq and Afghanistan, if this occurs again then we will have failed because then we will have taken our collective experience, the wisdom that we have gained from doing this hard stuff, and decided that it is never going to apply again and dropping it as opposed to preserving it and nurturing it and continuing to train on it even while the need is not there. Is that not what armies are to be, the best in being ready for any mission that can occur no matter what it is? But for that we cannot just consign that mission to a couple of guys down at Fort Bragg or Civil Affairs branch. They can handle this military governance thing. We do not have to think about it. The Army has to continue to think about it as an institution long term because we cannot say that mission is not going to happen again. I suppose that is a fancy way of saying, "Ask me in five years." It is a way to say we have to be serious about capturing it now which I believe we are doing, but when the emergency is not there how then do we invoke it and continue to capture it and use it and learn from it. That is the test and that will be a test 5 years from now . . . 10 years from now.

Audience Member

I was just reading the new FM 3-07, *Stability Operations*, which is excellent, and in Chapter 5 they talk about "transitional military authority." If you look at the previous version of FM 3-07 it was not there, no analogous chapter. If you read FM 3-07, Chapter 5, much of it is a lift out of FM 27-5, *Military Governance*, that we published in 1940 before we were even in World War II. It was based on our progressive era experiences after the Spanish American War, and the occupation of the Rhineland after World War I. My question to you is, as a historian, a military historian, why does the Army forget its history? Why are we relearning what we once knew?

Dr. Stewart

That is an excellent question and one that I have continually wrestled with as an officer and as a civilian and as a historian because it . . . maybe I am predisposed to history, to think historically, to ask the question occasionally that is asked even in the Pentagon, has this ever happened before? Maybe I ought to look into this. If I could get people to ask one question when they are beginning to plan for an operation, that would be the one because at least it would get them thinking about the possibility that there might be information out there that might be discoverable and might be useable. The transition part, in particular, is interesting. Everyone talks about lack of Phase IV planning for Operation IRAQI FREEDOM, right? Oh, we did not do any Phase IV planning. I was in the 352d Civil Affairs Command at the time with planning for postoccupation operations in Iraq, and we had six phases in our particular plan. We had Phase IV yes, which was immediate postemergency reconstruction activities, but we went on in our plan and continually thought about Phase V, how do we begin to pull out of this and turn things over to Iraqi officials, and then Phase VI, close things out completely, sort of lock and head out the door as a civil affairs military government type organization? That entire plan with all six phases which dealt with transition was not used at all by CENTCOM when they decided to basically throw the plan out and start all over again. Not only do we not use history from Spanish American War, World War I, Rhineland occupation, and by the way they did not look at that report from 1920 when it was produced until 1940. So there was 20 years where they had collective amnesia. Not only did they not look at the far past, to look for appropriate historical examples, they do not even look at the more recent past, and the more recent possibilities, their own historical thinking and plans that they are generating now because they cannot be bothered to think about that old stuff. I remember briefing General Potter once from Special Operations Command on the Haiti Operations and I wanted to give him some sense of the problems that the Marines ran into in Haiti in the 1920s and 1930s. I had just started on my presentation full of historical beans and he says, "Look at those years, 1919-1938, that was 50 years ago; sit down." Maybe that is more Dick Potter than anyone else, but the fact is, he was not willing to listen to old historical experience because to him it had no application to today's current situation. If we could go back to inculcating in our officer corps, not just in the Army but in other services as well, some of the historical mindedness that was so much a mission, so much a part of TRADOC in the 1980s and 1990s. You remember Sir, a battle analysis and a tied year long process, a historical analysis as well as teaching by the Combat Studies Institute in an attempt to inculcate in every officer a sense that history is alive, that history is useful, that you need to learn it and know where to find it, and know it has continuing validity. Pretty important lessons. If we could go back to some of that perhaps that would help. Again you got me on my soap box. Thank you.

Audience Member (From the Blogosphere)

Sir, the military command and control structure is often alien and often too inflexible for many other government and nongovernment organizations to be successful in working together. How can we in the military adjust to accommodate without deteriorating our own effectiveness?

Dr. Stewart

A good question, especially that first word alien. To some people, and it is not just the uniform either though, that changes over time. To some people and organizations within our own government, let alone foreign governments, we come across as beings from another planet. It is not just how we look, but by the very way in which we use time. 1500. When you are talking to military folks be efficient, but remember always think of your audience before you begin to speak. Some of them may not know what 1500 is. Use civilian language occasionally. Acronyms, you cannot abolish acronyms from your vocabulary, God knows we learn in acronyms; we have acronyms in our soup every day it appears. But if we could learn that after using an acronym to any audience, military or otherwise, to then say what that acronym means we would go a long way toward other people beginning to understand the alphabet soup that we drop on them. So some of it is mindset, thinking about your audience before you go on and on, making sure that they understand what you are talking about so they feel included in the discussion and not simply being bombarded with alien terms and doctrinal terms; that would go a long way, I think, to reducing some of that alienness that you bring up. That is an important point but remember, when we are dealing with the interagency we are not dealing with other people in uniform. Even with the Navy sometimes I had trouble making myself understood and vice versa. But with other civilian agencies, they really are coming from a different organization. Take a moment, look them in the eye, talk to them using the English language which is a wonderful and marvelous tool of communication rather than military acronyms short, concise, and to the point that work in a certain environment. When you are briefing your patrol before a night operation, that is one thing. Learn to switch gears, change your language, change your approach, and speak English when you are speaking to the civilian audiences which you have to deal with in the interagency; that will be a good start. Thank you for your attention this morning.

Approaching Iraq 2002 in the Light of Three Previous Army Interagency Experiences: Germany 1944–48, Japan 1944–48, and Vietnam 1962–73

by

Lieutenant General (Ret.) John H. Cushman, US Army

The theme of this Symposium is "The US Army and the Interagency Process."

Since 2002, forces of the US Army have been engaged in an interagency mode in Iraq. The theme of my talk is that if the senior officers of the Army who were responsible in 2002 and 2003 for the Iraq effort had been students of Army history they would have understood the full dimensions of what the Army was getting into. And they would have been in a good position to stand their ground with a Secretary of Defense who, not understanding that situation, forced on them a faulty plan for the invasion of Iraq.

The US-led Coalition's two-month conquest of Iraq's armed forces in March-May 2003 was superbly done. But, in view of its wholly inadequate planning for postconquest conditions, the full Iraq invasion was the worst planned US military operation at least since the Spanish-American War. The United States has suffered grievously therefrom.

I will use three historical examples—Germany 1944–48, Japan 1944–48, and Vietnam 1962–73—to make my case. My primary sources are, for Germany *The US Army in the Occupation of Germany 1944–46*, by Earl F. Ziemke; for Japan, a 1949 Ph.D. thesis at Syracuse University *The Occupation of Japan: A Study in Organization and Administration*, by a later distinguished scholar, Ralph J. D. Braibanti; and, for Vietnam, my own experience and *Pacification: The American Struggle for Vietnam's Hearts and Minds*, by Richard A. Hunt.

United States planning for and conduct of the 1941–1945 war against Germany and Japan took place in an organizational structure quite different from that of Iraq 2002. There was no Department of Defense, simply a War Department, and a Navy Department. The Joint Chiefs of Staff were an ad hoc planning and directing arrangement, using existing very senior officers, created to parallel that of the British Chiefs of Staff Committee. Detailed planning and orders for execution were accomplished by the Army and Navy staffs, including the Army's Air Staff. Unity of command in overseas theaters was achieved simply by double-hatting. For example, in Europe,

General Eisenhower was the commander of all US Army forces, including their logistics, and also supreme commander of the allied (including US) forces. He had two separate staffs.

It was not until late in 1943 that it was finally decided that the Army in the field would be responsible for posthostilities governance of occupied territory. But in 1940, as war clouds began to gather, the Army War College prepared a draft military government manual (civil affairs and military government are related but distinct notions; think of military government as "civil affairs conducted on enemy territory," and civil affairs as "military government conducted in friendly territory"). At the same time, a War College committee prepared a manuscript on the administration of enemy territory.

Uncertain about which of its staff sections should have the responsibility, the Army General Staff wrestled with the problem. Within six months after Pearl Harbor an Army School of Military Government was up and running at the University of Virginia at Charlottesville, Virginia. Soon the Army had established a Military Government Division in the newly created Office of the Provost Marshal General.

In mid-1942 various civilian agencies of the government began to take note of what the Army was doing. As operational planning took place for the November 1942 invasion of North Africa, free-wheeling discussions began in the high levels as to how civil affairs/military government (CA/MG) were to be handled in that theater. At the direction of Assistant Secretary of War Robert Patterson, the Provost Marshal General and the Charlottesville school came up with a "synopsis" of the matter. It said that:

> • In the first phase, military necessity would govern and US armed forces would be responsible for CA/MG.

> • In the second phase, a civilian authority would probably supplant the military, but until then government of occupied territory would be in Army hands.

The announced gist of the synopsis was "to assert and maintain War Department leadership in military government and at the same time invite and employ a wide cooperation with other departments and agencies of the government."

Accompanied by a letter from Secretary of War Stimson, this Synopsis was widely distributed in the government. In October 1942 it was brought before a full meeting of President Roosevelt's cabinet.

There, quoting from Ziemke,

> Several members, who apparently would have liked larger roles for themselves and their departments, voiced suspicions; and the Secretary of the Interior Harold L. Ickes expressed outright alarm at what he saw as a germ of imperialism. The President

seemed to think it was a good idea but had doubts about the (Charlottesville school's) faculty.

That day the President wrote a memo to Secretary Stimson which said that the matter was something that should first have been taken up with him, that governing civilian territory was predominately a civilian task which required "absolutely first class men."

Again, from Ziemke:

> . . . the President's memo converted an interdepartmental squabble into a monumental misunderstanding and a dire threat to principle of unity of command. The Army doctrine that made the theater commander the military governor at least until hostilities ended was apparently unknown to the President and could not be fitted into his concept of military government . . . . (He) considered civil administration, no matter where it was conducted, a civilian responsibility and was totally unimpressed by the argument of military necessity.

Taken somewhat aback, Secretary Stimson sought to avoid precipitating a Presidential decision that could force the Army out of military government and create intolerable command problems in a theater of operations. Rather than respond to the President in writing, he made an oral report of the objectives of the Army school at cabinet the following week, disclaiming the Army's desire to control occupied areas after the war ended.

On November 8, just two days later, US and British forces landed in Algeria and Morocco, and matters became very real. Assuming that administration in French North Africa could be left entirely to local authorities, the President had assigned policy formulation and execution to the State Department and provision of relief supplies to the Lend-Lease Administration. But soon, again per Ziemke,

> Lieutenant General Dwight D. Eisenhower (theater commander) . . . protested that until North Africa . . . was secure, everything done there directly affected the military situation. His chief civil administrator, Minister Robert D. Murphy, could not be a member of the theater staff and at the same time be independently responsible to the State Department.

> The Chief of Staff, General Marshall, agreed and on 28 November informed Eisenhower that Murphy would not function independently and the State Department would not assume control of civil matters until the military situation permitted. The Secretary of State, Marshall said, was in complete agreement . . . .

> Marshall had rescued the principle of military necessity . . . the North African campaign, in its first weeks, had set a pattern for civil affairs and military government that would persist throughout the war . . . . Thirty thousand tons of civilian supplies were needed every month . . . and both the military and civilian agencies agreed that on the drive into Tunisia the Army would have to assume complete responsibility for civilian relief.

By February 1943 it had become evident that a more substantial arrangement for managing CA/MG was needed in the War Department. The Civil Affairs Division, with Major General John H. Hilldring its director, was established on 1 March.

General Hilldring was to report directly to the Secretary of War on "all matters except those of a military nature" and to represent the Secretary of War to outside agencies. For the future, War Department officials contemplated placing full responsibility for civil affairs in the staff of the theater commander "until such time as the military situation will allow other arrangements," and the Civil Affairs Division was charged with making certain that all plans to occupy enemy or enemy-controlled territory included detailed planning for civil affairs."

On April 10, 1943, the Joint Chiefs of Staff confirmed the Civil Affairs Division as "the logical staff to handle civil affairs in nearly all occupied territory." (Presumably an element of the Department of the Navy staff would handle civil affairs in some small part of occupied territory.)

Where would the people with the necessary expertise for the wide range of military government tasks come from? General Hilldring's solution was to staff the Charlottesville school, and a second school established at Fort Oglethorpe, Georgia, with expertise from other agencies or with people brought in from their civilian pursuits. He would meet in-theater needs, which were forecast into many thousands, by training officers and by direct commission of civilian expertise, including some into field grade. And he did just that.

The issue of when the Army would relinquish management of civil affairs was yet to be resolved. In March 1943 the President placed former New York governor Herbert Lehman's Office of Foreign Relief and Rehabilitation in charge of planning and administering US civil relief in liberated areas. The President remained convinced that civil affairs was a civilian job, and in June 1943 he proposed to put an Assistant Secretary of State in charge of a committee that would give central direction to all economic operations in liberated areas, with a subordinate in each theater, nominally responsible to the theater commander, who would receive his orders from the State Department.

The invasion of Sicily that summer demonstrated that divided command in the field would not work. The Army commander on the scene had both the resources and the ability to direct operations; a civilian did not. In November 1943 the President wrote Secretary Stimson: ". . . it is quite apparent that if prompt results are to be obtained the Army will have to assume the initial burden . . . until civilian agencies are prepared to carry out the longer range program."

So that is how it turned out in World War II for the "US Army and the Interagency Process" with respect to civil affairs and military government. In Washington the War Department was a key agency, but not necessarily dominant, in policy making; other agencies of the government had interests and much to offer. But in the field the theater commander's responsibility for executing CA/MG, within policy, prevailed in war and continued with little outside participation for a time even after hostilities ceased.

In December 1943, General Eisenhower was named Supreme Commander; SHAEF (Supreme Headquarters Allied Expeditionary Force) would be his staff. An interim planning staff had been busy; that month it published a Standard Policy and Procedure for Combined Civil Affairs Operations in Northwest Europe, which assigned full control of and responsibility for civil affairs and military government to the military commanders, from the Supreme Commander on down. Well before Normandy, a G5 Section, SHAEF, under a British lieutenant general, began to function.

In England the US structure for CA/MG had already been forming. A Civil Affairs Center at Shrivenham and another at Manchester would receive and further train the hundreds, then thousands, of officers and enlisted men gathered into the Army, trained, and sent from the United States. They would form these people into self-sustaining CA/MG detachments of various sizes which were then assigned to every command level, from army group to division. At each level a G5 section would direct their activities. Large or small, a detachment would carry out the essential CA/MG actions—government, public safety, public health, public welfare, utilities, communications, labor, transportation, resources, industry, commerce, agriculture, legal, fiscal, supply, and information.

The expertise for these tasks came largely from men brought in from civilian life, many of them directly commissioned as captains or at ranks as high as colonel. Thousands of others were trained from scratch by the US Army. Ziemke tells in rich detail how these teams received their direction and then operated as the Allies entered German territory.

As the end of the war came into sight, authorities at the highest level began to discuss how to bring a civilian into Germany's postwar government, possibly as General Eisenhower's deputy; Eisenhower was willing, and even had his own civilian candidate. The War Department's choice for that job was Major General Lucius D. Clay, the highly regarded director of the military production program. Clay, supported by President Roosevelt, was nominated for his third star. In April 1945 he reported for duty as Deputy Military Governor of the United States Army European Theater of Operations, Eisenhower's "other hat."

In July 1945, SHAEF dissolved. The unified command USFET (US Forces European Theater), headquarters in Frankfurt, Eisenhower in command, came into being. The Office of Military Government, United States (OMGUS) was created, with General Clay in charge. Clay began to reduce its military strength, to bring civilians in, and to bring about the orderly transfer of government to German control.

In March 1947 General Clay succeeded General Eisenhower as theater commander and military governor. In 1949 he was replaced by John J. McCloy, named the US High Commissioner for Germany.

*   *   *

The occupation of Japan was markedly different from that of Germany.

On August 10, 1945, the Japanese made their first offer of surrender. On September 1 the Eighth US Army began to enter Japan unopposed. The next day General Douglas MacArthur, as the Allies-designated Supreme Commander Allied Powers (SCAP) received the surrender aboard the USS *Missouri*. The substance of an Initial Post-Surrender Policy (approved by the President on September 6) was provided to him on August 29. That policy read:

> Although every effort will be made, by consultation and by constitution of appropriate advisory bodies, to establish policies for the conduct of the occupation and control of Japan which will satisfy the principal Allied Powers, in the event of any difference of opinion among them, the policies of the United States will prevail.

As SCAP MacArthur was nominally responsible to the 13-member (including the United States, the United Kingdom, the USSR, and China) Far Eastern Commission (FEC) in Washington, DC. However, the FEC, which established an Allied Control Council in Tokyo, did not meet until February 1946 and had minimal influence on policy formulation.

Major General Courtney Whitney, in his book *MacArthur: His Rendezvous with History*, tells of MacArthur "pacing up and down the aisle of his C-54" as he dictated en-route to Japan. The

> . . . notes I took formed the policy under which we would work . . . First destroy the military power; then build the structure of representative government; enfranchise the women; free the political prisoners; liberate the farmers; establish a free labor movement; encourage a free economy; abolish police oppression; develop a free and responsible press; liberalize education; decentralize the political powers. . . .

MacArthur went right to work. Until September 2, 1945, his one headquarters in Manila, SWPA (Southwest Pacific Area), was a US unified warfighting command; it also served MacArthur in his role as commander of US Army Forces Pacific, including Sixth and Eighth Armies and other Army engineer, logistical, and administrative commands. On that date, GHQ SWPA became GHQ SCAP, responsible for the occupation of Japan. If MacArthur was to carry out his program, he must bring in substantial new expertise for management of occupation affairs, and his headquarters must adjust its working methods.

The solution, arrived at in the next six months, was to create within a single head-quarters two interconnected staffs, one for Japan's governance and one for military matters. General MacArthur, as SCAP, would be the ultimate authority for the former function. As concurrently commander of the US Far East Command (and directly commanding the US Army element of FEC), he would be the ultimate authority for the latter. Depending on the issue involved, the same officer would often serve on one staff or the other, his paperwork arriving for decision at the appropriate authority, or MacArthur himself. (I am told by my neighbor in Washington, Lieutenant General Edward L. Rowny, US Army, Retired, who served on MacArthur's staff in those days, that this was indeed how it worked, and very well.)

On September 6, 1945, General MacArthur received these instructions from the Joint Chiefs of Staff:

> The authority of the Emperor and the Japanese Government to rule the State is subordinate to you as Supreme Commander for the Allied Powers. You will exercise your authority as you deem proper to carry out your mission. Our relations with the Japanese do not rest on a contractual basis, but on unconditional surrender. Since your authority is supreme, you will not entertain any question on the part of the Japanese as to its scope.

Dr. Braibanti, in his Syracuse University doctoral dissertation gives many examples of how General MacArthur used this authority with great skill to bring about the transformation of Japan, including its adoption of a new constitution.

MacArthur decided right away to leave in place the Japanese structure of government with its prefectures and subordinate echelons. Annex 8, 28 August 1945, of Operating Instructions No. 4, 15 August 1945, provided that MacArthur would

> . . . issue all necessary instructions to the Japanese Emperor or to the Imperial Government and every opportunity would be given the Government and the Japanese people to carry out such instruction without further compulsion. If necessary, however, (he) will issue appropriate orders to (US) Army and Corps commanders . . . to secure compliance with (SCAP) instructions.

Eighth Army deployed a total of 46 military government teams, one team at each prefecture, each team with its subordinate detachments. As was the case in Europe, each team encompassed the full range of governmental functions. Operating for the most part under I Corps at Kyoto and IX Corps at Sendai and their assigned division and supervised by army, corps, and division G5s, this was the field structure that, along with the Tokyo SCAP establishment, carried out the occupation of Japan.

*　　　　　　*　　　　　　*

In 1962 President Kennedy's ordered increase in the US advisory effort and troop support began to arrive in Vietnam. In 1973 the last US troops were withdrawn and the

Congress forbade any further US military involvement. Throughout those years, US authorities sought the right recipe for civil-military action through which the Government of Vietnam (GVN) could take back the countryside from the communist Viet Cong.

I grappled with this problem myself, in 1963–64 when I was senior advisor to the 21st ARVN (Army of the Republic of Vietnam) Infantry Division in charge of a four province area in the deepest regions of Vietnam's delta, its headquarters in Bac Lieu. President Diem had recently launched his ultra-ambitious strategic hamlets program, which called for fortifying hamlets, and occasionally for moving people into new hamlets. But the strategic hamlet program had tried to do too much too fast and was in a shambles.

The fundamental reality was that out there in the countryside there were two governments competing for the loyalties and control of the same population. One was the GVN, the official government of Vietnam, with its province chiefs, district chiefs, and village and hamlet chiefs, its tax collectors, its armed forces down to civil guard companies, self-defense corps platoons, and hamlet militia, its schools and information machinery, and so on. The other was the Viet Cong with its own structure of province and district and other chiefs, its own tax collectors, its own schools, entertainers, and propaganda squads, and its own armed forces down to the hamlet militia, farmers by day and fighters by night. The VC side had its own doctrine, that of revolutionary war, and in the countryside it was winning.

I decided that our advisory team would work with the division commander and his four province chiefs to develop a doctrine of our own that would reverse the situation and win back the countryside. I was fortunate in having two very good colleagues, Lieutenant Colonel Robert M. Montague, a deputy senior advisor, and Richard Holbrooke, later famous but then a brand new foreign service office who was working in my area as a USAID representative.

Working with a grizzled old major on the division staff, among other Vietnamese officers who spoke pretty good English, we developed the "oil spot" concept. Major Yi, who gave the concept its name, told us that it was how the French had operated in Algeria.

Pacification by "oil spot" meant that we would start at the fringe of an area under GVN control and, using a *civil-military* organization and *civil-military* action, we would patiently pacify one contested hamlet at a time. We were rather successful; when I left Vietnam in April 1964 we had a division school in operation and systematic hamlet pacification underway in each province. (Those interested in the details can read about it in my article, "Pacification Operations in the ARVN 21st Infantry Division," *Army Magazine*, March 1966, and also in pages 108–116 of *Strange Ground: Americans in Vietnam 1945–1975, An Oral History*, by Harry Maurer, Henry Holt and Company, New York, 1989.)

The 21st ARVN Division initiative was not replicated elsewhere. It was later melded into a joint GVN–MACV (Military Assistance Command, Vietnam) program called HOP TAC (Victory), the execution of which then suffered from faulty management and, with coup after coup, from government disarray. As ARVN and local province forces were increasingly buffeted by the growing Viet Cong, by mid-1965 the Republic of Vietnam was losing the war.

President Johnson then committed US Marines, an Army air cavalry division, and two airborne brigades to save the situation. Embarking on a "strategy of attrition," General Westmoreland, COMUSMACV (Commander, US Military Assistance Command, Vietnam), asked for more. In July the President announced that he would send 44 combat battalions to Vietnam, increasing the US military presence to 125,000; it would triple by end-1966. President Johnson, meanwhile, was casting about for ways to improve the progress of pacification. Various avenues were explored; none satisfied him.

On March 26, 1966, the President (per Richard Hunt)

. . . appointed Robert W. Komer as special assistant for supervising pacification support from the White House. . . Komer's powers were substantial. He was authorized to draw support from the secretaries of state, defense, treasury, agriculture, and health, education, and welfare, from the administrator of USAID and from the directors of CIA and USIA. . . . The President made it clear that . . . Komer 'will have access to me at all times'. . . .

Komer handpicked a small group of people experienced in pacification to work for him. Lieutenant Colonel Robert Montague was his executive officer . . . (also) Richard Holbrooke of State. . . . Komer set out to solve problems, prodding officials in Washington and Saigon . . . earned the nickname 'Blowtorch.'

In September 1967, having just taken command of the 2d Brigade of the 101st Airborne Division, which had been alerted to deploy by air to Vietnam in December, I visited Vietnam with the division commander in an orientation party. In Saigon I spent an evening with Robert Komer, who was living out the consequences of his appointment 18 months earlier. Since May, Komer had been there as Deputy COMUSMACV for Civil Operations and Revolutionary Development (CORDS), with the rank of ambassador and with authority to manage all US participation in the GVN-US pacification effort. Bob Montague was his executive assistant.

Bob Montague and Dick Holbrooke, both brilliant men, had with their accomplished colleagues in Washington created CORDS—a Cadillac version of the Model T pacification machinery that the three of us had built in Bac Lieu three years before. [When President Nguyen Van Thieu in 1968 set up the GVN's Central Pacification and Development Council (CPDC) with a staff that reported to the Prime Minister and kept close ties with CORDS, he appointed our old 21st ARVN Division commander, Major General (then Colonel) Cao Hao Hon, to take charge of it.]

Richard Hunt:

> CORDS unique feature was to incorporate civilians into a . . . single chain of command that consolidated control of all pacification support. (Komer) exercised command of all pacification personnel from Saigon to the provinces (and down into the districts). . . . CORDS interleaved civilian and military personnel throughout its hierarchy. . . . Of the province senior advisors roughly half were civilians, and half were military, although the less secure provinces and districts tended to have a military head. . . . Civilians wrote the performance reports of their military subordinates, and army officers evaluated the Foreign Service officers under them. . . . The CORDS staff, called MACCORDS . . . functioned as a regular staff section under the MACV chief of staff. . . .

> General Westmoreland . . . transferred the responsibility for advising and supporting the RF/PF (Regional Force companies and Popular Force platoons commanded by the province chief in his capacity as sector commander) from the J3 (Operations) section in MACV to a directorate within CORDS. No single change was more important to the eventual course of pacification. It allowed CORDS to increase substantially the number of advisors and at last gave the pacification program access to forces that could provide sustained local security.

Commanding the 2d Brigade, 101st Airborne Division, north of Hue during and after Tet 1968, I watched province and district chiefs and their RF/PF units weather the intense fighting of that period, supported by their advisors and the CORDS structure and by the ARVN 1st Infantry Division and by my brigade. The system rebounded. When in March 1970 I returned to Vietnam's delta as senior advisor to the Commander, ARVN IV Corps/Military Region 4, pacification in the region was thriving and CORDS was going strong. (See my "Senior Officer Debriefing Report" of 14 January 1972, Headquarters, Delta Regional Assistance Command . . .)

*   *   *

Thirty years later as the Joint Chiefs of Staff and the Commander, US Central Command, were addressing the Iraq situation, Vietnam was a distant memory and CORDS had been forgotten.

From my January 15, 2007, paper "Planning and Early Execution of the War in Iraq: An Assessment of Military Participation" (http://www.westpoint.org/publications/cushman/ForArmyWarCollege.pdf).

> At CENTCOM there was little post-hostilities planning; Mr. Rumsfeld's key principals had told General Franks to 'leave Phase IV [the post-Hussein-defeat phase] to us.' Mr. Rumsfeld himself waved off help offered by the State Department. ARCENT set up a post-hostilities planning cell, but not until Secretary Rumsfeld named retired Army Lieutenant General Jay Garner to take charge of an Office of Reconstruction and Humanitarian Assistance two months before the invasion did real planning begin.

Garner would in effect work for the Secretary of Defense, creating divided in-theater command. Hastily collecting a staff, he deployed to Kuwait just days before D-Day.

As operations began, plans for constituting key ministries of an Iraqi post-Hussein national government and for putting in place provincial governments were essentially unformed. Provisions ensuring that there would be an Iraqi army and police force did not exist. The troops were not told what to do when the Iraqi Army was defeated. Post-hostilities operational concepts were not developed and made known. PSYOPS plans and capabilities were rudimentary at best. Gaping holes remained.

We know the rest of the sad story. In 2003 I had written an article for the November US Naval Institute *Proceedings*. Its title, "President Bush Deserved Better." Excerpts:

On 8 September, a somber President George W. Bush told the nation that accomplishing his goals in Iraq would take a lot more money and a lot more time than the public had been led to believe. He deserved better from his military. . . .

(W)hile Secretary Rumsfeld and the Chairman of the Joint Chiefs of Staff must be held primarily accountable for the second phase ineptness and the resulting problems, I also hold General Franks accountable. The Rumsfeld–Franks partnership extolled by Secretary Rumsfeld during the major combat phase had failed thereafter—and at a huge cost. . . .

Only in January 2003 did Secretary Rumsfeld appoint retired Army Lieutenant General Jay Garner as head of the Office of Reconstruction and Humanitarian Assistance to deal with postwar Iraq. It was late already, but that month, General Franks should have said something like this to Army General John Abizaid, his newly arrived and highly capable Arabic-speaking deputy commander:

> The postwar planning effort is not going well. If it falls short, I will be responsible. So I am going to tell Secretary Rumsfeld that you will plan the second phase, which will begin seamlessly as soon as we defeat Saddam Hussein's army. As my deputy, you will take over General Garner's operation and become the temporary military governor of Iraq. Get busy planning for a military-civilian operation, basing your organization on the solution applied by General Creighton Abrams that was successful—but too late—in the Vietnam War.
>
> We will find a civilian to play the part of Robert Komer, who first headed the Civil Operations Rural Development Support effort in Vietnam. You will make your own estimate of post-victory conditions, which will be chaotic. Here is one idea I want you to consider. There are 18 provinces in Iraq. Organize 18 province teams under three regions, one of which will be Baghdad and vicinity. I will get the Army War College to name 18 smart students to do full-time planning and to stand by to move to their province seats with their teams, complete with communications and local security from US troops.
>
> Next week, I will move you and a small planning staff to Carlisle Barracks, near Washington, DC, for convenient interagency planning. Start

gathering data on Iraq and doing research on former occupations of countries. Be ready to brief me on your concept by mid-February. I will fight for sufficient resources.

I ended the article with "General Franks should have acted along these lines and insisted that Secretary Rumsfeld accept his approach. Think of the difference it would have made."

Panel 4—Interagency Case Studies

A Mile Deep and an Inch Wide: Foreign Internal Defense Campaigning in Dhofar, Oman, and El Salvador

by

Mr. Michael J. Noonan
Foreign Policy Research Institute

Good morning. It is a great pleasure to be here today and an honor. It is my first time to be here at Fort Leavenworth, especially as a soon to be former captain. It is always nice to be able to talk to your betters in uniform. The title here of my presentation—I do not mean this in derisive way or in a dismissive way, "A Mile Deep and an Inch Wide" really refers to, as we will see in the case studies, a very small footprint approach to foreign internal defense where you do not have a large buildup or a large presence of troops. This is a possible model for what some people call phase zero in operations. Before I start all that I would like to say a line from Mark Twain. "The three most difficult things are, climbing a wall that leans towards you, kissing a woman that is leaning away from you, and addressing an audience that knows more about your subject than you do." And certainly in the case of Dr. John Fishel, I am sure he has probably forgotten more about El Salvador than I have probably read.

Like some of the other presenters I am also trained in the dark arts of political science, coming before an August crowd of military historians (Slide 1). At FPRI (Foreign Policy Research Institute) a lot of my colleagues are trained historians. As a political scientist I promise not to put any two by two tables in this presentation. Political scientists are generally an Aaron Simpson in the crowd. We tend to be lumpers. These kind of political scientists like to dichotomize things and there are lumpers and splitters. We like to aggregate things and then slice it up. Whereas I think historians like to split things up. I think in this presentation perhaps we can look at these discrete cases and get to more lessons learned issues as discussed before. A second caveat is that this research is in progress and this is part of a larger project. So this part is largely gathered from secondary source information with some primary source information. Due to the 30-year rule as was explained yesterday, there is more information coming out on the Dhofar case, and I have been able to assemble a lot of that as well as do some archival research in the United Kingdom, but this is sort of an initial macro view of these cases. The same is true of the case with El Salvador. George Washington University has an archive of information that deals with El Salvador that is fairly complete, but there are other resources out there that I am trying to gain access to, going into the historian's office at the John F. Kennedy Special Warfare Center and School. I think only one academic *[inaudible]* has gained access to this information for his dissertation on Central American strategy. The third caveat is that I think my experience in Iraq may have affected my analysis on this so I certainly welcome any push back. But my experience in Iraq was much different than the two cases that we are going to be talking about today. These were much smaller, much more discrete, and the advisors on the ground were on a much longer string without the type of support that my team and I were able to get.

The first question is, why look at these two cases? (Slide 2) I think the first and obvious point is the contemporary operating environments. You have al-Qaeda and affiliated movements, whether one thinks that this is an existential threat or more of a strategic nuisance. In 2001, discussions of al-Qaeda, we talked about this movement being spread across 60 countries and could tap into a sea of a billion Muslims throughout the Muslim world stretching from Morocco to Southeast Asia, from Central Asia down to the eastern littoral of East Africa. The second point gets to discussions of irregular warfare versus major combat operations. I do not know if anybody has seen it, but I recommend you take a look at Colonel Andrew Bacevich. He has a piece in the latest *Atlantic* where he talks about that in the United States Army there are two schools of thought. I think he oversimplifies, but there are definitely strands of truth in each one of the schools. One school is pushing for irregular threats in the future saying that obviously because the enemy gets a vote, the enemy is not going to organize conventionally against us. Then we will see a continuance of irregular threats in the future as opposed to another school that thinks we should wait. The Army should really focus on big picture stuff and major combat operations. Colonel Gian P. Gentile at the US Military Academy is considered the archetype of this school. Some people say that John Noggle is the archetype of the other school. So this sets up a Harry Summers-Andrew Krepinevich type of debate between people like Noggle who thinks that an irregular future is the path and people like Colonel Gentile who say let's hold on a second and not lose sight of things that actually could be an existential threat to the country. Finally on this point, we have seen since the Long War began you had Operation ENDURING FREEDOM–Philippines, Combined Joint Task Force–Horn of Africa, and the Pan Sahel Initiative working with armies in the Sahel region south of the Sahara in Africa going against groups like Abu Sayyaf in the Philippines, the Salafist Group for Preaching and Combat in Algeria moving down into the Sahel regions. It takes advantage of geopolitical dead space where lack of formal governments allows for an infestation or breeding like in the tribal areas of Pakistan today. These cases look at historical harbingers of those types of environments. The second point here, resource constraints in an economy of force as I believe one of the speakers said yesterday, I think it was Lieutenant General Caldwell, the defense budget is not going to be getting bigger over time therefore the supplemental is funding so many activities and operations today that as we move forward, particularly depending on what happens in November, there will be resource constraints and an optempo issue with the Army that will cause us to . . . I have little faith that there is going to be any sort of large Iraq or Afghanistan types of operations in the future. However, in order to shape the international environment, case studies like El Salvador and Dhofar might give us an alternative path. Lastly, these two cases deal with very politically sensitive operating environments which cause the type of force employment that we will discuss later on.

For internal defense you can see here from the DOD dictionary, I will give you a moment to read the definition here (Slide 3). As you can see this is a subset of counterinsurgency. It is not a big picture. During publication 3-07.1 *Joint TTPs (Training, Techniques, and Procedures) Foreign Internal Defense* tells us that there are three varieties of foreign internal defense. Indirect—which is basically training and equipping of foreign forces in a nonpermissive environment; direct—which is a direct support but not combat support for host nation governments; and finally, combat operations where you are actually going out on combat operations with the host nation in order for them to gain time to build up their own force structure to handle the lawlessness, insurgency, or subversion within the society.

Somebody said yesterday there is an obligatory Clausewitz slide. Well, in talking about counterinsurgency in foreign internal defenses, this is my obligatory David Galula slide (Slide 4). I will give you a moment . . . these are just some points that I find interesting from Galula. Now, in counterinsurgency operations there are basically two approaches. There is the enemy-centric approach where you are trying to attrit the enemy down and then there is a population-centric approach where you are working by, with, and through the host nation in order to address the political, economic, and social conditions on the grounds. As Galula quotes the Chinese General, "Counterinsurgency is kind of an 80/20 split." Only 20 percent of it is military and you can debate these numbers, but the important fact is that the other 80 percent, or whatever percentage you assign to that, is the main part of his operations, the political, economic and cultural underpinnings of the grievances on the ground that is feeding the insurgency in that country.

The advisors and advising, basically there are two approaches for advisors or advising (Slide 5). First, you have materiel advisory and support where you are basically providing weapon systems and training on those weapon systems in order to increase the host nation's capabilities. And secondly, there is nonmateriel advising and support, small unit training, and developing cohesion. A number of scholars such as Stephen Peter Rosen wrote articles in the *Washington Quarterly* back in 1982, particularly on insurgency environments. The second form of advising is much more important because all armies overseas for the most part see themselves as sand. They want to be armies; they want the equipment that makes them have a martial spirit and martial outlook. They want such things as artillery, tanks, and other things. Dealing with my Iraqi counterparts this would always be the crux of the matter. They wanted the big toys that made them more of a military unit. Seeing Coalition M1 Abrams and Bradleys or Strykers—you could see the glint of jealousy or envy in their eyes and them wanting these things. But in an insurgency environment you had a need for the more important things—radios, boots, good weapons, and good small arms in order to engage against the insurgents. Lastly, on selection, there is a big debate within the literature about what is more important, whether it should be a functional knowledge—expertise and small unit tactics—or whether you need a deep cultural base in order to relate with the forces that you are working with, by, and through on the ground. We will get into this in the two case studies. Secondly, on selection for advisors, this is the volunteer versus the volun-told dichotomy. I think it is a big issue today with the transition teams in the Army. The team that I was with was largely a Reserve Component unit. We had three Active Component soldiers on our team. The teams that replaced us were Active Duty soldiers and there was some debate there among them about why people were there. In the Army today there is a big debate, and we will get to this later on, about the people that you select for advising and what type of backgrounds and personalities those people have. Finally, you have an organizational cultural essence issue. This gets back to the discussion before about an irregular or a major combat operation focus of the force. Organizational essence was what Morton Halperin came up with in his book *Bureaucracy in American Foreign Policy*. It says the organizational essence is the dominant outlook of the profession itself. Talking about Vietnam, one of the things that he said was that the creation of Special Forces was the biggest threat to the Army's organizational essence since the split off of the Air Force in 1947. This dominant outlook within the profession itself plays back on to things like who should be advisors or whether the advisor mission should be an important mission or not.

The first case we will deal with is what I call the Dhofar War, probably more technically the Dhofar Rebellion, in Oman (Slide 6). Oman had an interesting background for this part of the Middle East. It had a treaty of friendship with the United Kingdom dating back to about 1798. So it was not a colonial possession of the United Kingdom, unlike the Trucial States which are the modern day United Arab Emirates or Southern Arabia with Aden that later became the People's Democratic Republic of Yemen, and then later just became Yemen when North Yemen and South Yemen united. The Dhofar Rebellion—an important historical antecedent to this was Sultan Sa'id bin Taymur who rose to power in 1932. He had a very backward looking approach so Oman had been stuck in the 17th century. He was a very tyrannical leader. He did not allow education. He did not allow foreign travel. He did not allow people to smoke cigarettes in the kingdom which originally was called Muscat. The Dhofar region which is the southern region to the southeast had always been an occupied portion of the country. The people there are different from those in the north. They have an Ethiopian or Somalian background. The language on the ground is different. They spoke a version of Aramaic and they were looked down upon by the north. There is an old Omani proverb or saying that says if you come across a Dhofari and a snake on the road, kill the Dhofari first. So they were looked down upon by the Arabs in the northern parts of Oman. Starting in the 1950s and 1960s the Sultan had built a palace down in Salalah which is in the southern part of the country. He had married a local Dhofari woman. You can see from the topographical map there, it is a very rugged terrain. There is basically only one road that leads up from Salalah to Durmat that reaches the rest of Oman. He did not allow any kind of agriculture throughout that coastal plain that you can see in front of the Jebel Dhofar which is a mountain range that runs down to the Yemeni border.

At its highest part it is about 3,000 feet. As you can see, it is a very rugged terrain that is perfect for guerilla activity. Oman at this time had a population of about 700,000 people and Dhofar had a population of about 50,000. Ten thousand of the Dhofaris were Jebeli living up in the mountains, and in the 1960s a group emerged called the Dhofar Liberation Front which is a Marxist organization that wanted development and aid for this part of Oman. The Sultan had really applied his resources to Muscat in the Arab portions of the country so this rebellion starts to break out. It starts out very slowly going after some oil concessions and other things that were exploring up toward Saudi Arabia and then eventually leads on to going after the Sultan's Armed Forces. So in 1965 the Sultan invites the British in and they do some initial operations from the Trucial States going through Dhofar that were very ineffective and small and did not last for a long time. The Sultan's Armed Forces, however, at this time were very small, there were a few thousand soldiers drawn mainly from Arabs in the north, and Baluchis from Pakistan. Until 1958 the Sultan had been in possession of Gwadar in Pakistan until he gave up that possession. The British Army had assigned seconded officers to the Sultan's Armed Forces. The commander of the Sultan's Armed Forces was a British officer and all the way down to battalion level you had as many as three British Army captains for every company in the Sultan's Armed Forces, but it is a very small organization. So from 1965 to 1970 is a Phase I of the counterinsurgency operations in Dhofar. This period is an abject failure. One of the interesting things about the Dhofar case is that it is two separate periods. It is a nice case to look at because there are discrete time frames that can show a lot of progress depending upon approaches of counterinsurgency. The first portion of operations, from 1965 to 1970, were very ineffective because it was a very attrition-based strategy. They tried to go after the Dhofar Liberation Front which in 1968 becomes the People's Front for the Liberation of the Occupied Arabian Gulf (PFLOAG). This organization was supported by the People's Democratic Republic of Yemen which had gained independence from Britain in 1967, and they were armed fairly heavily from the Soviet Union and China and they were much better equipped than the Sultan's Armed Forces. They had Kalashnikovs (AK-47), they had RPKs (Kalashnikova light machinegun), and they had RPGs (rocket propelled grenades), whereas the Sultan's Armed Forces was equipped with Lee-Enfield rifles and Bren guns. Phase II is 1970-1972. In 1970, the commander of the 22d Special Air Service Regiment, looking for work for his regiment, goes to Colonel Johnnie Watts, goes to Dhofar and does a hasty assessment on the ground and he comes up with a five part plan for winning the war there. This plan is based on civil reorganization, agriculture and economical development, intelligence gathering, physiological warfare, and military operations and training of local forces. However, because Sultan Sa'id bin Taymur was still in power, he really wanted to keep this operation very low key. He did not want an acceleration of British support there. So his son, Said bin Qabus launches a coup d'état against his father in the summer of 1970 and takes power. The son trained at Sandhurst, had served in the Cameronian Rifles, and had done a tour in the United Kingdom looking at modernization issues. With his rise to power he decides to change the approach on the ground there. The first thing he does is to address the social, political, and economic concerns. He offers general amnesty to all the subjects who opposed his father. He incorporates Dhofar formally into Oman as the southern province. He decides to provide effective military opposition to the rebels who did not accept the amnesty offer. He started a vigorous nation-wide program of developments, and finally he started diplomatic initiatives with the aim of having Oman recognized as a genuine Arab state and to isolate the People's Democratic Republic of Yemen. At this point he invites the 22d SAS (Special Air Service) to come into Oman. They start to do civil action programs, and they began to raise local civilian regular defense groups called Firqats which is Arabic for company. Those groups start to work in the Jebel. I will speed things up here since I am running out of time. In any event he takes a very population-centric approach between the SAS and with the Sultan's Armed Forces. They start working with the Jebelis. They start building wells and doing things like cattle inoculation. They do a cattle drive to bring cattle to market there. And in combination with the Sultan's Armed Forces they began to push the supply lines of the PFLOAG back toward Yemen with a series of defensive lines—the Hammer Line, the Hornbeam Line, and Damavand Lines. In 1973, the Iranians come in because of the broader geopolitics of the region. In the 1960s, the British Prime Minister declared that the British were going to start to leave the Gulf. However, you have this very important geostrategic position of Oman on the Straits of Hormuz. The Iranians wanted to ensure that they had a peaceful neighbor across the straits. The Jordanians start to send in troops. Oman is admitted into the Arab League, and eventually into the UN. As these population-centric and more effective military operations take place, you have increasing numbers of surrendered enemy personnel, and the insurgency itself peters out. In 1975, there is an offensive and basically that breaks the back of PFLOAG which at this time had dropped the Occupied Arabian Gulf portion from its name and basically went into a small handle and more of a brigand situation there.

British involvement there, you had about 300 seconded officers and air crew, field surgical team, etc. (Slide 7). You had the British Army training teams which were Special Air Service squadrons that served there, and then you had British contractors as well who worked with the Omanis, particularly pilots as defense expenditures increased and the Sultan's Armed Forces were modernized.

I will try to do this fast and furious here with the El Salvador case (Slide 8). El Salvador at the time was a country of about 5.4 million people, very high population density, about 239 people per square kilometer, the highest population concentration in North or South America, and basically a one commodity economy of coffee which basically was responsible for about 80 percent of GDP (gross domestic product). The problem, however, is that 60 percent of the population owned no lands and were basically seasonal migrant workers on the coffee plantations. About 4 percent of the population owned about 65 percent of the lands. Beginning in 1932 as the Great Depression set in and the economy there got worse, you had a despotic form of government take place between the landowners and the military in the country. In 1932, 30,000 peasants are killed in an uprising by Augustin Farabundo Marti. Fast-forwarding, 1969, you have a war with Honduras which is important because the El Salvadoran forces really saw themselves, as I said before, as a standing military force and saw the external threat as more important than internal strife. In 1977, FMLN which is the Farabundo Marti National Liberation Front begins kicking off minor operations throughout El Salvador due to human rights violations by the military. General Carlos Romero becomes President. Because of these human rights violations the Carter administration cuts off all aid to El Salvador. However, in 1979, with the Iranian Revolution, with the rise to power of Daniel Ortega and the Sandinistas in Nicaragua, the Carter administration decides to start backing the El Salvadorian government again. Because of these death squads about 30,000 people were killed and this fueled the FMLN insurgency in El Salvador. With President Reagan coming into office, he begins to seize three important geopolitical things in Central America at the time. He wanted a base in Honduras, he wanted to roll back the Sandinistas in Nicaragua, and he wanted to ensure that the government in El Salvador did not fall. So he begins to introduce advisors there.

Because Vietnam was very much in the short-term memory at the time, Congress and the administration worked out an arrangement where there would only be 55 advisors on the ground. This is from 1981 and these numbers shift over time (Slide 9). Special Forces soldiers from 7th Special Forces Group and other personnel came in to work with six El Salvadorian brigades. Five of the teams were Army or Special Forces teams and one team was a Marine Corps team working with an El Salvadorian brigade. In order to fight the insurgency however, they had to increase the size of the El Salvadorian forces. In 1979, there were 11,000 soldiers in the ESAF (El Salvadoran Armed Forces) and this number grew to about 50,000 soldiers by the end of the 1980s. The *USA Today* version of the strategy there was KIS, this is from Tommie Sue Montgomery, "Keep it simple, sustainable, small, and Salvadoran." In other words, we did not want them to get deeply involved in El Salvador. The military focus on Fulda Gap in Europe ensured that the military itself would not pay a lot of attention to this, and this space and top cover from the Embassy working with the Mil Group allowed these advisors to work with the El Salvadorans first to build capacity for the military and then to work on things like human rights and civil development in order to defuse some of the underlying causes of the insurgency. Over time, with the election of Jose Napoleon Duarte, and then with the election in 1989 of Alfredo Christiani who was with the National Republican Alliance, a very right-wing organization, they were able to build up the El Salvadoran forces to get to a draw situation with the insurgents on the ground which allowed, later on, a political resolution to the crisis finally with a peace treaty in 1992.

We can talk about this in the Q & A session because I know I am drastically over time (Slide 10).

Thank you for your patience.

Up Front Points

- **Political science background**
  - **"Lumper" vice "splitter"**

- **Research in progress**

- **Deployment as advisor may bias analysis**

Slide 1

Why Look at El Salvador and Dhofar?

- Contemporary operating environment
  - Al-Qaeda and affiliated movements
  - Irregular warfare vs. major combat operations
  - OEF-Philippines, CJTF-Horn of Africa, Pan-Sahel

- Resource constraints and economy of force

- Politically sensitive operating environments

Slide 2

Foreign Internal Defense

- "Participation by civilian and military agencies of a government in any of the action programs taken by another government or other designated organization to free and protect its society from subversion, lawlessness, and insurgency."

  http://www.dtic.mil/doctrine/jel/doddict/data/f/02204.html

- *JP 3-07.1 Joint Tactics, Techniques and Procedures for Foreign Internal Defense (FID)* tells us that there are three varieties: indirect, direct, and combat operations.

Slide 3

Insurgency According to David Galula

- "...an insurgency is a protracted struggle conducted methodically, step by step, in order to attain specific intermediate objectives leading finally to the overthrow of the existing order."

- "Promoting disorder is a legitimate objective for the insurgent. It helps to disrupt the economy, hence to produce discontent; it serves to undermine the strength and the authority of the counterinsurgent. Moreover, disorder—the normal state of nature—is cheap to create and very costly to prevent." Asymmetry of resources.

- "The insurgent, having no responsibility, is free to use every trick; if necessary, he can lie, cheat, exaggerate. He is not obliged to prove; he is judged by what he promises, not by what he does. Consequently, propaganda is a powerful weapon for him. With no positive policy but with good propaganda, the insurgent may still win."

  David Galula, *Counterinsurgency Warfare: Theory & Practice*

  (New York: Praeger, 1964)

Slide 4

Advisors and Advising

- Materiel advising and support
  - e.g., Weapons systems

- Non-materiel advising and support
  - e.g., small unit training, develop cohesion

- Selection
  - Functional knowledge vs. cultural knowledge
  - Volunteer vs. "voluntold"
  - Overall organizational culture or "essence"

Slide 5

Dhofar War, 1965-1975

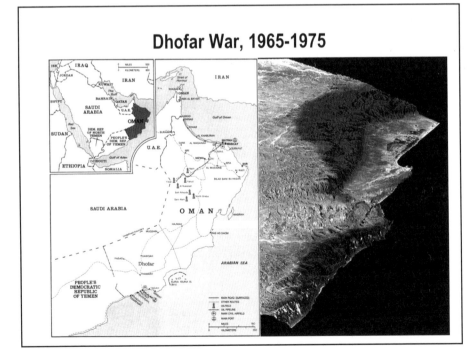

Slide 6

British Involvement

- "Seconded" officers and NCOs to the Sultan's Armed Forces (SAF)

- British Army Training Team (BATT; Special Air Service Squadron)

- Contractors (pilots, etc.)

Slide 7

El Salvador, 1980-1992

Slide 8

U.S. Military Advisors to El Salvador Armed Forces (ESAF), March 1981

- 6 MilGroup Staff
- 5 Mobile Training Team pax (admin, log, personnel)
- 6 Naval Training Team pax
- 14 helicopter training and maintenance pax
- 15 small unit training pax (3 teams)
- 10 Operational Planning and Assistance Team pax (2 teams)

Slide 9

The Future of FID Campaigning?

- Iraq and Afghanistan and American Grand Strategy

- The Future of Army Special Forces (Indirect vs. Direct)

- The Army and Combat Advisors
  - Internal vs. External Transition Teams
  - Advisory Corp?

Slide 10

Lieutenant General (Ret.) John H. Cushman
Mr. Michael P. Noonan

Moderated by Mr. John J. McGrath

Mr. McGrath

Okay, we will open it to the general public. Any questions or comments?

Audience Member

Mike, you talk about, in both of your cases, the role of Special Forces (SF) and British SAS (Special Air Service) and Army Special Forces Group in El Salvador. Can you talk a little bit today about the role of Special Forces in the Foreign Internal Defense mission and why perhaps we have abandoned it?

Mr. Noonan

That is a very interesting question. I think a lot of it depends on who you talk to in the Special Forces community. I gave a talk down at the Asymmetric Warfare Group last year and I pushed on this. It appeared that Special Forces was a functional area until 1986 when we created a branch. Then there is a line of argument advanced by people like Richard Downy and Anna Simons and others and Hy Rothstein who is also a Special Forces veteran, saying that they conventionalized their force and that they were more interested in how to deal in more direct action and special reconnaissance missions and less focused on foreign internal defense and their other core mission which is unconventional warfare. I think there is a lot of push back even within the Special Forces community today. It really depends on who you talk to. Some people, like Colonel Dave Maxwell, will say no, we have not abandoned this. We actually do a lot of foreign internal defense missions. We work a lot with the Iraqi or Afghani Special Forces elements. But then there are so many anecdotes out there that say hold on a second. Hy Rothstein calls it "Delta envy." Working in Afghanistan they wanted to do more direct action type stuff, more door kicking stuff and have backed away from indirect things. I think one of the interesting things was that the Special Operations Command handed off a large part of its civil affairs capability back into the Army Reserve and only kept a small . . . they expanded it, there used to be only one civil affairs battalion on Active Duty. Now there is a group headquarters, but because of the promotion processes the more kinetic realm of Special Operations Forces has been ascended. A few things got bumped up like the Special Missions Units, the JSOC commander, the Joint Special Operations Command which is responsible for these classified elements. It used to be a two-star command. Then General Stanley McChrystal was given a third star and was told that this was going to be a temporary situation. Then his replacement, Admiral McRaven was also given a third star. On top of that, Special Forces officers, because they are in their own distinctive branch cannot not rotate back out to more conventional units in order to gain rank. If you look at the most recent brigadier generals appointment lists, people on there with an SF or SOF background are mainly guys coming from this more kinetic community. So there is big debate on that, but what I am seeing at least, anecdotally, leads me to believe that kinetic is the way to get ahead and has dominated the culture a bit.

Audience Member

They certainly have no interaction with the training effort at Fort Riley and SOCOM does not support the Marine side of training advisors either. So that is an interesting development from that respect.

Mr. Noonan

It is also an interesting element that as they have become a branch they have so many LNOs (liaison officers) that have to be assigned outside their headquarters that people that get outside that tract are told they are persona non grata back in the Special Forces community, particularly some of the people who worked in Asymmetric Warfare Group.

Audience Member

My question is for General Cushman. Looking at your case studies it strikes me that, we did a lot of talking yesterday about trigger events, and it strikes me that of your three case studies they are all in established institutionalized crisis—the end of World War II and the hot part of the Cold War, whereas Iraq was a response to a new set of emergencies or perhaps one that had just been recognized. I wonder if you could talk a little bit about that as an issue of whether the length of time between September 11th and the invasion of Iraq may have played a role in perhaps the lack of consideration being given to postwar outcomes and long-term expectations of the changing security situation.

LTG Cushman

I think that the amount of time between 9-11 and the invasion of Iraq had very little to do with the decision as to whether it would be a theater commander's job or the job of some civilian agency as it turned out to be. It just had nothing to do with that. In my opinion, the lack of comprehension by Mr. Rumsfeld and the other senior people around him, caused the follow-on posthostilities work to be done by the theater commander. They just did not remember their history. There is an even longer time since World War II, but the history is there and they just did not pay attention to it. I think it was the job of the institution. The Army as an institution was handed over by the Chief of Staff to keep that history alive. We all know what happened with respect to Vietnam. All the lessons learned in Vietnam disappeared. I was here at Leavenworth in 1953, 1973, and 1976 and we could not get out of the Vietnam experience fast enough. We quickly decided to go back into warfare European style. Did that answer your question?

Mr. McGrath

General Cushman, could I ask an aside to that? Do you think force structure might be a factor because the World War II Army was large total mobilization, and even during Vietnam there was a draft Army much larger than the 2003 Army.

LTG Cushman

Well, I think that the size of the Army was too small. General Shinseki, when he retired, said, "Beware a twelve division strategy with a ten division Army." The Army was allowed to shrink and it should have been increased. That is a decision by management of the Department of Defense.

Mr. McGrath

So could the host conflict military government considerations be factored in because they wanted troops to do it?

LTG Cushman

Well, I think that if they used the troops that are there, even with the size of the Army that existed in 2003, properly employed, would have made a huge difference. The fact is there were no instructions to the forces that finished the invasion with the possession of Baghdad. The troops were not told what to do. They let the looting just go on. There were no plans to secure any government facilities. There were no plans for anything you properly would plan for. There was just a total void of effective planning. Now some planning was done. Third Army was quickly moved out of the sphere and pulled back to Kuwait. V Corps under General Sanchez was put in as the occupying force totally unprepared for the mission they were assigned. And of course, the 1st Cavalry Division was not deployed. They did not deploy the 1st Cavalry Division and should have. That would have made a difference. Planning was just dismal for the follow-on phrase, and the one person to blame most for that is General Franks because he had no conception. He was happy to get rid of that job. He had no conception of the historical significance of World War II and Vietnam, in my opinion.

Audience Member

Sir, this is for General Cushman. Do you see the current doctrine placing the military in a support role to the interagency and nation building? Do you see that as a failure? If so, how could that be improved?

LTG Cushman

Does current doctrine place the US military in a support role? Is that the current doctrine?

Audience Member

That is in the Joint Publication 3-0 and 5-0. It says that we will support interagency for defense support, public diplomacy and also for nation building such as in Iraq and Afghanistan.

LTG Cushman

I think it is a bad doctrine. I had no idea that was the doctrine. Of course, all I know is what I read in the newspaper. I am sorry I have not seen that one in the paper.

Audience Member (from the blogosphere)

The question is how do you manage competing agendas and how in your specific case studies did they do it?

LTG Cushman

Say that question again. What are the competing agendas?

Audience Member (from the blogosphere)

In your case study when you did the comparison with Vietnam and then to Iraq. How did the management of the competing agendas, I guess probably in your specific case, potential of the service Chiefs versus the JCS versus General Franks, the agendas that were in competition?

LTG Cushman

Well, one agenda was—as I see the question—one agenda was that this is going to be a short war and we turn it over to the Iraqis without any problem and go on home. And the other agenda was that this is going to be a posthostilities war that has to be taken care of and planned for. The agendas were managed in such a way that the first agenda won out. And I do not know how the person that made the final decision on that was the President, given advice by people he had put into positions of authority. He listened to his Secretary of Defense and what I say is that the military, the uniformed military with their experience in history that they should have understood, should have said, "This is a bad idea." My whole pitch is that they did not have an agenda and they should have had one that said, "Do not do it this way. Do it the way the history tells us to do it." And that is to give the theater commander responsibility for posthostilities planning. As far as I can tell, nobody in a position of authority, the four-star arena responsible for Iraq, ever made that pitch. Someone may have thought it, but they did not make it.

Mr. Noonan

Just real quickly, in El Salvador and in Dhofar, one issue was that, in El Salvador in particular, all of the advisors assigned there were pretty much fluent Spanish speakers. That being said, there were still agendas being played by the units that they supported. By keeping the advisory mission so small although the 55 limit was breached on many occasions because they would rotate in units from Panama and Fort Bragg to get above that cap for short periods of time because they kept the advisor and the Mil Group so small and the aid that they distributed was about a billion dollars over the decade, but by keeping it small they were able to use that as an excuse for things the El Salvadorans wanted, but they did not think was fitting into the campaign plan and put that off. So the small part of it was the flexibility of that and the size of the organization allowed them to offset some of the competing agendas there and in Oman, again, it was a small group. Because there were British officers commanding the Omani units and because of the SAS's background and their intervention force mystic surrounding them since the post Second World War period you had a lot more synergy and unity of effort on the ground so in that 1970-1975 period they were really on the same page with the Sultan, whereas in the 1965-1970 period if you go back into some of the archives, I looked through some of Peter Thwaites records at King's College, Sultan Bin Taimur would get so down into the weeds that he would talk about imposing taxes on officers who brought hunting rifles into the country. He was that far into the weeds on issues. So that change in leadership was a perfect case because you have these two distinct periods and in the second period you have somebody that was bought into and is totally on the same page with those people so that there was not really that many agendas that needed to be pushed aside there.

Panel 5—PostWar Germany

(Submitted Paper)

Role and Goal Alignment: The US Military-NGO Relationship in PostWorld War II Germany[1]

by

Major Tania M. Chacho, Ph.D., US Army[2]
United States Military Academy

History has hailed the US experience in the reconstruction of Germany in the aftermath of World War II as a "success story" that epitomized the capabilities and vision of a democracy in a postconflict environment. The economic recovery of Europe (enabled to a large degree by the Marshall Plan), along with the development of West Germany as a liberal democracy, provided convincing evidence that the reconstruction and stabilization efforts of the United States and her allies achieved success. During the course of this reconstruction effort, the US military found itself working closely with many nongovernmental organizations (NGOs) to restore the basic necessities of daily life to the German (and other former Nazi-occupied) people. This paper examines the intricacies of this military-NGO relationship, and presents several salient characteristics that defined the nature of the interaction. Of these, the leadership and direction provided by the US government emerges as particularly critical in achieving an integrated effort. This guidance served to mitigate many of the organizational and logistical challenges faced by those working to alleviate the humanitarian crisis and set the stage for the eventual reconstruction of Germany. Overall, the humanitarian relief and reconstruction work carried out by the US Army and nongovernmental organizations in postwar Germany provides testament to the value of the relationship developed between these two entities, and the circumstances under which they developed these ties offers interesting insights regarding the ability of two different organizational entities to work in the same geographic space to achieve their goals.

This study will begin with a brief overview of the general state of thinking regarding crisis-response and developmental environments, followed by an examination of the situation in postwar Germany, and in Europe in general, to set the context for the relationship. Next, a discussion of the goals and objectives of both the US military and the various NGO actors establishes the conditions for "success" in the humanitarian mission that both undertook. The third section will explore the interaction itself, identifying the characteristics that marked the relationship. The penultimate portion of the paper investigates the challenges that organizations on both sides faced, and the conclusion explores the role of US governmental leadership in setting goals and forging the structure around which the military-NGO relationship could develop. Ultimately, this paper seeks to investigate the salient characteristics of the relationship between US military forces and NGOs that resulted in the successful handling of the humanitarian crisis and set the stage for reconstruction efforts in postWWII Germany.

Immediate and Long-Term Efforts: From Crisis-Response to Development

Much of the existing literature focusing on postwar environments emphasizes the need for capacity building and places the challenge in a developmental context.[3] Yet often, it is the effectiveness of the preceding immediate humanitarian efforts to save lives and alleviate human suffering that sets the stage for these societal building activities. Such humanitarian relief takes place as soon as hostilities cease (or in some cases, while the security situation is still tenuous), and scholars view these efforts as largely short term and usually unsustainable in scope.[4] PostWorld War II developments, such as the Geneva Conventions of 1949 and the additional protocols of 1977, have articulated the rights of victims of armed conflicts to receive assistance and protection with the purpose of satisfying their immediate needs,[5] thus lending further credence to the short-term nature of the involvement. However, often the initial response is just a stop-gap measure, intended to mitigate the suffering in order to allow a more sustainable, long-term effort to get underway.

The concept of separate military and civilian domains within the humanitarian and reconstruction environments is also a prevalent theme of the literature, both during the World War II era and today. Recent writings have emphasized the concept that the military has "taken on new and significant political roles"[6] that place it within the realm of what NGOs regard as civil space. Missions in the 1990s and early 21st century have seen the military engaging in tasks such as providing shelter for displaced persons, supervising the return of refugees, organizing and monitoring elections, and supporting civilian reconstruction—many of the same functions performed by their nongovernmental counterparts.[7] This overlap has created tensions and misunderstandings about the propriety of military involvement in these types of situations.

Yet these debates are not new. Recent years have seen larger numbers of civilian relief workers and organizations engaged in postconflict operations, and a better articulated concept of humanitarian space has emerged. Yet many of the issues and challenges that exist today between soldiers and civilian aide workers were also present in the postWorld War II period. Each perceived their roles to be distinctly different and separate, and this self-identification of roles persists in each organization today. The US military in 1945 had just waged four years of high-intensity combat and was not eager to pick up additional missions as an occupying force—missions that the Department of War believed would be best honchoed by the Department of State.

But we must take care not to overstate any parallels present between postwar Germany and more recent humanitarian and reconstruction efforts. Certain specific elements of the postwar situation in Germany may mirror more modern situations, yet their very context is different due to the vast changes in the international environment over the past 60 years. As many recent commentators have observed, parallels between current US operations in Iraq and those in postwar World War II Germany are difficult to make, because the contrasts tend to outweigh the similarities.[8] The danger in attempting analogies is that the investigation often creates links out of context, potential ignoring critical shaping aspects in a desire to capitalize on similarities that make sense of the situation and potentially illuminate the way forward.

But if, as the saying goes, history "rhymes, but never repeats," perhaps a micro approach—in this case, examining the specific dynamics of a carefully defined aspect of the military-NGO relationship—may offer some insights into the elements necessary for a fruitful partnership. It is in this vein that this study moves forward.

Context: The Crisis Situation in PostWar Germany

Almost six years of war left Germany utterly and completely devastated. By the time of her unconditional surrender to Allied forces in May 1945, experts estimated that more than 20 million Germans were homeless or without adequate shelter, a food shortage and coal crisis was looming, and a further 5.2 million displaced persons (mostly liberated civilians and prisoners of war) were moving throughout SHAEF-held territory.[9] Of these issues, the lack of food was perhaps the most acute. And the situation was indeed grim: American officials estimated that the caloric intake of German civilians living in the British and US-occupied zones in the fall of 1945 was only 1250 a day (on average), compared to 3,000 calories a day in Great Britain and between 3,000 and 4,000 calories a day in the United States military.[10] Some deemed even this figure too high: a report by Colonel Joe Starnes the week that the war ended indicated that the "average basic ration is less than 1,000 calories."[11] A typical week's ration for a German citizen in May 1945 consisted of the following: "bread, 3 pounds; meat, 4 ounces; butter and fat, 2 ounces; sugar, 7 ounces; macaroni and spaghetti, 5 ounces; potatoes, 6 pounds."[12] This added up to just under 1,000 calories a day. By the fall of 1945, the infant mortality rate approached 65 percent in many places, according to the US Deputy Military Governor General Lucius Clay.[13] He further noted that "by the spring of 1946, German observers expect that epidemics and malnutrition will claim 2.5 to 3 million victims between the Oder and Elbe."[14]

Yet losses in the rest of Europe were also severe, and the hardships suffered there (many at the hands of the Nazis) created a situation in which many of the Allies had little desire, inclination, or capacity to assist the German population. In the US, the Morgenthau Plan—a punitive "pastoralization" of Germany to ensure that she never again took hostile action—initially gained favor among governmental leaders as the preferred method of administering the occupation, and reparations were a key component of this strategy.[15] But even if the desire to help was present, the French, British, and other Europeans had very little to offer the German people, as they themselves struggled to feed and shelter their own citizens. The displacement of millions of Europeans, caused both by the war and by the migrations of those that wished to position themselves on different territory as the lines between the Soviet east and the democratic west became clear, added to the humanitarian crisis.

The people of Europe faced a daunting task just meeting one of their most basic needs: food. Due to the devastation of the war, sufficient food was simply not available in theater, save for that brought in by Allied armies for use by their troops.[16] Worries about the ability of European farmers to harvest adequate crops in the fall of 1945 became compounded by a worldwide food shortage that shrunk the available supply and added to the crisis.[17] In the United States, President Truman's Cabinet Food Committee worked with the multinational Combined Food Board allocated resources in an attempt to mitigate hunger in Europe.[18]

Yet the intent of this aid was primarily for the newly liberated areas of Europe, not for the German people. Furthermore, the Allied government-supported agencies supplying aid to refugees, displaced persons, and those in the newly liberated territories did not view assistance to German citizens, seen as the perpetrators of the war, as part of their mandate. Eventually, in February 1946, President Truman did approve the creation of a Council of Relief Agencies Licensed for Operation in Germany (CRALOG) and allowed this umbrella organization to operate under the direction of the US Military Government.[19] Yet CRALOG's creation came nine months—and one winter—after the cessation of hostilities in May 1945.

And the worldwide food situation did not improve over the next few years. By 1947, a report by the Cabinet Food Committee spelled out the story in convincing detail: because of poor crops abroad, because of a sharp drop in US corn production, Europe now faced a food shortage of 4.5 million tons in grain alone.[20] By 1947, the Office of Military Government, United States (OMGUS) had the responsibility for supplying food to the displaced persons in Germany, a population of roughly 552,000,[21] which further strained the availability of food for ordinary German civilians.

The humanitarian crisis that Germans faced from 1945–1948 was one of unprecedented magnitude.[22] And on top of the physical hardships endured by the German people, they also faced worldwide censure and condemnation for their role in the war. This left many people in Allied nations unable to identify with their plight, or even to demand that they be forced to continue to live at bare sustenance levels.[23] Goodwill toward Germans was not readily apparent in many quarters, and this reluctance to aid a former enemy further complicated the humanitarian situation.

Yet some segments of American society did view the German people with compassion, and they worked to provide a means by which humanitarian aid could flow to ordinary German citizens. American voluntary agencies wanted to provide help to the war-stricken population for a variety of different reasons, ranging from pure humanitarianism, to a desire to help ethnic brethren and relatives, evidenced by German-

Americans. Religious organizations, particularly those with members of their faith in Germany (Lutherans, Catholics, and Mennonites, for example) were also interested in rendering assistance, as were organizations such as the American Friends Service Committee (AFSC) that had a record of operating in Germany, having helped German children in the aftermath of World War I.[24] Nongovernmental agencies were critical in shaping public perception of Germany, helping it to overcome the negative image of being the country of Nazism.[25] Once this stigma was no longer at the forefront of American minds, the need to provide aid for a starving people became more readily apparent. Such was the environment in which the US Army and NGOs performed their work.

"Success" in the Reconstruction Effort

Setting the criteria for "success" in any endeavor that involves multiple actors is always difficult, which is perhaps why in the realm of postconflict reconstruction, a clear definition is often absent. Most early researchers who examined the US occupation of Germany have evaluated it as successful, noting that America achieved her objective of an independent, liberally democratic, economically viable, and nonaggressive West German state.[26] Later scholars challenged this overall positive assertion, arguing that certain occupation policies contained notable shortcomings.[27] Still, the overall impression of American reconstruction activities in Germany during the postwar period remains favorable, at least in retrospect. At the time, there was much more angst about the course taken, and much more uncertainty about the eventual outcome.[28]

The US government initially set its criteria for postwar success through a series of directives and agreements, beginning in the spring of 1945 while the war was still ongoing. First, in April 1945, Joint Chiefs of Staff (JCS) 1067 set the policy for the occupation, although scholars generally believe that a discrepancy existed between official policy and the views of those who carried it out.[29] JCS 1067 was, in fact, a harsh document, intent on producing the "hard" peace favored by Henry Morgenthau and devoid of any mention of humanitarian assistance for the German people. In fact, the paper provided General Eisenhower with the "Basic Objectives of Military Government in Germany" as follows:

> a. It should be brought home to the Germans that Germany's ruthless warfare and the fanatical Nazi resistance have destroyed the German economy and made chaos and suffering inevitable and that the Germans cannot escape responsibility for what they have brought upon themselves.

> b. Germany will not be occupied for the purpose of liberation but as a defeated enemy nation. Your aim is not oppression but to occupy Germany for the purpose of realizing certain important Allied objectives. In the conduct of your occupation and administration you should be just but firm and aloof. You will strongly discourage fraternization with the German officials and population.

c. The principal Allied objective is to prevent Germany from ever again becoming a threat to the peace of the world. Essential steps in the accomplishment of this objective are the elimination of Nazism and militarism in all their forms, the immediate apprehension of war criminals for punishment, the industrial disarmament and demilitarization of Germany, with continuing control over Germany's capacity to make war, and the preparation for an eventual reconstruction of German political life on a democratic basis.

d. Other Allied objectives are to enforce the program of reparations and restitution, to provide relief for the benefit of countries devastated by Nazi aggression, and to ensure that prisoners of war and displaced persons of the United Nations are cared for and repatriated.

Furthermore, paragraph 5 of JCS 1067 clearly stated that Military Government officials were to restrict themselves to promoting the production and maintenance of only those indigenous goods and services "required to prevent starvation or such disease and unrest as would endanger the occupying forces."[30] Planners envisioned no mass scale relief efforts to assist the German population; instead, they directed their attention to the care of displaced persons and the imposition of order within the US zone of occupation. Thus, JCS 1067 did not supply any stated humanitarian relief and reconstruction objectives of the US as an occupying force.

Yet those responsible for running the military occupation received additional guidance from the Potsdam Agreement, issued in August 1945. In the few months that passed between the end of the war in Europe in May and Potsdam in late July-early August, doubts had emerged about the ability of the communist USSR and the democratic West to sustain their wartime partnership into the postwar era, and the document reflected some of the strategic maneuverings between the Allies. Ultimately, under the rubric of economic unity, the West used the Potsdam communiqué to consolidate their three zones and create a capitalistic and democratic West Germany.[31] But more importantly from a developmental standpoint, the spreading East-West divide and developing Cold War that became clear at Potsdam eventually allowed the US to shift its policy regarding the long-term viability of the German state. This marked the starting point of American leadership's strategic turn away from the Morgenthau plan and their move to an acceptance of a "softer" attitude toward defeated Germany. Secretary of State James Byrnes officially announced this change in American attitude in a speech at Stuttgart on September 6, 1946, when he emphasized (among other things) the need for an improved level of industry for Germany.[32] As Franklin Davis notes, from an American policy standpoint, the Stuttgart statement "oriented the Occupation Army away from a repression of German militarism and dispelled once and for all any concept of the Army in Germany as a force to exploit the purely military values of the victory in Europe."[33] So approximately 16 months after the formal cessation of hostilities, the role of the occupying US forces finally formally transitioned into one that could engage in relief operations for the benefit of the German people.

In addition, from a legal perspective, there existed support for use of the military in a limited humanitarian role. In fact, and some law experts argued that international law dictates that an occupying force provide adequate feeding of civilian populations under their control, and thus the care of the German civilian population should ultimately become a mission of the US military.[34] Military doctrine of the time tends to support this view, and by 1946, Defense Department officials argued before Congress that international law obligated them to import and distribute food to prevent "disease and unrest" among the German population.[35] The pacifying effects that an adequate caloric intake provided the occupied citizenry was something that the military recognized and appreciated, as ultimately it facilitated the job of preserving the security environment.

Initially, then, the relief mission was thus an implied task required to complete the stated goal of maintaining order. But while military involvement in providing food aid to the German population may have originated with these practical operational considerations, evidence suggests that military leaders in theater were among the first to recognize and call for the need for an expanded effort to avert a humanitarian disaster. For example, the Military Governor of the US Zone (Germany), General Clay, writes in his memoirs that, "From the first I begged and argued for food because I did not believe that the American people wanted starvation and misery to accompany occupation. . . ."[36] Clay further notes that in November 1945 he "made a hurried trip home" to discuss the food shortages with government officials and "ask personally for their assistance in increasing the food supply."[37] While Clay's efforts were largely unsuccessful (due to the world grain shortage and the desire to ensure that Allies received preferential treatment over former enemies), they do indicate a military awareness of the critical humanitarian need.

So as the mission of the US military morphed by default and necessity into one that included a formalized humanitarian role beyond that required to maintain the peace, the definition of "success" changed along with it. By taking on this role, the US occupying forces now had to measure their efforts against a new standard. Given this, a contemporary definition of relief "success" may prove useful here: Francis Fukuyama describes the first phase of successful nation-building as the United States solving immediate problems of physical infrastructure through the infusion of security forces, humanitarian relief, and technical assistance.[38]

This contemporary definition also provides a useful bridge to a discussion of the goals of humanitarian relief organizations involved in postwar Germany. In December 1945, a group of American voluntary relief agencies formed a united body—CRALOG —that received formal recognition from President Truman in February 1946. While each voluntary aid organization maintained its own individual mission and mandate, through participation in CRALOG, those operating in Germany agreed to a standardized set of four goals: "to provide a channel through which interested Americans could send relief to Germany; to avoid duplication among the agencies; to provide liaison with government and military authorities as well as German welfare agencies; and to provide facilities for other agencies and individuals, not members of the Council, to participate in this humanitarian undertaking."[39]

These generalized goals created the framework under which voluntary aid organizations operated. Yet some also opted for a more specialized, "niche" approach. For example, the Cooperation for American Remittances to Europe (CARE) had a stated mission of selling food remittances to interested individuals, groups, and organizations for designated beneficiaries. This person-to-person aid (with its striking similarities to micro-finance ventures promoted today) appealed to many Americans who wanted reassurance that their donation made a tangible difference to an individual recipient's life. As General Clay explains:

> The physical and psychological effects of this aid were immense. Much larger quantities of bulk food, largely grain, brought in with appropriated funds, lost their identity through processing before they reached the consumer. He knew something of the huge extent of this aid, but it remained impersonal. On the other hand, when a CARE package arrived, the consumer knew it was aid from America and that even the bitterness of war had not destroyed our compassion for suffering.[40]

Voluntary aid organizations often reported on the amount of food and supplies provided from the monies raised from donors, and this became a measure of their "success." Although many recognize now that this is a crude guide, at best, since it focuses on quantifiable rather than qualitative measures, it was nonetheless an accepted standard of the time. Using these criteria, CARE's Executive Director reported in February 1948 that "since the actual operation commenced, we have supplied between 52 and 53 million dollars worth of food and other packages."[41] This relief represented a vast impact on the lives of many Germans and kept many from starving during the initial rough postwar years. As such, it warrants mention in a discussion of the criteria of voluntary relief agency "success."

Characteristics of the Military-NGO Relationship

Postwar interactions between civil relief workers and military authorities occurred in an environment marked by clear (if often debated) policies and regulation, coordinated by a Presidential-appointed central authority. The government mobilized philanthropy, like other aspects of national life, in the interests of "efficiency and speedy victory."[42] The President's War Relief Control Board, established by Executive Order of President Roosevelt on 25 July 1942, oversaw the overall coordination of the efforts of public charities with regard to the provision of relief, reconstruction, or welfare arising from the war or its immediate aftermath (the order was valid until six months after the termination of hostilities, "unless revoked by Presidential order.")[43] Not only was the Board active throughout the war, but President Truman saw fit to have the functions of the Board continue into peacetime, and on 16 May 1946 he renamed it the Advisory Committee on Voluntary Foreign Aid.[44]

The development of the War Relief Control Board created some controversy. It essentially forced the mergers of many independent agencies, citing efficiency concerns, and made licensing decisions among ethnic-orientated agencies with different viewpoints that also caused contention.[45] The Board sought to clearly identify aid agencies with the United States government and its citizens, and so required charitable organizations to change their names to include "American" in their title to identify the source of the aid.[46] So, the French Relief Fund became American Relief for France; the Queen Wilhelmina Fund changed it's moniker to American Relief for Holland, and so forth.[47] Such partisan identification clearly linked the aid organization to the foreign policy objectives of the US government, but this identification was one that most made willingly,[48] in an atmosphere of national mobilization. This wholehearted identification with the aims of the national government is a unique feature of the time, and a response to the call for a national wartime effort. The ideological appeal was clear-cut and well-defined, which aided the government in its appeal for support and its establishment of a centralized framework for aid work.

The United States government was also instrumental in founding an international coordination effort for humanitarian relief. On 9 November 1943, 16 months after President Roosevelt established the War Relief Control Board, the United Nations Relief and Rehabilitation Agency (UNRRA) began its operations. A collaborative effort involving 44 Allied nations, the purpose of UNRRA was to "plan, co-ordinate, administer or arrange for the administration of measures for the relief of victims of war in any area under the control of any of the United Nations through the provision of food, fuel, clothing, shelter and other basic necessities, medical and other essential services"[49] Although UNRRA eventually operated in occupied Germany (primarily operating displaced persons camps), the organization did not render assistance to ethnic Germans.[50]

This distinction was critical, and Richard Wiggers sees it in the overall context of a lack of American desire to care for a population seen not only as the enemy, but also as perpetrators of horrible crimes against humanity.[51] Slowly, as the dire nature of the situation became clear through independent reports, US relief agencies and others began to advocate greater amounts of assistance for the German people. Yet President Truman remained initially reluctant, noting that it was difficult to feel "great sympathy" for the Germans, "who caused the death of so many human beings by starvation, disease and outright murder."[52]

Public pressure, coupled with an increasing understanding of the criticality of the situation in Germany and the emerging geopolitical outlines of the Cold War, slowly changed the minds of US leaders. Notably, however, this did not occur until about 12 months (and one harsh winter) into the occupation period. The mindset of US leadership toward the Germans at the cessation of hostilities (and thus the beginning of the postconflict period) was one of retribution and punishment, and as mentioned above, this influenced the guidance given to those in charge of the occupation. But some historians have argued that all along, the American Occupation Forces did "frequently perform their duties at variance with the policies expressed by the civil authorities at home."[53] This dissonance between stated government policy and actions of the military government on the ground in Germany is a critical aspect of the postconflict relief operation. Acknowledgement that the military performed tasks designed to alleviate human suffering suggests recognition that those soldiers on the ground were perhaps best poised to assess and address the operational situation.

In 1946, Congress authorized emergency aid for Germany, Japan, and Austria under the Government Aid and Relief in Occupied Areas (GARIOA) program. This money came with Congress' stipulation that OMGUS use the funds only to import "food, petroleum, and fertilizers" to prevent "disease and unrest" in Germany.[54] In 1946, GARIOA provided almost $9 million in aid and supplies.[55] Military authorities worked with aid organizations (both foreign and local national) to distribute this assistance and, in this way, developed their relationship.

Prior Planning: Immediate Humanitarian Needs

As indicated above, discussions regarding the role of charitable organizations in postwar Europe began long before the surrender of the Third Reich. Indeed, a long war (and a mobilized public) provided ample time and opportunity to engage in planning regarding the postwar challenges that the European continent as a whole would face. American agencies paid time and attention to both the immediate humanitarian needs faced by the European people (many in newly liberated territory in the wake of the advancing Allied armies) and to the longer-term reconstruction challenges that these war-torn societies would encounter.

The President's War Relief Control Board set the priority of effort. Although it came under "continued and increasing pressure from persons and organizations who desire to assist in the reconstruction of damaged towns, institutions, monuments, etc. in Europe," the Control Board took the position that "in view of the present conditions in Europe, private relief resources should be utilized for the direct relief of human suffering rather than for reconstruction."[56] While some might argue for a simultaneous rather than a sequenced effort, the Control Board's determination was unambiguous. Letters from private citizens and organizations reached the White House and State Department and other government agencies, requesting guidance as to how to proceed in reconstruction efforts for Europe. Citing the Control Board's policy, government officials replied that this was premature, and that the sole focus of current relief efforts should be on the alleviation of human suffering.[57]

The human aspect of the postwar situation was indeed critical, and many offered ideas for providing and facilitating this relief as well. For example, a national committee headed by Henry J. Kaiser organized a United National Clothing Collection in the spring of 1945—a unified effort on the part of United Nations Relief and Rehabilitation Administration (UNRRA) and all the voluntary war relief agencies of the United

States.[58] The campaign's publicity department (it was a very well organized effort) issued a release to inform Americans that the nationwide drive would occur between April 1-30, 1945, to "secure the maximum quantity possible of good used clothing for free distribution to needy and destitute men, women, and children in war-devastated countries."[59] This drive was strictly to benefit the newly liberated areas of Europe, so it did not encompass Germany.

Yet after the war, other organizations did take the needs of the German people into account. For example, concern over the immediate humanitarian needs spawned the creation of the Cooperation for American Remittances to Europe (CARE) by "some twenty leading American charitable organizations."[60] Created "largely at the insistence of the President's War Relief Control Board," CARE had a stated mission of selling food remittances to interested individuals, groups, and organizations, for designated beneficiaries throughout Europe.[61] By June 1946, CARE had concluded a formal agreement with OMGUS that specifically delineated the obligations expected of both sides. On paper, the relationship appeared to be an unequal one: CARE listed 4 points under their obligations, and in return articulated 12 items (some with subcomponents) under the paragraph titled, "Exemptions and Facilities Accorded to CARE."[62] This was a reflection of the environment: OMGUS administered the US Occupied Zone, and thus were the entity responsible for executing the occupation plan's rules, including those dealing with humanitarian assistance.

Yet aid and economic assistance was not a business that the military desired to remain involved in beyond what was necessary for as an emergency, stop-gap, measure. During the war, in February and March 1945, President Roosevelt sent Judge Samuel Rosenman to investigate Europe's reconstruction needs. The Rosenman Report identified the need to remove the military from responsibility for civilian supply as soon as operationally feasible, citing concerns that the continuation of such involvement would "mitigate against ultimate economic recovery."[63] The report noted that General Eisenhower, SHAEF Commander, had made the same recommendation regarding the termination of military responsibility for civilian supply in the newly liberated territories—namely, that it occur at the earliest practicable date.[64] The Army's eagerness to get out of the humanitarian business did come with the recognition that initially, they were the only organization who could perform some essential tasks. For example, the Rosenman report readily acknowledged that war created conditions in which the "only effective medium for the initial provision of civilian supplies" is the Army.[65]

Soon after the Rosenman report, President Truman instructed the Secretary of War to plan for the termination of military responsibilities for shipping and distributing relief supplies for liberated areas of Europe as soon as the military situation permits.[66] He also clearly indicated that he expected a relationship between military and civil authorities to develop:

> In addition, I think that the general policy of the Army, upon such termination, should be to continue to assist the national governments involved and the appropriate civil-

ian agencies of our own Government and UNRRA to the extent the military situation permits. This should include, where possible, and to the extent legally permissible, the transfer of supplies which are in excess of essential military requirements.[67]

Thus, initial relief planning efforts from both the US governmental and voluntary agencies had expectations of close interaction and working relationships developing between the military and NGOs.

Prior Planning: Reconstruction Efforts

While the focus of this study is immediate humanitarian efforts, it is worth mentioning that longer-term reconstruction planning also occurred—and not just in the governmental realm. Private organizations and agencies engaged in strategic planning regarding reconstruction on the European continent. For example, the Director of the International Relations Board of the American Library Association penned a letter to the Secretary of State in May 1945 expressing that the Association was "anxious to renew and extend their relationships with European libraries," including the restoration of the "flow of books, pamphlets and serial publications from the United States to Europe, and from Europe to the United States."[68] He further notes that the American Library Association "has now in stock piles in the United States nearly half a million dollars' worth of carefully selected books and periodicals (purchased with Rockefeller Foundation money)" with which to bring European libraries up-to-date in "American publications and American scholarship and thought."[69] The identification of the need to provide the necessary materials to resume educational opportunities speaks to the recognition of a long-term effort to rebuild war-torn societies, and is thus firmly in the realm of reconstruction activities.

For Germany, as well as for the rest of Europe, the ultimate reconstruction effort occurred with the announcement of the Marshall Plan (or the Economic Recovery Program, as it was formally known) in 1947. This also signaled the end of the emergency relief effort and the transition to foreign aid for development.

Goal Alignment

During the humanitarian assistance phase, several characteristics of the relationship between the US military and voluntary aid agencies quickly became evident. First, thanks to the establishment of the President's War Relief Control Board, there existed an alignment of goals between aid agencies and governmental policies. The recognition of this commonality was important in that it established the framework for operations within the theater of operations. Of course, the path was not smooth, and disagreements arose regarding the best means to achieve the end.

For example, the Control Board regulated "appeals to the public for funds and other contributions for foreign relief"[70] and thus arguably stifled the free speech of agencies that would have otherwise asked the public for donations. And by setting the conditions under which private relief agencies could operate, the Control Board ensured that it controlled the means by which each goal was tackled. For example, it ". . . has refrained from giving its approval for campaigns to raise funds for the restoration of cities abroad. It has taken the position that the solicitation of funds for war relief at this time should be confined to emergent projects, that is, for the direct relief of human suffering through provision of food, clothing, medicaments, et cetera."[71]

NGOs' reaction to this control was overall rather muted. For example, CARE documents from the time show that the organization developed within the system emplaced by the Control Board, and sought to align their goals with those of the Board in order to achieve success by emphasizing the areas of overlap. CARE emphasized that by design, they filled a "need for an individual to individual and group to group package service" that the American voluntary relief agencies "were not equipped to handle."[72] Like any good business plan, the CARE concept sought to find a niche to fill. Once it identified this, its supporters then lobbied an initially reluctant government to grant it recognition and allow it to operate in pursuit of this goal.[73] In doing so, it did not change its stated mission, nor did it alter its vision regarding the essential functions of this mission. The organization's concept of providing a method for interested Americans to purchase remittances to send food overseas to designated or undesignated individuals and groups remained intact. The organizers of CARE marketed their plan effectively and, in doing so, helped the US government to see that their goals aligned.

Another compelling force for goal alignment was CRALOG. The requirement to register with CRALOG prior to operating in theater ensured that the ends pursued by voluntary agencies coincided, with the intent of reducing inefficiencies and duplication of effort. Of course, the means by which each agency decided to pursue the goal was still matter of discussion.

Coordination and Deconfliction of Effort

Once the actors established agreement on goals, or ends, a discussion over means (procedures and processes) naturally ensued. The State Department handled many inquiries from citizens and groups regarding providing assistance to the civilians liberated in Europe. The standard response was that the provision of supplies was a military responsibility and that the interested party should contact the President's War Relief Control Board to determine how best to render assistance.[74] Although the provision of supplies was indeed a military responsibility at the end of the war, it was one that the Department of War wanted divested as soon as possible.

In fact, interagency disagreements permeated the discussion of the means of execution for the postwar relief aid plan. A central issue was who held ultimate responsibility for the provision of civilian supplies, and this was a matter of intense debate between the Secretary of War and the Secretary of State. President Roosevelt directed the

Secretary of War to provide initial relief supplies in newly liberated areas necessary to "avoid disease and unrest."[75] In a letter dated 21 May 1945, President Truman directed that the War Department cease taking responsibility for these relief efforts "as soon as the military situation permits" and after consultation with the State Department.[76] What ensued was a series of letters back and forth between the Secretary of State and the Secretary of War, discussing the terms of the handoff arrangement. State expressed reluctance to take on the mission immediately, as they had very few resources in theater that would allow them to effectively oversee the effort. For its part, the military wanted to relinquish this mission—and its ensuing financial obligations—as rapidly as practicable.[77] Ultimately, the military wound up retaining responsibility for these aid functions longer than they felt necessary given the President's directive, and they did so under protest.

Initial military plans for the occupation of Europe overlooked the need for coordination with voluntary aid organizations. For example, the Basic Preliminary Plan for Allied Control and Occupation of Germany 1944–45 does not include any nongovernmental organizations in an extensive matrix of those organizations with which military divisions or headquarters sections should coordinate.[78] This oversight persisted into the administration of the military government and occupying force in Germany. Indeed, as late as the fall of 1946, many in the US government opposed allowing voluntary aid organizations to even enter into Germany. A case in point is the reaction of the Office of Political Affairs as they expressed nonconcurrence with a proposal from the American Friends Service Committee (AFSC) to allow 10 Americans into a town in Bavaria to provide relief to displaced persons and to facilitate relations between them and the German population.[79] The Director of Political Affairs indicated his concern with this plan, noting:

> The door, admitting the entrance of relief personnel, cannot be readily closed once it is opened. The British opened the door at an early stage and now find themselves with more than 600 non-German volunteer workers in the British Zone working in the interests of the German population.[80]

The worry about "opening the floodgates" of relief workers is clear, and the Director goes on to note that "every non-German individual who comes into the Zone occupies space and utilizes facilities which are at a great premium and definitely needed by the German people themselves."[81]

The discussion about the best means to use to execute the relief plan produced many challenges for both the relief organizations and the military. For example, CARE initially struggled to obtain authority to purchase surplus "10-in-1" rations from the Army, which the organization needed in order to establish their individual food package plan. These rations could feed 10 combat soldiers for 1 day or 1 combat soldier for 10 days—a total of 30 meals or 45,000 calories.[82] The military initially wanted to sell these surplus rations to UNRRA, apparently to dispose of them in the "simplest way and in one transaction."[83] CARE ultimately achieved success here, and managed to purchase 2.8 million to successfully launch their remittance program and also inaugurate their operations as a private nonprofit agency.[84]

Scholars of humanitarian assistance often cite a clear recognition of areas of responsibility as a key element necessary for the successful execution of the aid mission.[85] Yet this was not immediately achieved in US governmental operations in postwar Germany, and still the organizers found ways to overcome this difficulty. As noted above, disagreements most certainly existed, and interagency and interorganizational debates challenged smooth transitions and timelines. Yet as this occurred, the force on the ground, in theater, with resources—the US military government—stepped in to address the gaps created (and often self-inflicted). As a recent RAND study on US nation-building experience in Germany notes, "The US Army's focus on "getting things moving" was key to minimizing humanitarian suffering and accelerating economic recovery in its zone in the immediate aftermath of World War II."[86]

Reliance on Local National Groups (Aid Organizations)

Another key characteristic of the relationship between the US military and American-based voluntary aid organizations was their reliance on another actor; namely, German relief agencies. In fact, the lessons learned by the War-Torn Societies Project from its experience of working in war-torn societies[87] have a familiar ring: "Local solutions and responses to rebuilding challenges are often more effective, cheaper and more sustainable" than any of those imported by foreign counterparts.[88] The Military Governor, General Lucius Clay, fully recognized and appreciated this distinction:

> While German welfare agencies deserved high praise for their work in the winter of 1945–46 with meager resources, there was a growing consciousness of their need for help from the United States. I was convinced that German organizations were competent to distribute supplies and that United States aid sent directly to these agencies would prove most effective.[89]

In retrospect, General Clay felt even more strongly about the role played by German organizations, "While it is true that without American food, bought with American money, loss of life in Germany would have been appalling, the major relief burden was carried by the German state governments and private welfare organizations."[90]

Fortunately, in postwar Germany, there was a network of private welfare organizations that the Allies could work with to provide this emergency relief. The organizations did have to undergo a vetting process to ascertain that they did not have ties to the Nazi Party, but then could operate under the direct supervision of local German authorities. Many were religious organizations, and they worked with the administrators of the Lander (State-level authorities) to distribute aid, particularly when it began arriving from abroad.[91]

Understanding and Appreciation of Roles

Mutual recognition and understanding go a long way to reduce friction, and this was evident in the relationship between US soldiers and American civilian aid workers in Germany. The gap was not a wide one to breech, since conscription in the US ensured that most aid workers had a relative, neighbor, or friend serving in uniform. Furthermore, many voluntary aid organizations were led by former members of the US military. For example, CARE's Executive Director during the immediate postwar period was former Army Major General William Haskell.[92] The benefits derived from this rather widespread understanding of military had the effect of minimizing any potential institutional culture divides between civilians and soldiers working in Germany.

Clearly outlined expectations also assisted the smooth progression of the military-NGO relationship. On behalf of OMGUS, General Clay signed a concise two-page agreement with the American Council of Voluntary Agencies that captured the understanding of the necessary procedures that both sides would follow.[93] Notable in this agreement was the liaison functions that the voluntary relief agencies performed, which provided evidence of the close nature of the cooperation between the military and these organizations.

The CARE program operated in a similar close fashion with the military government, which provided "general supervision of the CARE program" and received the attachment of three CARE representatives "to observe and assist in the operation of the CARE program."[94] Such close collaboration, so problematic today, was a feature of the postwar landscape in 1946 Germany.

Challenges

The mutual understanding of roles also allowed both organizations to overcome challenges. For example, perhaps one of the greatest concerns for the aid organizations operating in Germany was the living and working conditions for the volunteers themselves. Many did without rations, living on bread and a meager ration of "one or two sardines on a small triangle of cheese"[95] and contracted illnesses. According to a Quaker relief worker, the "lack of food and the consequences of not getting enough" created a "constant worry."[96] These conditions made arrangements for military provisions all the more necessary. The military agreed to provide "billets, mess, and transportation"[97] but, in return, could set limits on the number of aid workers allowed in country and receiving this type of support. This unequal relationship was not accepted by all agencies; for example, aid workers with the American Friends Service Committee (AFSC–Quakers) "lived on the German economy and shared to considerable degree the hardships of the German,"[98] as noted above. This conscious choice created a unique role for the AFSC, and it differentiated their efforts from those of other aid organizations.

A final thought in relation to roles deals with budgets and donor accountability. As Barakat and Chard point out, in the nonprofit arena, a donor culture of "financial accountability tied to the management of short, fixed-term budgets by means of measurable indicators of expenditures" often exists and presents challenges to those agencies operating in a postconflict environment. And although the agencies operating in post-World War II Germany did so long before this analysis, there is substantial evidence to indicate that donor concerns were very similar in the late 1940s. For example, the precise CARE method of ensuring that donor remittances went to the named individual involved the "proper presentation of credentials" and then CARE transmitted a receipt of delivery to the donor.[99] The need to formally notify the donor that his or her aid money was converted to a food package that reached the intended friend or relative was clear.

In fact, if CARE could not find the designated beneficiary, they promised to notify the donor and refund his or her money.

Conclusion

Times have undeniably changed since the postWorld War II era. In the humanitarian and reconstruction arena, nongovernmental organizations have expanded their work, and their definition of the international environment in which they conduct business. Concepts of impartiality, neutrality, and independence have become critical to the operations of most aid organizations, and in some cases, this has created clear lines between military and nongovernmental actors.

Can the characteristics of the relationship be duplicated, or are they a product of a specific time period and set of circumstances? A powerful argument exists for the uniqueness of the experience, with its specific configuration of social, political, and economic factors creating an environment conducive to such communication, cooperation, and coordination of efforts. But to what degree do leaders affect the political (and arguably the economic and social) sphere, thus shaping and ultimately creating this environment? If we accept the notion that individual actors (leaders) can influence and affect the political environment, then perhaps lessons learned from a particular type of environment can provide useful insights into the conditions that breed successful operations, from both a governmental and a nongovernmental perspective.

The skill sets that both soldiers and aid workers possess are critical to the humanitarian relief endeavor. The understanding of each other's mission, and a willingness to search for aligned goals and common ground, also assists the relief work on the ground. An ability to communicate with each other further facilities the relationship, although it cannot (and will not) overcome different points of view regarding the means of mission accomplishment. Context matters, and drawing analogies imprecisely can create imperfect solutions. But when viewed on a microscale, certain elements of a "successful" relationship in a humanitarian relief environment can perhaps provide guidance to those looking to improve such interactions in the future. Whether or not there should be a normative basis for the relationship is a question that deserves discussion. It is in this spirit that this study offers these thoughts regarding the interaction of the US military and NGOs in Germany during the postconflict phase of World War II.

Works Cited

Abrisketa, Joana. "The Right to Humanitarian Aid: Basis and Limitations," in *Reflections on Humanitarian Action: Principles, Ethics and Contradictions*. Edited by Humanitarian Studies Unit, Transnational Institute. London, Sterling, VA: Pluto Press, 2001.

Agreement for United Nations Relief and Rehabilitation Administration, November 9, 1943. Pamphlet No. 4, *Pillars of Peace*. Documents Pertaining to American Interest in Establishing a Lasting World Peace: January 1941–February 1946. Carlisle Barracks, Pa.: Army Information School, May 1946. Articles 1 and 2.

Backer, John H. *The Decision to Divide Germany American Foreign Policy in Transition.* Durham, NC: Duke University Press, 1978.

_____. *Winds of History: The German Years of Lucius DuBignon Clay.* New York: Van Nostrand Reinhold, 1983.

_____. "From Morgenthau Plan to Marshall Plan," in *Americans as Proconsuls: United States Military Government in Germany and Japan, 1944–1952,* Robert Wolfe, ed. Carbondale: Southern Illinois University Press, 1984.

Balabkins, Nicholas. *Germany Under Direct Controls: Economic Aspects of Industrial Disarmament 1945–1948.* New Brunswick, NJ: Rutgers University Press, 1964.

Barakat, Sultan, and Margaret Chard. "Theories, Rhetoric, and Practice: Recovering the Capacities of War-Torn Societies." *Third World Quarterly* 23, No. 5 (October 2002).

Benjamin, Daniel. "Condi's Phony History: Sorry, Dr. Rice, Postwar Germany Was Nothing Like Iraq" posted August 29, 2003, http://www.slate.com/id/2087768/ (accessed January 22, 2008).

Bramwell, Anna C., ed. *Refugees in the Age of Total War.* London: Unwin Hyman Ltd, 1988.

Browder, Dewey A. *Americans in PostWorld War II Germany.* Lewiston, NY: Edwin Mellen Press, 1998.

Cecil, Robert. "Potsdam and Its Legends." *International Affairs* 46, No. 3 (June 1970).

Clay, Lucius D. *Decision in Germany.* Garden City, NY: Doubleday and Company, Inc., 1950.

Curti, Merle Eugene. *American Philanthropy Abroad.* New Brunswick, NJ: Rutgers University Press, 1963.

Davis, Franklin M. *Come As Conqueror: The United States Army's Occupation of Germany, 1945–1949.* New York: The MacMillian Company, 1967.

Department of State, *Foreign Relations of the United States, 1945*, Vol. 3, European Advisory Commission; Austria; Germany.

Dobbins, James, John G. McGinn, Keith Crane, Seth G. Jones, Rollie Lal, Andrew Rathmell, Rachel M. Swanger, Anga R. Timilsina, "America's Role in Nation-Building, From Germany to Iraq." Santa Monica, CA: RAND, 2003.

Frederiksen, Oliver J. *The American Military Occupation of Germany: 1945–1953.* United States Army, Europe: Historical Division, 1953.

Fukuyama, Francis. *State-Building: Governance and World Order in the 21st Century.* Ithaca, NY: Cornell University Press, 2004.

Genizi, Haim. *America's Fair Share: The Admission and Resettlement of Displaced Persons, 1945–1952.* Detroit MI: Wayne State University Press, 1993.

Gimbel, John. *The American Occupation of Germany: Politics and the Military, 1945–1949.* Stanford, CA: Stanford University Press, 1968.

Goedde, Petra. *GIs and Germans: Culture, Gender, and Foreign Relations, 1945–1949.* New Haven, CT: Yale University Press, 2003.

Gulgowski, Paul W. *The American Military Government of United States Occupied Zones of Post World War II Germany in Relation to Policies Expressed by its Civilian Authorities at Home, During the Course of 1944/45 Through 1949*, Frankfurt am Main: Haag und Herchen Verlag, 1983.

Jennings, Ray Salvatore. "The Road Ahead: Lessons in Nation Building from Japan, Germany, and Afghanistan for Postwar Iraq" *Peaceworks* No. 49. Washington, DC: United States Institute of Peace, April 2003.

Kaplan, Fred. "Iraq's Not Germany: What a 60-Year-Old Allen Dulles Speech Can Teach Us About Postwar Reconstruction," *Slate*, posted October 17, 2003, http://www.slate.com/id/2089987/ (accessed January 22, 2008).

Killick, John. *The United States and European Reconstruction: 1945–1960.* Edinburgh, UK: Keele University Press, 1997.

McCleary, Rachel M., and Robert J. Barro, "Private Voluntary Organizations Engaged in International Assistance, 1939–2004." *Nonprofit and Voluntary Sector Quarterly Journal* , December 2007.

Maresko, Deborah. "Development, Relief Aid, and Creating Peace: Humanitarian Aid in Liberia's War of the 1990s." *OJPCR: The Online Journal of Peace and Conflict Resolution,* 6.1 (Fall 2004), 94–120.

Marrus, Michael R. *The Unwanted: European Refugees in the Twentieth Century.* New York: Oxford University Press, 1985.

Olick, Jeffrey K. *In the House of the Hangman: The Agonies of German Defeat, 1943–1949.* Chicago and London: The University of Chicago Press, 2005.

Peterson, Edward N. *The American Occupation of Germany: Retreat to Victory.* Detroit, MI: Wayne State University Press, 1978.

Rupieper, Hermann J. Review of *Humanitare Auslandshilfe als Bruke zu atlantischer Partnerschaft: CARE, CRALOG und die Entwicklung der deutsch-aamerikanischen Beziehungen nach ende des Zweiten Weltkrieges* by Karl-Ludwig Sommer, *American Historical Review*, Vol. 106, No. 5 (December 2001): 1,828.

Schoenberg, Hans W. *Germans from the East: A Study of their Migration, Resettlement, and Subsequent Group History Since 1945.* The Hague: Martinus Nijhoff, 1970.

Schor, Heney Clark, and Harmon L. Swan. "Simultaneous Surveys of Food Consumption in Various Camps of the United States Army." Chicago: Department of the Army Medical Nutrition Laboratory, 1949.

Stiefel, Matthias. *Rebuilding After War: Lessons Learned from the War-Torn Societies Project.* Geneva: UNRISD/PSIS, 1999.

Truman, Harry S. *Memories, Volume I: Years of Decisions.* Garden City, NY: Doubleday, 1955.

"Waste Less." *Time Magazine*, 6 October 1947.

Weyerer, Godehard. "CARE Packages: Gifts from Overseas to A Defeated and Debilitated Nation." *The United States and Germany in the Era of the Cold War, 1945–1990: A Handbook, Volume 1: 1945–1968.* Edited by Detlef Junker. Washington, DC: Cambridge University Press: German Historical Institute, 2004.

Wiggers, Richard Dominic. "The United States and the Refusal to Feed German Civilians after World War II." *Ethnic Cleansing in Twentieth-Century Europe.* Edited by Steven Bela Vardy and T. Hunt Tooley. Boulder, CO: Social Science Monographs, 2003.

Williams, Michael. *Civil Military Relations and Peacekeeping,* Adelphi Paper 321. London: International Institute for Strategic Studies, 1998.

Winslow, Donna. "Strange Bedfellows: NGOs and the Military in Humanitarian Crises." *The International Journal of Peace Studies* Vol. 7, Issue 2 (Autumn/Winter 2002).

Woolley, John T., and Gerhard Peters. *The American Presidency Project* [online]. Santa Barbara, CA: University of California (hosted), Gerhard Peters (database). Available from World Wide Web: http://www.presidency.ucsb.edu/ws/?pid=16287.

Wriggins, W. Howard. *Picking up the Pieces from Portugal to Palestine: Quakers Refugee Relief in World War II: A Memoir.* Lanham, MD.: University Press of America, 2004.

Ziemke, Earl F. *The US Army in the Occupation of Germany, 1944–1946.* Washington, DC: US Army Historical Series, Center of Military History, 1975.

Zink, Harold. *The United States in Germany, 1944–1945.* Princeton, NJ: D. Van Nostrand Co., 1957.

Primary Sources

CARE Collection. Manuscripts and Archives Division. The New York Public Library. Astor, Lenox and Tilden Foundations.

General Records of the Department of State (RG 59), National Archives Microfilm Publication 840.48, Records of the Department of State Relating to the Problems of Relief and Refugees in Europe Arising From World War II and Its Aftermath, 1938–1949, Rolls 14–15, National Archives at College Park, College Park, MD (NACP).

OMGUS Entry 25, Records of US Occupation Headquarters, World War II, RG 260 390/40/19/7, NACP.

Records of US Occupation Headquarters, World War II, RG 260 390/40–5/17, Entry 2, Records of the US Group Control Council (Germany)—Records Relating to the Basic Preliminary Plan for Allied Control and Occupation of Germany 1944–45, NACP.

Notes

1. The views expressed in this paper are those of the author. They do not necessarily reflect the official policy or position of the Department of Defense, the US Army, or the United States Military Academy at West Point.

2. Please contact before citing: Tania M. Chacho, Department of Social Sciences, USMA, West Point, NY 10996; (845) 938-7758, Fax (845) 938-4563; tania.chacho@usma.edu

3. See Sultan Barakat and Margaret Chard, "Theories, Rhetoric, and Practice: Recovering the Capacities of War-Torn Societies," *Third World Quarterly*, Vol. 23, No. 5 (October 2002): 817-818.

4. See Deborah Maresko, "Development, Relief Aid, and Creating Peace: Humanitarian Aid in Liberia's War of the 1990s," *OJPCR: The Online Journal of Peace and Conflict Resolution*, 6.1 (Fall 2004): 94–120 (2004), 102. Maresko offers a definition of "relief aid" as "any provision of aid during an emergency that is meant to attend to a person's immediate requirements for survival or recovery, which include food, clothing, housing, medical care, necessary social services, and security when a person is faced with circumstances beyond her or his control."

5. See Joana Abrisketa, "The Right to Humanitarian Aid: Basis and Limitations," *Reflections on Humanitarian Action: Principles, Ethics and Contradictions*, ed. ited by Humanitarian Studies Unit, Transnational Institute (London, Sterling, VA: Pluto Press, 2001), 55.

6. Michael Williams, *Civil Military Relations and Peacekeeping,* Adelphi Paper 321 (London: International Institute for Strategic Studies, 1998), 14.

7. See Donna Winslow, "Strange Bedfellows: NGOs and the Military in Humanitarian Crises," *The International Journal of Peace Studies* Vol. 7, Issue 2 (Autumn/Winter 2002).

8. See Fred Kaplan, "Iraq's Not Germany: What a 60-Year-Old Allen Dulles Speech Can Teach Us About Postwar Reconstruction," *Slate*, posted October 17, 2003, http://www.slate.com/id/2089987/ (accessed January 22, 2008).

9. Earl F. Ziemke, *The US Army in the Occupation of Germany, 1944–1946* (Washington, DC: US Army Historical Series, Center of Military History, 1975), 275, 283–284.

10. Heney Clark Schor and Harmon L. Swan, "Simultaneous Surveys of Food Consumption in Various Camps of the United States Army," (Chicago: Department of the Army Medical Nutrition Laboratory, 1949), 56. See also Military Government of Germany, "Monthly Report of the Military Governor, US Zone," No. 4, "Monthly Report of the Military Governor," 20 November 1945, Record Group (RG) 94, OpBr/B1174, National Archives at College Park, College Park, MD (NACP).

11. Ziemke, 283.

12. Ibid, 274.

13. Lieutenant General Lucius D. Clay, Office of the Deputy Military Governor to John J. McCloy, Assistant Secretary of War, War Department, 5 October 1945, RG107, E180/B26, NACP; Hans W. Schoenberg, *Germans from the East: A Study of their Migration, Resettlement, and Subsequent Group History Since 1945* (The Hague: Martinus Nijhoff, 1970), 32, 38.

14. Ibid.

15. The simplification of the harsh treatment argument into "Morgenthau" has come under valid critique by at least some current historians. For a nuanced study of the perils of the reductive statement the "Morgenthau Plan," see Jeffrey K. Olick, *In the House of the Hangman: The Agonies of German Defeat, 1943–1949* (Chicago and London: The University of Chicago Press, 2005).

16. See "Prospectus for Cooperation for American Remittances to Europe, Inc.," CARE Collection, Box #1, Folder: Prospectus for CARE. Manuscripts and Archives Division, The New York Public Library, Astor, Lenox and Tilden Foundations.

17. Paul W. Gulgowski, *The American Military Government of United States Occupied Zones of Post World War II Germany in Relation to Policies Expressed by its Civilian Authorities at Home, During the Course of 1944/45 Through 1949* (Frankfurt am Main: Haag und Herchen Verlag, 1983), 283.

18. John Gimbel, *The American Occupation of Germany: Politics and the Military, 1945–1949* (Stanford, CA: Stanford University Press, 1968), 54–55 and Harry S Truman, *Memories, Volume I: Years of Decisions* (Garden City, NY: Doubleday, 1955), 468–469.

19. Richard Dominic Wiggers, "The United States and the Refusal to Feed German Civilians after World War II," *Ethnic Cleansing in Twentieth-Century Europe*, eds. Steven Bela Vardy and T. Hunt Tooley (Boulder, CO: Social Science Monographs, 2003), 282.

20. "Waste Less," *Time Magazine*, October 6, 1947.

21. Oliver J. Frederiksen, *The American Military Occupation of Germany: 1945–1953,* (United States Army Europe: Historical Division, 1953), 74 and 78.

22. Michael R. Marrus, *The Unwanted: European Refugees in the Twentieth Century* (New York: Oxford University Press, 1985), 296–299.

23. See Ziemke, 104.

24. Haim Genizi, *America's Fair Share: The Admission and Resettlement of Displaced Persons, 1945–1952* (Detroit MI: Wayne State University Press, 1993), 55.

25. See Hermann J. Rupieper, Review in the *American Historical Review* 106, No. 5, 1,828.

26. See Harold Zink, *The United States in Germany, 1944–1945* (Princeton, NJ: D. Van Nostrand Co., 1957); Gimbel, *The American Occupation of Germany;* Edward N. Peterson, *The American Occupation of Germany: Retreat to Victory* (Detroit MI: Wayne State University Press, 1978); John H. Backer, *The Decision to Divide Germany American Foreign Policy in Transition* (Durham, NC: Duke University Press, 1978); John H. Backer, *Winds of History: The German Years of Lucius DuBignon Clay* (New York: Van Nostrand Reinhold, 1983); Dewey A. Browder, *Americans in PostWorld War II Germany* (Lewiston NY: Edwin Mellen Press, 1998).

27. Petra Goedde, *GIs and Germans: Culture, Gender, and Foreign Relations, 1945–1949* (New Haven: Yale University Press, 2003), xv.

28. In August 2003, members of the Bush administration drew similarities between the uncertainty of the outcome of Germany in 1945 and that of Iraq in 2003. See Daniel Benjamin, "Condi's Phony History: Sorry, Dr. Rice, Postwar Germany Was Nothing Like Iraq" Posted August 29, 2003 http://www.slate.com/id/2087768/ (accessed January 22, 2008).

29. Gimbel, 5. See also Department of State: Foreign Relations of the United States, 1945, vol. 3, European Advisory Commission; Austria; Germany, p. 484.

30. SHAEF Food and Agriculture Section, Economic Control Agency, G5 Division, *"The Food Position in Western Germany as of 1 June 1945,"* 3 July 1945, RG332, ETO,SGS/B57, NACP; see also John H. Backer, "From Morgenthau Plan to Marshall Plan," *Americans as Proconsuls: United States Military Government in Germany and Japan, 1944–1952,* ed. Robert Wolfe (Carbondale: Southern Illinois University Press, 1984), 157.

31. See Robert Cecil, "Potsdam and Its Legends," *International Affairs* 46, No. 3 (June 1970), 462.

32. Franklin M. Davis, *Come As Conqueror: The United States Army's Occupation of Germany, 1945–1949* (New York: The MacMillian Company, 1967), 186.

33. Ibid.

34. See Wiggers, 275.

35. Ibid.

36. Lucius D. Clay, *Decision in Germany* (Garden City, NY: Doubleday and Company, Inc., 1950), 263.

37. Ibid.

38. Francis Fukuyama, *State-Building: Governance and World Order in the 21st Century* (Ithaca, NY: Cornell University Press, 2004), 100.

39. Genizi, 56.

40. Clay, 277.

41. Memorandum from Paul Comly French, Executive Director, CARE, dated February 20, 1948. CARE Collection, Box #1, Folder "CARE—Organization" CARE records. Manuscripts and Archives Division. The New York Public Library, Astor, Lenox and Tilden Foundations.

42. Merle Eugene Curti, *American Philanthropy Abroad* (New Brunswick, NJ: Rutgers University Press, 1963), 452.

43. Executive Order No. 9205, *Establishing the President's War Relief Control Board* (July 25, 1942). John T. Woolley and Gerhard Peters, *The American Presidency Project* [online]. Santa Barbara, CA: University of California (hosted), Gerhard Peters (database), http://www.presidency.ucsb.edu/ws/?pid=16287.

44. See Anna C. Bramwell, ed., *Refugees in the Age of Total War* (London: Unwin Hyman Ltd, 1988), 96.

45. Curti, 454.

46. See Bramwell, 96; Curti, 453–455.

47. Curti, 453.

48. See Rachel M. McCleary and Robert J. Barro, "Private Voluntary Organizations Engaged in International Assistance, 1939–2004," *Nonprofit and Voluntary Sector Quarterly Journal* (December 2007): 20.

49. Agreement for United Nations Relief and Rehabilitation Administration, November 9, 1943, Pamphlet No. 4, *Pillars of Peace,* Documents Pertaining to American Interest in Establishing a Lasting World Peace: January 1941–February 1946 (Carlisle Barracks, PA: Army Information School, May 1946), Articles 1 and 2.

50. Wiggers, 281–282.

51. Ibid., 287.

52. Ibid., 281.

53. Gulgowski, 10.

54. Nicholas Balabkins, *Germany Under Direct Controls: Economic Aspects of Industrial Disarmament 1945–1948* (New Brunswick, NJ: Rutgers University Press, 1964), 101.

55. John Killick, *The United States and European Reconstruction: 1945–1960* (Edinburgh, UK: Keele University Press, 1997), 76.

56. Department of State, War Problems–Special Division Letter, dated September 12, 1945, signed by Edwin A. Plitt, General Records of the Department of State (RG 59), National Archives Microfilm Publication 840.48, Records of the Department of State Relating to the Problems of Relief and Refugees in Europe Arising From World War II and Its Aftermath, 1938–1949, Roll 15, NACP.

57. See, for example, Department of State Airgram to US Ambassador in Brussels, dated February 22, 1945, General Records of the Department of State (RG 59), National Archives Microfilm Publication 840.48, Records of the Department of State Relating to the Problems of Relief and Refugees in Europe Arising From World War II and Its Aftermath, 1938–1949, Roll 14, NACP.

58. Department of State, Plitt letter, RG 59, 840.48, Roll 15, NACP.

59. General Information from the Publicity Department, United National Clothing Collection, 100 Maiden Lane, New York 5, New York, General Records of the Department of State (RG 59), National Archives Microfilm Publication 840.48, Records of the Department of State Relating to the Problems of Relief and Refugees in Europe Arising From World War II and Its Aftermath, 1938–1949, Roll 14, NACP.

60. Cooperative for American Remittances to Europe, Memorandum dated January 8, 1946, CARE Collection, Box #1, Folder: Development—Committee on Cooperatives, Manuscripts and Archives Division, The New York Public Library, Astor, Lenox and Tilden Foundations

61. Ibid, 3; Agreement Between the Deputy Military Governor of the United States Zone of Occupation in Germany and Cooperative for American Remittances to Europe, Inc., Records of US Occupation Headquarters, World War II (RG 260), Office of Military Government, US Zone (Germany)-CARE file, Entry 81, NACP.

62. Agreement Between Deputy Military Governor of US Zone and CARE, RG 260, Entry 81, NACP.

63. Report prepared by Rosenman Mission London, England, April 15, 1945, General Records of the Department of State (RG 59), National Archives Microfilm Publication 840.48, Records of the Department of State Relating to the Problems of Relief and Refugees in Europe Arising From World War II and Its Aftermath, 1938–1949, Roll 14, NACP.

64. Ibid. General Eisenhower identified this date as May 1, 1945 for France.

65. Ibid.

66. President Harry S. Truman letter to the Secretary of War, dated May 21, 1945, General Records of the Department of State (RG 59), National Archives Microfilm Publication 840.48, Records of the Department of State Relating to the Problems of Relief and Refugees in Europe Arising From World War II and Its Aftermath, 1938–1949, Roll 14, NACP.

67. Ibid.

68. Keyes D. Metcalf letter to the Honorable Edward R. Stettinius, Secretary of State, dated 14 May 1945. General Records of the Department of State (RG 59), National Archives Microfilm Publication 840.48, Records of the Department of State Relating to the Problems of Relief and Refugees in Europe Arising From World War II and Its Aftermath, 1938–1949, Roll 14, NACP.

69. Ibid.

70. Department of State Airgram to US Ambassador in Brussels, dated February 22, 1945, NACP.

71. Ibid.

72. Memorandum from Paul Comly French, dated February 20, 1948, CARE Collection, Manuscripts and Archives Division, The New York Public Library, Astor, Lenox and Tilden Foundations.

73. See for example, Letter from Wallace J. Campbell, Chairman, Committee on Cooperatives, to Secretary of State James F. Byrnes, dated October 18, 1945, General Records of the Department of State (RG 59), National Archives Microfilm Publication 840.48, Records of the Department of State Relating to the Problems of Relief and Refugees in Europe Arising From World War II and Its Aftermath, 1938–1949, Roll 15, NACP.

74. See, for example, Telegram from the Reverend Steven H. Fritchman, to Matthew J. Connelly, Secretary to the President, dated April 24, 1945, General Records of the Department of State (RG 59), National Archives Microfilm Publication 840.48, Records of the Department of State Relating to the Problems of Relief and Refugees in Europe Arising From World War II and Its Aftermath, 1938–1949, Roll 14, NACP.

75. See Rosenman Mission Report, dated April 15, 1945, RG 59, 840.48, Roll 14, NACP.

76. President Harry S. Truman letter to Secretary of War, dated May 21, 1945, RG 59, 840.48, Roll 14, NACP.

77. See Letters Between the Department of War and the Department of State, General Records of the Department of State (RG 59), National Archives Microfilm Publication 840.48, Records of the Department of State Relating to the Problems of Relief and Refugees in Europe

Arising From World War II and Its Aftermath, 1938–1949, Roll 14, NACP. See Rosenman Mission Report, dated April 15, 1945, RG 59, 840.48, Roll 14, NACP.

78. This matrix did include a caveat, noting that it does not list "all agencies with which coordination may be required." See HQ US Group CC—Sub-Plans (Annexes) to Preliminary Basic Plan Allied Control and Occupation of Germany—Tab D, Records of US Occupation Headquarters, World War II, RG 260 390/40–5/17, Entry 2, Records of the US Group Control Council (Germany)—Records Relating to the Basic Preliminary Plan for Allied Control and Occupation of Germany 1944–45, NACP.

79. American Friends Service Committee Proposal for Aid to German Expellees (undated), OMGUS Entry 25, Records of US Occupation Headquarters, World War II, RG 260 390/40/19/7, NACP.

80. Memorandum from Sumner Sewall, Director, Internal Affairs and Communications Division, "Aid to German Expellees by AFSC," dated November 12, 1946, OMGUS Entry 25, Records of US Occupation Headquarters, World War II, RG 260 390/40/19/7, NACP.

81. Ibid.

82. CARE's Financial Requirements Memorandum, dated January 8, 1946, CARE Collection, Box #1, Folder "Development-Committee on Cooperatives," CARE records, Manuscripts and Archives Division, The New York Public Library, Astor, Lenox and Tilden Foundations.

83. E.D. Kuppinger, Department of State, Memorandum, dated October 17, 1945. General Records of the Department of State (RG 59), National Archives Microfilm Publication 840.48, Records of the Department of State Relating to the Problems of Relief and Refugees in Europe Arising From World War II and Its Aftermath, 1938–1949, Roll 14, NACP.

84. Agreement on 10-in-1 Rations dated 2–11–46, CARE Collection, Box #1, Folder "Development-Committee on Cooperatives," CARE records, Manuscripts and Archives Division, The New York Public Library, Astor, Lenox and Tilden Foundations.

85. Ray Salvatore Jennings, "The Road Ahead: Lessons in National Building from Japan, Germany, and Afghanistan for Postwar Iraq," *Peaceworks* No. 49 (Washington, DC: United States Institute of Peace, April 2003).

86. James Dobbins, John G. McGinn, Keith Crane, Seth G. Jones, Rollie Lal, Andrew Rathmell, Rachel M. Swanger, Anga R. Timilsina, "America's Role in Nation-Building, From Germany to Iraq*"* (Santa Monica, CA: RAND, 2003), 22.

87. Matthias Stiefel, *Rebuilding After War: Lessons Learned from the War-Torn Societies Project* (Geneva: UNRISD/PSIS, 1999), 16–19.

88. Barakat and Chard, 827.

89. Clay, 276.

90. Clay, 279.

91. Hermann J. Rupieper, Review of *Humanitare Auslandshilfe als Bruke zu atlantischer Partnerschaft: CARE, CRALOG und die Entwicklung der deutsch-aamerikanischen Beziehungen nach ende des Zweiten Weltkrieges by* Karl-Ludwig Sommer, *American Historical Review* 106, No. 5 (December 2001): 1,828. See also Godehard Weyerer, "CARE Packages: Gifts from Overseas to a Defeated and Debilitated Nation," *The United States and Germany in the Era of the Cold War, 1945–1990: A Handbook, Volume 1: 1945–1968*, ed. Detlef Junker (Washington, DC and Cambridge University Press: German Historical Institute, 2004), 524.

92. William Haskell, Executive Director of CARE, Letter to Robert Patterson, Secretary of

War, dated September 9, 1946, CARE Collection, Box #1, Folder "Contracts" CARE records, Manuscripts and Archives Division, The New York Public Library, Astor, Lenox and Tilden Foundations.

93. Memorandum: Agreement from German Delegation, American Council of Voluntary Agencies to Office of Military Government (US) Re: Relief Operations by American Voluntary Agencies, OMGUS Entry 112, Records of US Occupation Headquarters, World War II, RG 260 390/40/19/7, NACP.

94. Alden E. Bevier, Chief, Public Welfare Section, Information Memorandum: "CARE Food Package Program," dated 2 July 1946, CARE Collection, Box #13, Folder "Germany – US Zone," CARE Records, Manuscripts and Archives Division, The New York Public Library, Astor, Lenox and Tilden Foundations.

95. W. Howard Wriggins, *Picking up the Pieces from Portugal to Palestine: Quakers Refugee Relief in World War II: A Memoir* (Lanham, MD: University Press of America, 2004), 151.

96. Ibid.

97. Memorandum: Agreement from German Delegation, American Council of Voluntary Agencies to Office of Military Government (US) Re: Relief Operations by American Voluntary Agencies, OMGUS Entry 112, Records of US Occupation Headquarters, World War II, RG 260 390/40/19/7, NACP.

98. Clay, 277.

99. See "Prospectus for Cooperation for American Remittances to Europe, Inc.," CARE Collection, Box #1, Folder: Prospectus for CARE, Manuscripts and Archives Division, The New York Public Library, Astor, Lenox and Tilden Foundations.

Panel 5—Post-War Germany
(Submitted Paper)

Between Catastrophe and Cooperation: The US Army and the Refugee Crisis in West Germany, 1945-50

by

Adam R. Seipp, Ph.D.
Texas A&M University

The day after Thanksgiving, 1960, a group of American and West German dignitaries arrived at the refugee camp in Heidingsfeld, bringing with them quantities of turkey for a proper American feast. Colonel Jack Dempsey from Leighton Barracks in nearby Würzburg told the assembled crowd that he was "grateful that he was stationed in the miracle city of Würzburg, which sank into ashes 15 years ago yet stands again and has become one of the most beautiful cities in Europe." More boldly, Dempsey suggested that the American day of giving thanks mirrored the situation of the Eastern European refugees living at Heidingsfeld. After all, he reminded his audience, what was Thanksgiving but a day when "a group of European immigrants gave thanks for the end of a difficult time?"[1]

The fact that this speech happened at all, and that it took place fifteen years after the end of the war, highlights the extraordinary set of crises that befell Central Europe after 1945. The war generated millions of refugees, many of whom would never return home. At the same time, the Cold War created an institutionalized emergency across the region and led to the creation of a network of military installations on both sides of the Iron Curtain. In this paper, I will demonstrate that these two developments have to be understood together.

This paper has three arguments. First, that the United States Army played a critical and little appreciated role in the management of the postwar refugee crisis in Germany. Second, that US strategic priorities often conflicted with the needs of German and international agencies and organizations charged with managing refugees, a conflict that escalated as the Cold War standoff in Central Europe worsened. Finally, that troops on the ground in occupied Germany often acted on their own initiative in refugee affairs, sometimes directly contradicting or contravening US policy on the matter.

We cannot understand the history of the American military presence in West Germany without examining the refugee problem. Histories of the Cold War US Army and histories of West Germany have tended to gloss over this critical issue.[2] The refugee problem complicated American efforts to establish bases in post-sovereignty West Germany because many of the spaces suitable for quickly establishing such posts were already occupied by the displaced. At the same time, West German responses to the American presence were in turn conditioned by these living reminders of the costs of war and the looming Cold War division of Europe.

Perhaps the best reason for the relative absence of refugees in histories of the period is one of sources. Simply put, the Army and the American military government (OMGUS) had good reason to move refugees onto the books of other agencies, both German and international. Refugees proved a logistical nightmare, one that the Army was more than happy to shift elsewhere. The result was a lack of clear authority and confusion that sometimes mirrored the chaotic conditions that created the refugee flows in the first place.

Several distinct streams of refugees existed in Germany after 1945. Large numbers of Germans from across the ruins of the Third Reich found themselves homeless, a problem exacerbated by Germans fleeing the Soviet occupation of the eastern zone of the country. Even before the end of the conflict, ethnic Germans began to flee their homes in Eastern Europe, beginning a bloody exodus from countries recently subject to the Nazi empire. Between legal and extra-judicial expulsions, nearly 14 million "expellees" made their way west, often under terrible conditions. More than 2 million of these refugees ended up in American occupied Bavaria alone.[3] As OMGUS attempted to establish viable German institutions to manage civilian affairs in occupied Germany, one of the first acts was to create state-level refugee agencies tasked to feed and care for the growing stream of refugees.

At the same time, occupied Germany played host to a new kind of refugee. More than 5 million people, many of whom were survivors of German slave labor or concentration camp facilities, were classified as "Displaced Persons" (DPs). These DPs enjoyed the protection of the new United Nations and could, in theory, depend on the international community for their lodging and provisioning.[4]

The changing relationship between American authorities and refugees can be seen in the administrative structure established to deal with them and their concerns. Under OMGUS, which lasted from 1945 until West German sovereignty in 1949, a Prisoners of War and Displaced Persons Division existed as a distinct functional division reporting to the Executive Office. German refugees did not formally concern OMGUS officials after the establishment of German agencies to oversee their needs. After 1949, when the Office of the High Commissioner for Germany (HICOG) replaced OMGUS, the calculus of refugee affairs had changed. Now, even with the DP problem still a major issue across West Germany, refugee affairs shifted names and, importantly, responsibilities. Now, the office of the HICOG Political Advisor had a Displaced Populations Branch, as did the Public Affairs Division and each of the Resident Officers in the field. This shift from Displaced Persons to Displaced Populations was more than semantic.[5] As West Germany emerged as a sovereign Cold War ally and frontier state, American concern shifted from the specific needs of the DP population to that of the overall refugee problem.

The remainder of this paper will examine this shift, and the implications that it had for Germans, Americans, and refugee populations, in the Bavarian district of Lower Franconia (*Unterfranken*). Within a few months of war in 1944 and 45, the district of Lower Franconia transformed from a sedate, rural part of Central Germany into a possible Cold War flashpoint. The region, about 12% of the landmass of Bavaria, occupied a critical junction along the emerging inter-German border. Unterfranken's 124 kilometer border with what is today the German state of Thuringia was, after 1949, an international frontier. That border fell squarely along the southern edge of the most famous strategic route in Cold War Europe: the Fulda Gap.[6] This space between the Vogelsberg and Spessart hills offered the easiest approach to Frankfurt, and both sides of the emerging Cold War sought the advantage along this corridor.

Lower Franconia became one of the most highly militarized districts in Bavaria, home to a division headquarters and a number of smaller facilities near or along the border. Two of the places where American basing policy and the refugee crisis collided were the district capital of Würzburg and the small town of Wildflecken about seventy kilometers northwest. In these communities, and in many others across Central Europe, refugees, refugee agencies, and the American military presence shared close quarters, with results that helped to shape West Germany's Cold War future.

Würzburg lay in ruins in 1945. On March 16, a massive RAF assault obliterated nearly 80% of the city center, killing about 5,000 people in 20 minutes. A month later, American troops crossing the Main River met stiff resistance and many of the remaining buildings fell during fierce fighting in the streets. This series of disasters, along with the low-level civil war in Franconia at the end of the war, created waves of displaced, terrified, and starving humanity. At the end of the war, Würzburg had a population of only about 6,000, bolstered by nearly 5,000 refugees. While the population rebounded to nearly 70,000 within two years, the number of refugees did not decrease markedly for almost five years.[7] Clearly, the refugee problem was going to be a long-term one, and local officials scrambled to find places to house this influx of extra mouths.

The American footprint in the city was tiny, with a small headquarters in one of the few intact buildings along the river. This meant that the occupation forces did not need the abandoned *Wehrmacht* airfield on the Galgenberg Hill above the city. The Bavarian Refugee Office, headed locally by an energetic and respected bureaucrat named Josef Winter, began housing some of the refugees in the buildings along the airfield, which had likely survived the bombing because of their relative distance from the center of the city. Between Galgenberg and a smaller camp at Heidingsfeld south of the city, refugees could be drawn out of the enormous construction zone in the city center.

There were significant formal contacts between the occupation forces and the mass of refugees in Würzburg and the region. In 1947, more than 6,700 refugee children from around Unterfranken were treated to Christmas parties hosted by the occupation government. The Americans made much of the fact that most of the refugees involved came from places under Soviet domination, noting that the entertainment for these parties included "Ukrainian Cossack dances, Baltic choirs, Polish dances, folk music, and the showing of the colorful garments of the different nationalities." On a more prosaic level, the budget of the Würzburg Military Post indicates that the Americans helped to finance the recovery of local industries by paying for a range of materials to be delivered to DP camps and other refugee installations. The glassmaker Otto Weigan, for instance, provided 10 DM glass eyes for inhabitants of local DP camps.[8]

To the north, in the tiny farming community of Wildflecken, a very different story unfolded. Wildflecken had been a rural backwater until 1936, when the German Army identified it as an ideal spot to conduct training in hilly and wooded terrain. They built an 18,000 acre training facility there, with a ring of buildings on a hill above the village. During the war, the remote location made it an ideal site for a hidden armaments plant, which included hundreds of slave laborers. The town and the nearby base escaped bombing, and Wildflecken's war ended with the arrival of a lone American jeep in early April 1945.[9]

The Americans left a small garrison in Wildflecken, largely to guard the munitions plant and demolish the network of bunkers that ringed it. In taking charge of the base, they also inherited supervision over the workers living therein. Between May and October, the Army began the process of consolidating a warren of facilities for laborers into a few large camps, divided by putative ethnicity. Wildflecken, with a tremendous amount of unused space, became one of the largest such catchment facilities in Europe, with nearly 15,000 Poles in residence by the time the Army gave over control to UNRRA on 1 October 1945. The American Kathryn Hulme, the camp's Assistant Director, recalled that:

> When we entered the camp, Army was in control in the form of a Captain, but he took off at the end of the first week, leaving us a handsome large office equipped with mahogany desks empty of all documents, reports or even carbon copies of letters which might have given us a clue to what had gone on in the camp prior to our arrival.[10]

While the Army turned over formal control of the DP camps to UNRRA, it was forced to take a more active role in their maintenance than had been expected. Despite initial plans to provision the facilities from locally available stocks, the devastated German economy proved unequal to the task. As early as 1946, most DP calories came from occupying, and mostly American, supplies. "Preferential feeding for DPs," one report noted, "has always been US Army policy. It gave them 425 calories above the basic ration for Germans."[11] The strategy of clearly favoring DPs over both Germans and German refugees created enormous resentment among local populations. It also did not work particularly well on the ground, where conditions and personal behavior made the situation far more complicated.

Scarcity created competition between UNRRA and the Army, with the results typically favorable to the latter. UNRRA could claim preferential treatment, but small American garrisons in places like rural Franconia faced similar supply problems and were unlikely to yield to the demands of the refugee agency. Ephraim Chase, a young

American UNRRA employee at the DP camp in Dillingen, captured this problem nicely with regard to the perpetual problem of lumber:

> . . . often when our trucks arrived at the saw-mill to gather the products of the previous two or three days we would find that someone beat us to it. The XYZ Engineers, we would learn, appeared on the scene a day or an hour ago and removed every splinter of wood on the premises leaving a tell-tale notice behind to the effect that from now on everyone else is to keep out of here. At first we felt frustrated, but gradually we became inured and mastered the game of matching wits. Compromise and gentlemens' agreements were effected, and the supply of lumber resumed its steady flow. So that we shall have a quantity of lumber left over to meet emergency repairs during the winter months and for the camp shop to make simple furniture.[12]

DPs were also supposed to receive priority when it came to hiring for civilian labor. In mid-1947, US Army Labor Service Companies employed about 40,000 DPs.[13] While this number was high, there is significant evidence to suggest that many units, when given the option, chose to hire non-DP refugee labor. The company guarding the facility could easily employ Poles from the camp as civilian laborers but chose to hire almost entirely from the ranks of other refugees. In March 1946, twelve of twenty civilian workers came from either the Soviet zone or from Poland/Silesia. Only three came from the Rhön region.[14]

Despite a policy that officially favored DPs, many American officers found them, and their UNRRA supervisors, tiresome and untrustworthy. Despite wartime optimism on the part of the Allies, many Eastern European DPs had little desire to return home. By the time Eisenhower ordered an end to forcible repatriation in September, 1945, almost 2 million DPs remained in Western Europe, more than half of whom were Poles.[15] As relations between former allies deteriorated, Eastern Europeans came under increasing scrutiny as potential security threats. Nothing caused more problems between UNRRA and the Americans than the periodic camp inspections. OMGUS consistently emphasized the need to speed up repatriation. As a corollary, American suggestions or demands for camp administrators focused on making the camps physically safe but minimally comfortable in order to encourage repatriation without the risk of the sorts of epidemic diseases that might both create a public health nightmare and slow down repatriation. Relations between inspectors and UNRRA officials generally proceeded efficiently, if not warmly. Friction, however, was all but inevitable, spurred by the kinds of rumor and innuendo that acted as a motor for camp life.

In late 1945, the camp received frequent visits from one General Watson and his staff. UNRRA reports from the period suggest that Watson became an almost legendary figure among camp residents, prone to simply appear and begin questioning DPs and UNRRA officials. In September, rumors circulated that Watson believed that "all Poles should be treated as Germans." Whatever the conversation among the residents, Watson's relationship with the camp director, the Frenchman George Masset, did not go well either. In the very least, the General could be brusque, tone-deaf to the needs of the camp, and fixated on the idea of repatriation.

Watson tended to descend on the camp, leaving trouble in his wake. On an inspection in mid-September, he pointedly criticized the amount of litter in the camp, the frequency of bathing, and the existence of a Polish Committee, which he believed indicated that the camp ran along "Soviet lines." On several occasions, he told camp administrators that he expected repatriation to be sped up to 1,500 a week.

As might be expected, both the Wildflecken Poles and UNRRA expressed outrage at the tone and content of these inspections. Masset sent an angry letter to his superiors, reporting that the Wildflecken Poles now believed that the camp "is occupied more rigorously than were concentration camps during the war or German cities since the peace." How, asked Masset, did the Army expect sanitation to improve with limited resources and a hygiene situation that already was "more clean and healthy than many European cities of 15,000 inhabitants?"[16]

He took the additional step of writing to the UNRRA offices in Bad Kissingen unofficially to protest what he saw as outrageous behavior on the part of the American officers. Continuing to emphasize the connections between DP camps and concentration camps, Masset angrily noted that the General:

> [gave] hell to a DP worker who was dressed in the striped coat of Buchenwald! None of General Watson's commands and orders were made privately; at all times he was surrounded by a crowd of 200 to 300 Poles, many of whom understand English. You cannot imagine the state of this camp today.[17]

In 1950, a series of seismic shifts in the Cold War order changed the face of refugee policy and with it the relationship between the Army and refugee agencies. Truman's Troops to Europe decision and the beginning of the Korean War forced a reorientation of basing policy and created a vast need for usable space. At the end of 1950, there were 86,000 American soldiers in Europe. A year later, this number climbed to almost 232,000.[18] The demand for housing for these personnel and their dependents brought the refugee question to center stage.

In April, 1949, according to the city government, of 5,245 refugees living in Würzburg, 3,444 had some sort of private accommodation. 1,052 lived in the Galgenberg camp, while 749 were at Heidingsfeld.[19] Into this tense and difficult situation came the Americans. Following the reversal of the US drawdown in Europe, the US Army in particular needed facilities that could be made ready quickly. The most efficient way to accomplish this was to take over old *Wehrmacht* and *Luftwaffe* facilities, which served no purpose in a disarmed West Germany. The problem was that thousands of refugees were already living in these abandoned bases, supported by state-level bureaucracies like the Bavarian Refugee Office.

By early 1951, the Americans identified the Galgenberg facility as suitable for conversion to an American base. When local Refugee Office officials pointed out that there were more than 1,000 refugees living in the surviving barracks on the facility, the Resident Officer in Würzburg responded that money had already been budgeted and there was no way to delay the beginning of construction. By summer, virtually all of the building firms in and around the city were engaged in the largest postwar construction project in the region, the building of the Skyline barracks complex.

By October, the few refugee families left on the base received notice to vacate within ten days, awaiting the arrival of American families and the opening of a base school. The remaining residents either found homes and livelihoods in the private sector or moved to the Heidingsfeld facility, which remained open until the early 1960s.[20]

The goal of the *Land*-level refugee administrations was not to keep moving people from camp to camp. Their hope was to allow refugees the time to secure employment and find their own housing. At the same time, Federal money was available to help the *Länder* build new homes for West Germany's homeless. This process was achingly slow, making it imperative to keep camps operating as long as possible. In 1950, there were 9,635 refugees living in 35 camps administered by the Bavarian Refugee Office across Unterfranken. As settlement plans advanced, that number declined, slowly but surely, to about 5,000 at the end of 1952. About 3,000 families moved elsewhere in Bavaria, while others moved into newly constructed homes across the *Land*. Many of the apartment blocks built with funding from the Bavarian Refugee Agency remain in Würzburg, reminders of the haste and confusion of the city's conversion into a Cold War garrison town.[21]

The problems of conversion in Würzburg were serious, but nothing like the crisis over the future of the Wildflecken facility. While UNRRA turned over its operations to its successor the International Refugee Organization (IRO) in 1947, the camp remained in operation and, if anything, the pace of repatriation slowed. In January of that year, a team of Polish Army officers visited the camp to conduct a nationality screening. Fearing forcible repatriation, the inhabitants rioted and only the timely arrival of American troops saved the Polish officers from being beaten to death by an angry mob.[22]

At the same time, the Wildflecken facility came to the attention of the Bavarian Refugee Office. The DP camp occupied a small corner of the base, while the villages that had been cleared to build the camp in the 1930s had been continuously occupied by workers and remained reasonably intact. Beginning in 1946, the Refugee Office began to settle ethnic German refugees in the surviving houses in Werberg and Reussendorf. Werberg, now settled by farmers from Romania, Poland, and Czechoslovakia, emerged as a model community that seemed to highlight all of the virtues of thrift and hard work that the Bavarian Refugee Office wanted to publically display. "For expellees, the chance to have a place of their own is not common," wrote the *Main-Post* in 1950, "But here, thanks to their energy and drive a large number of refugees have found a new home on the grounds of the former base at Wildflecken." The senior refugee official for Lower Franconia, Anton Beck, expressed his confidence in the project. 'Despite the fact that Werberg has taken up nearly all of my time in the past few months . . . it brings me great joy to see the first loaf of bread baked, the first calf of the season, or a father's pride when he shows off a newborn resident before the town.'[23]

251

In July 1950, six days after the North Korean invasion of the South began, the IRO transferred jurisdiction over the last 600 DPs at Wildflecken to German authorities. Not only did the UN leave, but American policy toward refugees began to change in line with the unstated preferences of many of the troops on the ground. In the new security situation of Central Europe, DPs were a minor irritant compared with the potential problems of integrating millions of ethnic German refugees. The refugee issue now topped lists of concerns for the Americans in their vision of the continued development of Lower Franconia. When the mayor of Wildflecken went to Hammelburg to meet with the American Resident Officer in January 1951, he and the other assembled mayors were castigated for their obstruction on the refugee question. Many local mayors, the Resident Officer suggested, would rather 'foster division between locals and refugees than to lead the way toward integration.'[24]

The Americans had good reason to be interested in the improvement of local conditions and the resolution of outstanding refugee issues. As early as 1948, the Military Government informed local officials that they had approval to build a training facility on the grounds of the former *Wehrmacht* base above Wildflecken. The initial plans came with severe restrictions, notably that both the DP camp and local settlements could not be disturbed. Since even the reduced DP camp monopolized most military buildings on the site, and since the communities like Werberg and Reussendorf sat squarely in the middle of the proposed maneuver grounds, these restrictions were simply untenable. Six months later, the official requisition forms for the facility contained no such limitations.[25]

So began a nearly three-year conflict between local government and the American military over the fate of the communities around Wildflecken. While the Americans continued to promote the speedy clearance of the DP camp, they argued with local officials over the scale of the camp. At issue were the small towns dotted across the 18,000 acres of the facility. Dr. Maria Probst, a *Landtag* and later *Bundestag* delegate living in Hammelburg took special interest in the fate of these small towns, forcing meetings in Bonn over the issue and holding a series of rallies in the region to bring together farmers threatened with dispossession.[26]

The looming crisis over the base expansion had the effect of hastening the integration of those recent arrivals into the rural communities where they found themselves after the war. A 1950 meeting in Reussendorf issued a statement in the name of "old, new, and neighboring" inhabitants of the area urging the Bavarian government to distribute the territory inside the base to local farmers.[27] The local *Landrat* went even farther, darkly suggesting that the experience of dispossession might well unite the local population behind "politically extreme ideas."[28]

In the summer of 1951, the Bavarian government publicly suggested that the Americans concentrate their building program on Wildflecken. They expressed particular concern over the possibility of an American base in nearby Hammelburg, with a larger population and more refugees subject to resettlement. The *Münchner Merkur* applauded the move, suggesting that a base at Wildflecken would be minimally invasive, could take advantage of all the amenities built by the *Wehrmacht* during the construction program of the 1930s, would displace far fewer people, and, most importantly, would get rid of the DP camp. "Three groups would be helped: the Hammelburger who will not be made homeless, the displaced, who won't have to keep living in camps, and the public good, which will be saved from this smuggling center."[29]

The local government in Brückenau reacted furiously to the letter, pointing out, entirely correctly, that most of the luxuries (like a swimming pool and a modern town hall) didn't actually exist, that 55% of those who would likely be displaced by the base were, in fact, *Flüchtlinge*, and that Wildflecken was so close to the border that enemy agents could easily watch proceedings at such a facility. But the *Landrat* reserved his greatest indignation for the suggestion that Wildflecken was the center of a smuggling enterprise. "Coffee and cigarettes will no longer be smuggled by DPs from Wildflecken, since they all left four months ago. A 'smuggling center' is no longer there to be dissolved."[30]

Delicate negotiations over the precise demarcation of the new American facility proceeded through 1951, which added to the anxiety of residents who complained that they were being kept in the dark. In the end, even compromise was not enough to save the towns. Reussendorf had to be abandoned by its nearly 300 residents. Werberg, the model refugee town of a half-decade before, survived, but only after losing most of the farmland north of the settlement. While the town remained intact, it was no longer economically viable. By 1960, only 12 farms remained. Three years later, when the base expanded, the town vanished completely.[31]

As the towns faced the prospect of ruin, Bavarian government funds sponsored a crash building program designed to give resettled residents apartments in Wildflecken and in the community of Neuwildflecken on the grounds of the abandoned munitions plant. By October 1952, just as the cold rains of another Central European autumn arrived, the new town was passably ready. Despite the lack of a school or town hall, sixty new homes had been completed in Neuwildflecken, cause for a public celebration.[32] The invitation evoked the hilly country of the Rhön and the bucolic comforts of the new settlement. Such celebrations were common events in a country going through the early stages of an astounding economic transformation. For the citizens of Neuwildflecken, this was a transformative event, creating a new community in the shadow of an occupying army.

The story of the relationship between the Army and the refugee crisis in post-1945 West Germany is an important one. The Army had little or no interest in taking direct responsibility for refugees or refugee affairs and worked through German government agencies or the United Nations. Particularly in the case of UNRRA, which was dependent on the Army for supplies, this meant in practice a close working relationship that was often fractious. Efforts by the Bavarian government to manage their own refugee affairs consistently ran into difficulties because of the divided sovereignty arrangement in postwar Germany.

Relations between the Army, refugee agencies, and the refugees themselves mirrored the broader transformation of the American presence in Germany after the war. In the wake of the war, policy emphasized the importance of creating comfortable conditions to ensure the safe return of DP populations. As it became evident that DPs were likely to be a long-term presence, Army personnel on the ground began to turn on them, a shift that anticipated the larger change in focus that accompanied West Germany's emergence as a state at the end of the decade.

These relationships, marked by necessary cooperation and increasingly by frustrated disagreement, did more than decide the short-term fate of thousands of refugees in the years after the war. They shaped the emerging society of a Cold War frontier state and the American military presence that remained there for decades to come. While the Wildflecken training area was turned over to the *Bundeswehr* in 1994 and the last barracks in Würzburg are in the process of closing, the American presence there in the second half of the twentieth century shaped the lived experience of the Cold War for Americans and Germans. As this paper has argued, we cannot understand the creation of that Cold War order, or its centrality in the history of the divided century, without seeing in the context of the German refugee crisis. Through collaboration, negotiation, and confrontation, the interests of refugees, refugee agencies, and the US Army intersected in the limited space of places like the Fulda Gap. The result was a country, a region, and an Army transformed.

Notes

1. "Die Flüchtlingen schmeckte der Truthahn" *Main-Post*, November 25, 1960.
2. Harold Zink's classic *The United States in Germany, 1945–1955* (Princeton: Nostrand, 1957) has precisely one reference to the refugee issue. Jeffry Diefendorfer, Axel Frohn, and Hermann-Josepf Rupieper, eds., *American Policy and the Reconstruction of West Germany, 1945–55* (Cambridge: Cambridge University Press, 1993) has little more. Richard Merritt, *Democracy Imposed: US Occupation Policy and the German Public, 1945–49* (New Haven: Yale University Press, 1995) considers the role of later refugee flows, but has little about the larger issue in the postwar years. A recent exception to this is Sylvia Schraut, *Flüchtlingsaufnahme in Württemberg-Baden 1945–1949. Amerikanische Besatzungsziele und demokratischer Wiederaufbau im Konflikt* (Munich, 1995).
3. Among others, see Andreas Kossert, *Kalte Heimat: Die Gesichchte der deutschen Vertriebenen nach 1945* (Munich, 2008); Marion Frantzioch, *Die Vertriebene: Hemmnisse, Antriebskräfte und Wege ihrer Integration in der Bundesrepublik Deutschland* (Berlin, 1987); and

David Rock and Stefan Wolff, eds., *Coming Home to Germany? The Integration of Ethnic Germans from Central and Eastern Europe in the Federal Republic since 1945* (New York and Oxford, 2002).

4. Daniel Cohen, "Remembering Post-War Displaced Persons: From Omission to Resurrection" in Mareike König, Rainer Ohliger (eds.), *Enlarging European Memory: Migration Movements in Historical Perspective*, (Stuttgart: Thorbecke Verlag), 2006, pp.87–97; Atina Grossman, *Jews, Germans, and Allies: Close Encounters in Occupied Germany* (Princeton: Princeton University Press, 2007); Laura Hilton, "Prisoners of Peace: Rebuilding Community, Identity, and Nationality in Displaced Persons Camps in Germany, 1945–1952", PhD, Ohio State University, 2001; Anna Holian, "Between National Socialism and Soviet Communism: The Politics of Self-Representation Among Displaced Persons in Munich, 1945–51", PhD, University of Chicago, 2005; Wolfgang Jacobmeyer, *Vom Zwangsarbeiter zum Heimatlosen Ausländer: Die Displaced Persons in Westdeutschland, 1945–1951* (Göttingen, 1985); Kim Salomon, *Refugees in the Cold War: Towards a New International History of the International Refugee Regime in the Early Postwar Era* (Lund, 1991); Mark Wyman, *DPs: Europe's Displaced Persons, 1945–1951* (Ithaca, 1998).

5. Zink, 27 and 61. HICOG had a slightly different structure in the German states (*Länder*) administered by the United States. Württemberg-Baden had significantly fewer ethnic German refugees than Bavaria, so refugee affairs were not given as prominent a role.

6. Regierung von Unterfranken, *Unterfranken in Zahlen* (Würzburg: Regierung von Unterfranken, 1976), 6. I am using the 1976 edition because it was the last that used the names of administrative districts before the name and border changes that accompanied administrative reform in the 1970s.

7. Peter Moser, *Würzburg: Geschichte einer Stadt* (Bamberg: Babenberger Verlag, 1999), 289; Herbert Schott, *Die Amerikaner als Besatzungsmacht in Würzburg* (Würzburg: Freunde Mainfränkischen Kunst und Geschichte, 1985), 74; and Hans Steidle and Christine Weisner, *Würzburg: Streifzüge durch 13 Jahrhunderte Stadtgeschichte* (Würzburg: Echter, 1999), 223; Stephen Fritz, *Endkampf: Soldiers, Civilians, and the Death of the Third Reich* (Lexington: University Press of Kentucky, 2004).

8. "Christmas Parties Planned by WMP" *Würzburg Post-Argus*, November 26, 1947. Receipts in Staatsarchiv Würzburg (StaatWü), Rg. v. Ufr. 17224.

9. Gerwin Kellerman, *475 Jahre Wildflecken, 1524–1999* (Wildflecken 1999).

10. "History of a Polish Camp." United Nations Archives and Records Management Section (hereafter UNARM) S-0425-0006-17, p. 3.

11. Jane Perry Clark Carey, *The Role of Uprooted People in European Recovery* (Washington: National Planning Committee, 1946), 29.

12. Report of Ephraim Chase, November 1945. UNARM S-0436-0055-01.

13. Wyman, 113.

14. Fragebogen in StaaWü, Lra Bad Brückenau, 3700.

15. Wyman 69.

16. "Inspection of Camp by Brig. Gen'l Williams" and "The New Military Regime at Wildflecken", both September 17, 1945, UNARM 5-0436-0008-05

17. G. Masset to A.R.Truelson, September 14, 1945. UNARM 5-0436-0008-05

18. Thomas Leuerer, *Die Stationierung amerikanischer Streitkräfte in Deutschland: Militärge-*

meinden der US Armee seit 1945 als ziviles Element der Stationierungspolitik der Vereinigten Staaten (Würzburg: Ergon Verlag, 1997), 335.

19. Undated document appended to newspaper clippings about Galgenberg. Stadtarchiv Würzburg (StaWü), Flüchtlinge und Vertreibene.

20. "Niederschrift über Besprechung be idem Resident-Officer Mr. Sega," January 22, 1951, StaatWü, Rg. v. Ufr. 16497; "Lager Galgenberg wurde in zehn Tagen geraümt" *Main-Post*, August 29, 1951.

21. Report in StaatWü, Rg. v. Ufr. 16408, undated but 1952; Report on "The Ausländer in Bayern" from Bayer. Statistisches Landesamt, June 3, 1952; "Profile: Dr. Josef Winter," *Fränkisches Volkslbatt*, April 21, 1955.

22. Descriptions of the January 22 riot from Kathryn Hulme, *The Wild Place* (Boston: Little, Brown, 1953), 158–160 and in "Incident at K-7 (22 Jan 1947)," "Statement of Mrs. Nowicka," and "Report on Nationality Screening Incident" by K. Hulme, January 22, 1947 in UNARM 5-436-8-5, and "DP's in Camp Riot: Mob 'Soviet' Aides," *New York Times*, February 7, 1947.

23. "Flüchtlinge wieder auf eigener Scholle," *Main-Post*, 15 October 1946 and "Werberg, das Flüchtlingsdorf in der Rhön," *Main-Post*, 3 June 1950. I have left the word 'Heimat' untranslated. In German, it means something more than 'home,' generally referring to place with which an individual identifies. For a group of people who had recently fled or been pushed out of their place of origin, the question of attachment obviously had important political and cultural resonances.

24. Summary of remarks by US Resident Officer Miller, GaW, Gemeinde Wildflecken, 6.5.

25. Memo from MG Field Operations Division, 13 November 1948 and US Army Real Property Requisition, 27 May 1949. StaaWü, Lra Bad Brückenau, 4286 (III).

26. See, for instance, "Wohin mit den Bewohner des Truppenübungsplatzes?," *Fränkisches Land*, 23 April, 1952.

27. "Erhebungen" in StaatWü, Rg. v. Ufr. 17363; Minutes of meeting, 27 March 1950 in in StaatWü, Rg. v. Ufr. 17363.

28. "Beschlagnahme des ehem. Truppenübungsplatzes Wildflecken . . ." 8 November 1948, StaaWü, Lra Bad Brückenau, 4286 (III).

29. "Bayern weigert sich 21,000 Deutsche zu vertreiben," *Münchner Merkur*, 9 July 1951.

30. Letter from Landrat Baus, 10 July 1951. StaaWü, Lra Bad Brückenau, 4286 (II).

31. Shrenk, op. cit., 56–57, 115.

32. From StaatWü, Reg. v. Ufr, 17364.

Panel 5—Post-War Germany
Question and Answers
(Transcript of Presentation)

Major Tania M. Chacho, Ph.D., US Army
Adam R. Seipp, Ph.D.

Moderated by Curtis S. King, Ph.D.

Audience Member

You said in the beginning that you had indicated that the military was unwilling to help the refugees. Do you think that was more of a diplomatic issue? Did we have the resources to do that?

Dr. Seipp

Well, it is sort of a fine distinction because the displaced persons were clearly to be helped by the Army. As Major Chacho pointed out, that was something that was established early on. This left a tremendous number of refugees, an unknown number of refugees, outside of that circle. So what I am suggesting you end up with is that on the ground aid was being given more or less indirectly to these refugees, but it was the distinction between former enemies and former allies. So it was relatively easy to make the distinction in theory, but in practice it got much more difficult, particularly if you were trying to govern in a city that was 50% refugees. So 50% of the population had to be left to its own devices. I agree with you that if the decision had been made that we are going to help everybody this would have probably been a complete resource disaster, but the stark distinctions that were made on paper did not really hold up out in the field. So yes, it is a very good point.

Audience Member

This is a question for Major Chacho. Reference the United Nations (UN) and the various powers in Europe, I am wondering what was the headquarters relationship? You had SHAPE (Supreme Headquarters Allied Powers Europe) sitting on top of the western sector. Was there a UN office that plugged directly into SHAPE or worked in coordination with SHAPE and then pushed stuff down to the British sector, the French sector, and the American sector? I am just trying figure out how the command and control went for UN agencies and also how active was the UN in the Soviet sector?

MAJ Chacho

Those are great questions. I guess the short answer to what the command and control relationship looked like was that it was really a patchwork that occurred and initially since the sectors were divided up as you pointed out, the British, French, and US, there were aid agencies that plugged directly into each of those sectors. In my paper I have an interesting little vignette about the initial US reluctance to allow aid agencies into the US sector. There is a quote from the political advisor, so a State Department official, indicating, "Hey, we were concerned that once we opened the door, like the British next door had done, a lot of aid agencies would come in and start running around, and we did not really know what that command and control relationship looked like. So we were reluctant to actually do that." He had said that they had up to 600 people running around their sector . . . unbelievable. Can you imagine that? They cannot control these people who are out there administering humanitarian aid. From the UN perspective it was a little bit different and it evolved throughout. I really do not have very much information on the Soviet sector, but to say that there was even less UN presence in there than there was in the allied sectors which you know eventually emerged into West Germany. The information that I had from the National Archives just alluded to the fact that things were much worse in the Soviet sector without actually getting into specifics. I think part of that was due to the way the political situation developed . . . that information just was not readily available. So beyond just general statements there was not really much that they could do there. Then as far as the command and control relationship, it was interesting in that the UN also plugged in at different levels. There were liaisons there, but it was often just one or two people at various levels, and they found it actually more beneficial and just more operationally efficient to plug in at lower levels below SHAPE than to actually plug in at the headquarters itself because at the lower levels, you know at that time of course, the headquarters was in France, so to be able to plug in at the level of the Provincial Governors who were actually in control of the situation, they found it easier to actually facilitate the distribution of aid and that is just from what I found during this research and perhaps that there was eventually more of a relationship built. But the preponderance of information that I found was really pushed down to the operational-level, at the level that we could make a difference. There was the strategic agreement in place and then going down almost to the tactical level with that gap to use military terms, in that operational level as to where they were plugging in in-between. That might have developed beyond 1947, but quite honestly, I stopped in my research there so I am not quite certain how that relationship moved forward.

Dr. Seipp

Can I speak to this quickly? There is actually an interesting connection here between ideology and practice, and that is that the Soviets had a refreshingly direct way of dealing with the refugee problem as they tended to deal with most problems. The vast bulk of the refugees in the Soviet zone, of course, had fled from the Red Army, in advance of the Red Army. You do not really want to admit that you have created refugees or they were people who were kicked out by governments friendly to the Soviets who by definition are not refugees. So the Soviet saw themselves as having absolutely no refugee problem in Eastern Germany. There was briefly a refugee agency and some coordination with UNRA (United Nations Relief Agency) but it was designed to get people home. The refugee agency in East Germany was actually closed down in 1946. As of 1946, East Germany effectively legally had no more refugees. They had resettled people, but they had no refugees, whereas the Americans had adopted the famous "we come as conquerors, not as occupiers" mentality. Whether it was true or not, the idea was we are coming as a force of rebuilding and a source of liberation. The Soviets really did not quite have the same idea behind their occupation of Eastern Germany so that became really important in questions of relief in the rehabilitation.

Dr. King

I could briefly add, you touched upon my area of history of Russia and Soviet history that looking at it from the economic point of view they had no qualms about using the Morgenthau Plan and stripping East Germany of all the equipment that they could too.

Audience Member

You answered a little bit of what my was question was in reference to, the policy that the Soviet Union had for refugees. Were they adhering to the policy they had made with the alliance for repatriation of refugees or DPs (Displaced Persons)?

Dr. Seipp

This is a great question. I did not want to get into this part in the paper because it is a wee bit complicated, but one of the big problems, and I mean huge, was that, those of you who know your map of Eastern Europe after the war, a big chunk of Eastern Poland was annexed to the Ukraine and Poland was essentially picked up and moved westward absorbing a large chunk of what had been Eastern Germany. So overnight, millions of Poles became Ukrainians legally. The American policy had been that anyone who was a Soviet citizen could be forcibly repatriated. This fluctuated widely, but all of a sudden Poles were being reclassified as Ukrainians, and that riot I talked about in 1947 was when the Polish Army showed up and started saying, "You are now Ukrainian. You are going back." The Poles at Wildflecken rioted so this issue of who was a Soviet citizen, whether they wanted to be or not, or whether they knew it or not, was incredibly important in the question of the DPs and it was a major sticking point in trying to convince Poles to go home.

Audience Member

Thank you for two excellent papers. I have a question actually for both of you. You each mentioned using German aid organizations to help funnel some of the aid through because of their local connections and all that. Were the German aid organizations robust enough to handle this burden? I do not know what state they were in after the end of World War II and I am interested to hear. Secondly, and perhaps more importantly, to what degree in the post-war era were they trusted? Were they seen as being reliable agents who would account for this material, use it properly, and not just use it to enrich their friends or the Black Market?

MAJ Chacho

Certainly, I will address that first. It is very interesting. There was the political process that was going on, the de-nazification. So once that vetting process had gone through and you had the stamp of approval that you were a trusted German that did not have the connections with the Nazi Party that perhaps some of your neighbors did, my research showed that the relationship was actually fairly open and robust. The problem that the German aid organizations had was that they had a certain resource; they had manpower that was available and able and willing to work and be employed, but they had no further resources beyond that. So of course, what the United States did as the occupying force and in fact, the aid agencies coming in from the United States just included strict accountability procedures similar to what CARE had done in tracing each individual package. And it is interesting because CARE would spend more money trying to track a person with this package because they might have been relocated or they might have been displaced. This package would go all the way around. It might be worth five dollars and they just spent $55 trying to deliver it to the person to assure that that accountability was in place. I found that the same sort of degree of accountability was there which is not to say that there was not some siphoning off into the Black Market and there were certainly incidents of that that had come forward, but one of the benefits of having this military hierarchy in place in the military government was that there was a streamline procedure that if anything was found that was directly in violation of the military code that was in place, that was quickly shut down. The effects of that were so brutal upon that area that if they shut down the aid distribution in a certain town that could literally mean that hundreds were on the doors of starvation. There was an incentivized program to ensure that everything was distributed properly. It did not always work certainly. There were problems, but that is how I found the relationship ultimately wound up working out. I do not know if with regards to displaced persons if that was similar or not so I will defer to my colleague.

Dr. Seipp

The answer to the first question is it really depended on where you were. In Lower Franconia you had two very important and energetic refugee officers who managed to do with very limited resource a lot of good. But overall I think you can say that the efforts of the varying refugee offices really on into the 1950s until the Equalization of Burdens Law were hampered by consistent underfunding and most of their projects would end up on the face of it to failure. The second question is actually really interesting in a way that gets out some of the more, some questions that are applicable in other circumstances. The people who had the expertise to handle refugee affairs who were German tended to have been somewhere in what you might call the social work hierarchy of the Third Reich, and there were really very few parts of the Nazi dictatorship that were more affected by what you might call the criminal elements of German law, that is racial policy, than social work. If you think about it, one's existence as a human being was measured by categories that had not really existed before. So for instance, social workers were mowed down by de-nazification because these were people who had very specialized training, but they had been for twelve years enforcing some variant of German racial law. And in fact, if you look at the people who were important in refugee affairs they tended to be drawn from a kind of curious, and as far as I know absolutely unstudied, part of the vast Nazi bureaucracy, called the Reich resettlement office. Basically, if you lost your house because the German Army needed a base or if you had lost property to the German state, these are the people who resettled you and it was awfully easy to transform those people into refugee officials after the war. But it could not be social workers because social workers were among the first casualties of de-nazification. If you look at the civil affairs documents in the National Archives there is again and again complaints about we do not have any social workers we can work with because they are all in jail or they are awaiting de-nazification or there is absolutely no way they are going to get out of de-nazification. And it really points to just how difficult de-nazification and the process of political evaluation made doing refugee work.

Audience Member

(Audio too quiet. Cannot hear speaker.)

Dr. Seipp

For DPs the Americans were kind of late-comers. Belgium and France and then slightly later Australia were in dire need of agricultural and industrial labor after the war. They were more than happy to take in DPs that made it clear they did not want to stay. The Americans started taking DPs in 1948—the Displaced Persons Act—so late in the game by those standards. For ethnic German refugees it took longer. By 1949, when West Germany got sovereignty the Americans were now more willing to take in ethnic Germans as long as they had sponsors in the United States initially. And the other part of your question about getting the Germans in on it, there is a significant gap here into the early 1950s before the West German State passed a whole raft of absolutely revolutionary social legislation—the Equalization of Burdens Law that created a tax structure that was designed to help integrate refugees so this is really a ten year process just to get to the point where the West German State is saying we are now stake-holders in the permanent resettlement and economic integration of these refugees. And it actually takes decades to be accomplished.

Audience Member

Dr. Seipp answered one half of the question I was going to ask. I was going to ask of Major Chacho, the UN had only been set up that summer of 1945. Where did the personnel come from that made up this UNRA administration? Were they seconded military personnel or diplomats, and mainly, what nations were they drawn from? Thank you.

Dr. Seipp

Yes, they were not seconded military. As far as I know no one had come directly from any of the militaries except possibly the French. The largest numbers of UNRA employees, field workers, initially were French with a significant number of Britons and Americans. Most of the senior administrators were French because they tended to have more of the language skills and they could get there more easily. It is an amazing story because these people were set up, there were schools in the United States, Britain, and France and they were put in the field very fast. For instance, the woman I quoted, Kathryn Hulme, had been a tour guide in Europe before the war. During the war she worked in a shipyard in the United States. She got over there claiming to speak German and it is clear from the UN archives that she did not speak German so there was really no way for her to communicate. But if you read the reports there is a kind of blue sky effect. These people were there and they were being asked to invent everything. They more or less successfully did it and that is absolutely remarkable.

MAJ Chacho

I do not have too much more to add to that beyond them moving outside of the UN in that kind of population pool in voluntary aid organizations, the same types of things. Those who had been involved in some type of work, but often times their background was that they had organized some kind of town clothing drive and thought they could help. There was the need and if they raised their hand and said they were willing. Once that opened up in the later years, 1947 and on, they were on the next ship out there and they were landing and had discovered this type of environment. That is why it was so critical to actually have the relationship with the US military because one of the key concerns was if Germans are subsisting on about 1000 calories a day, where are aid workers going to get food? It was not like you can go down to the local market so they have to turn to the US military and work out some sort of agreement to allow them to eat in mess halls or to provide shelter and things like that.

Audience Member (Blogosphere)

We have a lot of programs dealing with JIIM (Joint, Interagency, Intergovernmental, and Multinational) and individual advancements and interagency postings. In your research did you find any place where officers that were involved with governance, humanitarian assistance, that they were rewarded later on and they became compelled or they were ostracized because they had this "stigma" of being humanitarian assistance type people? Soldiers have always filled the void, but I was wondering if there was any career impact that you were aware of?

MAJ Chacho

That is a great question, and quite honestly, I did not look at that specifically. I did not follow those officers and see what their career impact looked like. To a large degree the Army was going through a demobilization period then so people were looking to transition to civilian life and some of the officers that I have cited here found that that transition was made easier if they worked with a humanitarian aid organization. They naturally transitioned into that type of a role. But it does raise an interesting point regarding what the institution rewards or not. I cannot say with any degree of scholarly background and citation, but the impression I got from the research in the National Archives was that officers and the military at large were looking to divest themselves of this responsibility as quickly as possible, as evidenced by the Secretary of War's correspondence with the Secretary of State, "You take this. This is humanitarian. We will do military governance. We understand that we need to do that, but surely there has to be someone else out there who can step up and take on this other mission." That was the impression that I got from the institution writ large, but I do not have any specific examples as to whether it helped or harmed an officer's career which is very interesting especially given the current situation and environment we find ourselves in.

Audience Member

I am struck by the fact that you mentioned delivering Thanksgiving turkeys fifteen years after the war and we still have Palestinian refugees being supervised by the United Nations 60 years after 1948 so I do not think we should find fifteen years of UN refugee activities being unusual.

Dr. Seipp

Sure. And part of what happened in the 1950s, as the Cold War heated up, refugees were being publicized because the new refugees were coming from East Germany. Before 1961, when the Berlin Wall was built there was a wave of migration from the east and it obviously benefited both the Americans and the West Germans to highlight that people were leaving East Germany. In an odd way the second half of the 1950s was a great time for refugees in terms of public attention, but the camps would remain in constant operation until the early 1960s when, frankly, the Berlin Wall helped close off the taps and the refugees that were in the camps could then be parceled out, but yes, I think that is a great point.

Panel 6—The Interagency Process in Asia
(Submitted Paper)

Post-Cold War Interagency Process in East Timor

by

Major Eric M. Nager, US Army
United States Army, Pacific

Acknowledgments

In completing this project I wish to gratefully acknowledge the support of the Public Affairs Office, US Army, Pacific; especially that of Mr. David Hilkert, the Command Historian, who helped source, proof and edit my work. His interview with Colonel Uson inspired me to write this paper. I also want to thank Mr. Ken Gott of the Combat Studies Institute for his encouragement and pointing me in the right direction. Dr. Albert Palazzo of the Australian Land Warfare Studies Centre and Wing Commander Steve Kennedy of the Royal Australian Air Force provided Australian sources. Lastly and most importantly, I wish to thank my family and my civilian employer for their support of my military career, allowing me to be "twice the citizen."

Dedication

In memory of my Great Uncle, Howard McCormick, who served in the US Army, Pacific, during World War II.

Introduction

Since becoming the sole super power at the close of the Cold War, the United States is accustomed to taking the lead role in international military coalitions. As such, much interagency coordination is required between the US Army and US governmental agencies. But what of small, regional conflicts that do not directly impact US national security? In such a case, a US ally might take the lead in forming a coalition and the US Army, in a supporting role, then faces coordination with other external agencies. This type of "extra-agency" process is exactly what took place in late 1999 during operations in East Timor and it might well portend future operations with their attendant issues.

Background

As a brief history, the Island of Timor lies about 400 miles north of Darwin, Australia. Portugal colonized the eastern half in the early 1500's while the Dutch colonized the western portion along with the surrounding islands comprising present day Indonesia. From 1515-1975 East Timor was known as Portuguese Timor, a relatively neglected outpost. When a military coup in Portugal in 1974 forced them to abandon their colonies, it precipitated a short civil conflict in East Timor, culminating in a claim to independence on 28 November 1975.[1]

Before the world had a chance to recognize the prospective new nation, Indonesia invaded on 7 December 1975, citing concerns about communism and regional unrest. Indonesia annexed East Timor the following year over international protest and ruled with a heavy hand. Upon taking office in 1998, Indonesian President Habibe agreed to hold a referendum on self-determination in East Timor. That vote took place on 30 August 1999. The referendum passed overwhelmingly with over 98% voter turnout, yet it sparked a violent rampage by local militias backed by the Indonesian military that displaced over 300,000 civilians and destroyed East Timor's infrastructure.[2] The cry went out for international intervention. Thanks to a successful response, East Timor formally and finally declared independence on 20 May 2002.

The Coalition

Because of its proximity and good relationship with the Indonesian government, Australia was the natural choice to lead such a coalition. To be sure, coalitions have been the rule rather than the exception in military history. However, it is one thing to be a member of one and an entirely different matter leading one, for they require political, economic, and social objective coordination, as well as agreed strategic plans.[3] In this case there was not much time to act.

UN Security Council Resolution 1264 authorized forces to assist in the maintenance of peace surrounding the August referendum (see Addendum A). When violence broke out, the International Force East Timor (INTERFET) under Australian Major General Peter Cosgrove deployed on 20 September. Operation WARDEN was the first phase, whereby peace was restored, and the second was Operation STABILISE, maintaining peace. A total of 11,000 troops from 22 nations participated until 23 February 2000 when INTERFET handed the mission to the United Nations Transitional Authority for East Timor (UNTAET).[4]

Participating nations funded the INTERFET mission as did the United Nations through a substantial donation from Japan. This was in lieu of troops. Aside from Australia and the US the participating nations were Brazil, Canada, Denmark, Egypt, Fiji, France, Germany, Ireland, Italy, Jordan, Kenya, Malaysia, Mozambique, New Zealand, Norway, Philippines, South Korea, Singapore, Sweden, Thailand, and the United Kingdom.[5]

The INTERFET rules of engagement were threefold, simple and strong. The first goal was to restore peace and security in East Timor. Second was to protect and support the United Nations Mission to East Timor (UNAMET) in carrying out its tasks. Third was, within force capabilities, to facilitate humanitarian assistance operations.[6] Because of the scale of destruction wreaked by the militia following the referendum, this was a considerable task.

Australian Perspective

The 1998 Asian currency crisis sparked a government-directed internal review within the Australian Department of Defence. The situation plunged Indonesia, their neighbor to the north, into a financial crisis at the same time as they were transitioning to democracy and trying to stamp out long-running civil conflicts in Aceh, Irian Jaya, and East Timor. Geographically, Indonesia spans about 90% of Australian trade routes and has ten times the population so a stable Indonesia was clearly in Australian interests. Their defense planners correctly predicted that at least one of these three hot spots would require outside intervention in the near future and began retooling their military, which had not been deployed in any numbers since Vietnam.[7]

As the lead in the coalition, the Australian government had to devise its own interagency process. Their answer was to bring together representatives from the Department of Defence, Foreign Affairs, Trade, the Attorney General, as well as the Prime Minister. The Department of Defence formed the East Timor Policy Unit, and Foreign Affairs formed the East Timor Crisis Centre, both of which advised the central government. Seeing that there was limited interagency coordination below the level of cabinet minister, the government established in interdepartmental East Timor Policy Group to, according to Intelligence Officer John Blaxland, "help develop an integrated policy approach to the issues at hand, with senior officers from key government departments involved."[8] (See Addendum B for the current organizational chart of the Australian Department of Defence.)

This was all a new experience for the Australian government as it would have been for almost any government. INTERFET represented the first UN-mandated peace operation outside of Africa and not headed by the United States. Yet having participated in international coalitions before, the Australians recognized that a multilateral response garners greater international support.[9] How would they choose to shape their response in taking the coalition lead?

An interesting contrast is how the Australians perceived the US approach to military operations, especially in light of the US perspective of INTERFET that is shared later. First, they perceived the Americans as prepared to accept casualties. This stemmed from working with them in Vietnam. Second, they saw US tactics as "heavy handed." Third, they saw the US goal as achieving success through destroying the enemy's resistance, usually resulting in collateral damage. The Australian goal was to attain their political and military goals through "bending their opponent's will," as opposed to brute force.[10]

This approach the Australians call Information Operations. The term essentially means psychological operations backed by a competent military force in order to get the opposition in a "disadvantageous position." The goal is psychological rather than physical by trying to convey the idea to the opponent that "continued resistance is irrational."[11] As a self-described mid-level power, it seems a natural conclusion that Australian military planners arrived at such an approach. Interestingly, such an approach involves civil-military affairs, in which the US has great expertise.

The nature of Information Operations is an interagency approach since it brings the diplomat and the soldier together. In the United States, it is the equivalent of pairing the Department of Defense with the Department of State. Again according to Blaxland, "Australian defence planners and diplomats had to work together to reconcile military plans with foreign policy."[12] How they would coordinate with other members of the coalition remained to be seen.

What gave INTERFET teeth was the fact it was a peace-enforcement and not a peace keeping mission, under Chapter VII of the UN Charter. It allowed for "coercive" military use, yet there were 15,000 Indonesian troops on the island backing the local militia, versus about 5,000 ground troops for INTERFET. General Cosgrove boldly validated the Information Operations approach with a decisive show of force, landing the troops quickly and keeping supporting ships close to shore. He correctly identified that the militia would crumble once the Indonesian troops left. The latter had trained with the Australians, knew their capabilities, and knew they were serious about the mission. As a consequence, they withdrew within two weeks and the INTERFET mission reverted to peace keeping.[13] Sometimes extra-agency can mean coordinating with a potential adversary through joint training exercises.

US Perspective

The US mission within INTERFET was simply to provide the Australians the support they requested. This included four areas: communications, intelligence, civil-military operations, to include dealing with non-governmental organizations (NGO's), and heavy lift helicopter capacity. The US Army took responsibility for the first three. Since the US was not the lead nation in the coalition, it affected their assessments as to what equipment to bring for the mission since they had to wait for the Australian requests and the Australians could not anticipate everything they needed ahead of time.[14] From the start there was a different perspective on this operation. At the diplomatic level, the US suspended military to military relations with Indonesia during the operation.

The US Forces INTERFET (USFI) were commanded by Major General John Castellaw, USMC, and headquartered in Darwin, Australia. The US Forces East Timor

(USFET) were commanded by Colonel Randolph Strong and located in Dili, the capital of East Timor. At the height of operations, the US Army had 230 troops on the ground. The largest of the three missions was communications and USFI formed Task Force *Thunderbird*, made up of an element from the 11th Signal Brigade. Soldiers were spread in six locations across East Timor, with an additional element in Darwin to support USFI headquarters. The Australians provided force protection.[15]

Obviously the primary extra-agency coordination was with the Australian Department of Defense through their military. This extended to the intelligence mission as well. Six US Army personnel comprised Trojan Spirit II, downloading classified information from satellite. Another eight conducted counter intelligence. The remainder of the US intelligence personnel was integrated into the INTERFET C^2 staff.[16] The Australian military has a similar staff structure to that of the US.

The mission for the civil-military operations was to coordinate and facilitate the relationship between INTERFET, UN agencies and NGO's. The CMOC team consisted of 15 personnel from four nations (see Addendum C for CMOC organizational chart) with the US in the lead.[17] Not only did the US personnel have to adapt to Australian leadership, they had to integrate soldiers from other nations into the team. Expertise on the team included specialists in engineering, law, water purification, and power generation.

The US Civil Affairs teams worked primarily with humanitarian assistance, but they were largely on their own. There was an Information Operations cell at INTERFET Headquarters in Darwin, but it was ad hoc in nature, so there was little integration of Civil Affairs. Within the Australian military, Civil Affairs is a secondary assignment for artillery battery commanders and their forward observers.[18] This is why the US troops were invited to share their expertise, but it also made the mission more challenging.

The arrangement also made for odd working relationships relative to rank. The Australian battery commander in the CMOC was a major, and the senior US officer was a lieutenant colonel. Since the US Civil Affairs mission was coordination and not command and control, the US CMOC commander supported the Australian major at the regular INTERFET briefs. The US CMOC commander reported directly to Colonel Strong, the USFET commander.[19]

In his role as USFET commander, Colonel Strong wore several figurative hats. Perhaps his primary role was a diplomatic one as he represented General Castellaw at daily meetings with General Cosgrove to sort out issues with the US forces and how they were accomplishing the mission.[20] US soldiers did not strictly deal with the Australian Army, however. Their extra-agency coordination ran virtually the entire gamut of the Australian Department of Defence.

For example, when Colonel Strong and his troops arrived in Darwin prior to deploying to East Timor, the Australian Northern Command (NorCom) welcomed them. They provided facilities for US soldiers in newly renovated barracks, and the US unit had the honor of being the first organization to occupy a recently constructed bunker as a Command Post (CP). Meanwhile, the Royal Australian Air Force (RAAF) provided Post Exchange (PX), dining, barber, and other services on their nearby base.[21]

The assignment of liaison officers was a key military to military technique that facilitated good communication between key USFI and INTERFET staff. One aspect that made this easier was the years of experience conducting joint exercises between the US and Australian militaries. The US Army, Pacific conducts about 35 exercises per year as part of their Expanded Relations Program (ERP) with other Pacific nations. One example was Crocodile '99 that took place at the same time as the East Timor operation. Within the operation, five Australian signal officers had graduated from US programs, were familiar with US equipment, and this made them a perfect fit to work together with Task Force *Thunderbird*.[22]

At times alternate methods had to be found to accomplish the mission, as was the case with the Australian Navy. During the sealift of signal equipment from Darwin to East Timor, Her Majesty's Australian Ship (HMAS) *Jervis Bay*, a catamaran, was assigned but did not have the capacity to carry the largest items. Since they were also too heavy for available aircraft, other container ships were found.[23] Of course the extra-agency extended beyond the Australian Department of Defence.

The Australian government supplied all contributing nations with a computer in order to access the INTERFET Local Area Network (I-LAN) set up for the operation. This included over 20 computers along with hub, router, cable and other necessary equipment. The US AAR found that this was an unnecessary drain on already short resources, and that in the future participating coalition members can supply their own computers.[24] Nevertheless, this issue highlights the difference for US planners when they are not leading a coalition.

Another area of coordination was with the Australian medical community. Those concerned with mental health issues wanted to learn more from their American counterparts about the US Family Support system for deployed soldiers. At the time Australia had an informal program only, and it became apparent to US participants that other armies around the world are recognizing the need for such a support network.[25] Even when not in the lead, US coalition participation can be a teaching as well as a learning tool.

Extra-agency coordination is not merely confined to the government of the coalition leader. When Lieutenant Colonel Uson, the CMOC Commander, arrived in Dili on 3 October, his first meeting was with the UN Office for the Coordination of Humanitarian Affairs (OCHA). For the first time, the UN Secretary General designated OCHA as the lead NGO coordination agency.[26] Of course even when the US is in the lead of a coalition, such extra-agency coordination is necessary, but not being in the lead adds an extra layer.

The same holds true for NGO's, which are agencies unto themselves. One example of an issue that arose during Operation STABILISE came in the form of a letter from James Patterson of Medecins Sans Frontieres (Doctors Without Borders—see Addendum D). As of 19 October, the Australian 3rd Combat Engineer Regiment had moved into a hospital in Maliana, and the letter requested they move so the hospital would be seen as "neutral" by the local populace.[27] The letter came to the US officer in charge of the CMOC.

Such extra-agency coordination can also include entities within the nation in which the operation takes place. For example the CNRT was identified before the operation as an umbrella organization of Timorese communities. They were seen as a political organization by the UN and various NGO's who had to maintain their neutrality and therefore did not deal with them. Yet, the CNRT cooperated with the civil-military operations and collected over 30,000 names in the town of Christo Rei within a span of two weeks for the purpose of building a census.[28]

Perhaps no issue is more important to US planners, especially if they are not leading a coalition, than force protection and East Timor was no exception. Different nations have different standards for force protection, and these must be reconciled, sometimes under the stress of operations. From the US Commanding General briefing on 18 September, there were two components to US force protection. First, security for US forces was a US responsibility unless otherwise stated. Second, US personnel were not to go beyond the wire of secured areas unless the situation changed.[29] From a Civil Affairs perspective, the latter guidance presented practical problems.

According to one US member of the CMOC, much information had to be gathered from various NGO's and the local populace to determine conditions on the ground. However, the travel restrictions inhibited the Civil Affairs team travel such that they were reduced to collecting information from the Australian Civil-Military Operations teams that were freer to travel.[30] The irony is that these are the same teams that the US personnel were sent to augment!

Force protection has many aspects, the most obvious of which is physical. As noted, standards varied by nation and in some cases within different services of the same nation. As a condition of US support when carrying out their missions, US personnel insisted on US force protection standards. While not every coalition member agreed with these standards, they agreed to comply. It appears this issue was negotiated on the ground and not part of any extra-agency coordination. Until Dili was deemed secured, US soldiers worked ashore during the day and slept on ships anchored in the harbor by night.[31]

Force protection can also mean medical precautions, and again the US led the coalition in high standards. In one case flooding from a monsoon wiped out the Australian mess facility. Deeming the area to be unsanitary until properly cleaned up, US soldiers went to consuming MRE's while their allies reportedly suffered from gastro-intestinal problems. US soldiers also deployed with malaria pills, mosquito netting and other items to combat tropical diseases. During the exercise only one US soldier came down with such an illness while there were 191 other reported cases among coalition personnel.[32] Here it is only fair to note that in some of the AAR comments, US soldiers were at times bewildered by the severity of the force protection restrictions, but this hearkens back to the more traditional interagency process in conjunction with the US Department of Defense.

Overall Operation STABILISE functioned very smoothly, especially given the short time span under which the participants had to deploy. If there was one area of friction among the allies, it was the redeployment schedule. The Australians wanted the US to stay longer than their December pull out date. From a US perspective, the main criterion for exiting was that all non-US personnel were trained and could handle the functions performed by the US.[33] Even so, US soldiers had 179 days on their orders as a precaution for a longer operation. As it was, UNTAET took over in February 2000.

Australian Lessons Learned

As first time coalition leaders, the Australians learned many lessons about the interagency process. Of course these lessons can also apply to other nations. According to Dr. Alan Ryan, a Research Fellow at the Australian Land Warfare Studies Centre, "If Australia is to make an effective coalition contribution, foreign policy and defence planners need to work together to develop a clear-eyed appreciation of Australia's most likely partners; ensure adequate preparation and training for combined operations; and undertake a thorough re-evaluation of Australia's force structure."[34] The same applies for US extra-agency planning.

A specific lesson learned from the INTERFET operation was that the Australian Defence Headquarters developed a model to look at the requirements of future warfare. This model includes seven elements and was written about by Australian Brigadier General Steve Ayling and Ms. Sarah Guise of the Australian Department of Defence. The seven components are command and control; intelligence, surveillance and reconnaissance; tailored effects; force projection; force protection; force sustainment; and force generation.[35] Only a couple of these elements will be examined here as they pertain to extra-agency coordination.

For command and control, the Australian model emphasizes the importance of the national commander/force commander relationship. Each coalition participant nation should agree to predetermined limits, yet allow each nation to maintain control of its own troops. This is significant since it differs from the NATO model of unified command that the US is more used to operating under. The Australian argument is that their model is more appropriate for the Asia-Pacific arena since there are so many nations with differing political objectives.[36] And the more nations that participate lend more legitimacy to a given operation.

A main objective of the Australian model is to move away from the ad hoc nature of the INTERFET operation to a more permanent regional security framework. This includes more sharing of intelligence among coalition partners, staged deployment, if possible, versus direct deployment, and negotiating logistic and force generation requirements ahead of time.[37] This implies much extra-agency coordination with allies in peacetime.

Much self-examination came out of INTERFET on the Australian side. According to an Australian Defence White Paper, they have identified eight conditions for participating in future coalitions. First is the effect on Australian strategic, political and humanitarian interests, in addition to those of their allies. Second is a clear mandate, goals and an end point. Third is the achievability of the mission based on the given resources and circumstances. Fourth is the degree of international support. Fifth is the cost, including the effects on the Australian Defence Force (ADF) for other tasks. Sixth is the training benefit for the ADF. Seventh is the risk to personnel. Eighth are the potential consequences for Australian interests and relations.[38] These are criteria the US can apply as well.

US Lessons Learned

The US conducted its own analysis of Operation STABILISE. One reason for the success was that Australia is such a strong and capable US ally. According to David Dickens, director of the Centre for Strategic Studies at Victoria University in Wellington, New Zealand, the Australian government has a strategic decision-making capacity: "Canberra was capable of making strategic decisions, gathering and interpreting intelligence and staffing military and diplomatic organizations with experienced professionals."[39] In other words the agencies of the Australian government are internally efficient.

Another important factor according to Dickens is that the Australians had good defense relationships with other nations, especially the US. They knew whom to call in Washington and what to ask in terms of strategic and tactical intelligence, protection of communication sea lines, and strategic lift capability.[40] This made the US more willing to work with the Australians, to include assets outside of what the Army provided, such as the Aegis-class cruiser USS *Mobile Bay*. This ship was vital to the INTERFET show of force during the landing in East Timor.

From the American side it was a great advantage that they could seamlessly "plug into" the Australian intelligence system. Since Australia's military is fully compatible with NATO, it was a comfortable match. However, this does raise the question as to whether the US can realistically only support nations that feature such interoperability.[41] If they cannot, it might limit future operations.

There is little doubt that there is a wide technological gap from an operational standpoint between the US military and most other militaries around the world, so the question remains as to with how many nations the US can conduct future operations of a similar nature to East Timor. Dickens performed such an analysis as of 2002, dividing potential US coalition allies into various tiers. Only two nations, Australia and Great Britain, made it into the top tier. Does this mean that East Timor was a one-time rare event? His recommendations to increase the number of top tier nations is for the US to maintain close military cooperation with like-minded nations and to develop and nourish more such relationships.[42]

Another lesson for US military planners is opportunities such as that presented in East Timor represent great potential for non-combat arms Army officers. Major General William Ross, then the Commanding General of the US Army Signal Corps, called Colonel Strong's assignment as USFET commander a "history making event" that could lead to more such opportunities.[43] All Army officers, regardless of branch, must be prepared.

From a troop on the ground perspective, perhaps Colonel Strong's own words provide the best example of US lessons learned. According to Strong, "I think we learned the lesson that it is okay to be in a support role to an ally. The US doesn't always have to be in the lead."[44] With small, regional conflicts in various parts of the world seemingly more the norm than the exception, the US might have more opportunities to test Colonel Strong's words.

Analysis

In September of 1999, there was no Operation IRAQI FREEDOM to occupy the bulk of US Army planning and assets, yet the Army only sent a couple of hundred soldiers to East Timor. Vital US interests were not at stake, yet there was value in participating in and supporting the multinational effort approved by the UN in order to further national goals and objectives in a time of peace. With the US Army now focused on the Middle East in time of war, planning for similar operations takes on greater importance.

Normally when the US leads a coalition, a close working partnership is necessary between the Army and US government agencies in attaining national goals. This has happened often enough in recent years that a good working relationship and procedures are fairly firmly in place and institutionalized in such Army officer training as the Command and General Staff College. When such coalitions are built, primary coordination with allied government agencies is done through the US government and is thus invisible to the Army. However in the case of a coalition led by another nation, direct extra-agency coordination with allied government agencies and the Army comes into effect.

The case of East Timor, while not common enough to establish trends, nevertheless identified at least six issues associated with Army operations that require such cooperation. The first of these is the possibility that the lead nation in the coalition might be taking the lead for the first time, as was the case with Australia. The Australian government did an admirable job of bringing together an interagency task force of various cabinet ministers, yet by their own admission there was limited coordination below that level. While the Australians performed exceptional self-analysis of their performance after the fact in order to institutionalize some of their procedures, the next coalition might well be led by another nation.

The second issue, and key to being prepared for any deployment, is a needs assessment prior to the operation. When the US takes the lead, that process is relatively clear. In the case of East Timor, US Army planners had to wait for the Australian needs assessment, communicated through their Department of Defence. Again, since they were new to leading such an operation, the needs assessment took time to compile and the short timeframe in which to deploy exacerbated the problem for US planners. This presented dilemmas as to what equipment to bring.

The third issue raised by Operation STABILISE is the interoperability and capability of the lead nation's military. Fortunately in this case the Australian military is extremely capable and operates on the NATO model. Through many joint exercises and training conducted prior to East Timor, the US and Australian Armies had a good working relationship and were familiar with each other's strengths. This allowed for Australia to invite US Army Civil Affairs soldiers in order to augment an area in which they are not as organically strong. However as pointed out by David Dickens, the Australian army is one of the very few in the world that is as compatible with the US, and this points all the more to the importance within the US Army of continuing to conduct similar training and nurturing relationships with other potential coalition leader nations.

Whether we recognize it or not, even when leading a coalition, the US Army already performs extra-agency coordination with such entities as the UN, NGO's and agencies within the host country of a given operation. The difference and key fourth issue from the East Timor operation is that such extra-agency coordination is no longer directly between those agencies and the US Army, but the US Army becomes a conduit for coordination with the agencies of the lead nation. This is an important distinction and it might take more such operations for US planners to get used to and feel comfortable with it.

Even in coalitions that the US leads, force protection has been and will continue to be a major issue. This includes everything from physical protection such as body armor to sanitation and medical procedures. Standards will continue to vary by nation, but one critical difference in operations where the US is not the lead is that political decision makers might be even less inclined to accept casualties. There is no question that force protection issues negatively impacted the mission in East Timor that in all other regards was a great success. US soldiers confined behind the wire could not always get out and gather the information they needed. This will require better internal coordination in the future since the force protection standards were not imposed by the Australians.

The final issue is that of redeployment. In East Timor the US Army and Australian government had different ideas about when the operation was complete. When the US criteria for mission accomplishment were fulfilled, the Army had to negotiate with their Australian counterparts to let them go home. Ideally in future operations, this will be worked out ahead of time.

With the US Army stretched by current operations, now is the time to cultivate allies that can take the lead in future operations. Such cultivation will get the Army used to working with governmental agencies of other nations. This includes but is not limited to needs assessment, military capability and interoperability, extra-agency coordination with outside entities, force protection, and redeployment. The Army will know that it has arrived as an expert in the extra-agency process when that process can be moved from an ad hoc system as it was in East Timor, to an institutionalized one as it is when the US takes the lead in multinational coalitions.

Notes

1. US Department of State Website, Bureau of East Asian & Pacific Affairs, http://www.state.gov/r/pa/ei/bgn/35878.htm, (accessed 19 May 2008).

2. Ibid.

3. Alan Ryan, "From Desert Storm to East Timor: Australia, the Asia-Pacific, and 'New Age' Coalition Operations," Canberra, LWSC 2000, Study Paper 302, 14.

4. John Blaxland, "Information-era Manoeuvre: The Australian-led Mission to East Timor," Canberra, LWSC 2002, Working Paper 118, 2-3.

5. Steve Ayling and Sarah Guise, "UNTAC and INTERFET—A Comparative Analysis," *Institute for National Strategic Studies Pacific Symposium Papers* (March 2001), http://www.ndu.edu/inss/symposia/pacific2001/aylingpaper.htm (accessed 20 May 2008), 4.

6. David Dickens, "The United Nations in East Timor: Intervention at the Military Operations Level," *Contemporary Southeast Asia: A Journal of International and Strategic Affairs*, August 2001, Vol. 23, Issue 2, 216.

7. Eric Nager, Interview with Australian Wing Commander Steve Kennedy, RAAF, 29 May 2008.

8. Blaxland, 20, 21.

9. Ryan, 2, 8.

10. Blaxland, 9, 10, 18.

11. Ibid., 19.

12. Ibid., 22.

13. Dickens, 4, 5.

14. Bill McPherson, "The East Timor Tapes: An Interview with COL Randolph P. Strong, Commander US Forces East Timor (October-December 1999)," *Pacific Voice*, Special Edition, Spring 2000, 8-9.

15. Craig Collier, "A New Way to Wage Peace: US Support to Operation STABILISE," *Military Review*, January-February 2001, Vol. 81, Issue 1, 4.

16. Ibid.

17. Operation Stabilize CMOC AAR, Tab H.

18. Blaxland, 43.

19. David Hilkert interview with Colonel Jose Uson, 27 July 2006.

20. McPherson, 8.

21. Ibid., 10.

22. Collier, 7-8.

23. McPherson, 10.

24. AAR, Tab B.

25. Ibid., Tab D.

26. Ibid., Tab M.

27. Ibid., Tab U.

28. Ibid.

29. Ibid., Tab M.

30. Ibid., Tab D.

31. Collier, 6.

32. Ibid.

33. Hilkert, Uson interview and AAR, Tab I.

34. Ryan, 11.

35. Ayling & Guise, 5-6.

36. Ibid., 6.

37. Ibid., 7-9.

38. Ibid., 3.

39. David Dickens, "Can East Timor Be a Blueprint for Burden Sharing?" Washington Quarterly, Summer 2002, Vol. 25, Issue 3, 30.

40. Ibid., 31.

41. Ibid., 31-33.

42. Ibid., 39.

43. McPherson, 13.

44. Ibid., 14-15.

Bibliography

Ayling, Steve and Guise, Sarah. "UNTAC and INTERFET—A Comparative Analysis." *Institute for National Strategic Studies Pacific Symposium Papers* (March 2001), http://www.ndu.edu/inss/symposia/pacific2001/aylingpaper.htm (accessed 20 May 2008).

Blaxland, John. "Information-era Manoeuvre: The Australian-led Mission to East Timor." Canberra, Land Warfare Studies Centre, 2002, Working Paper 118.

Collier, Craig. "A New Way to Wage Peace: US Support to Operation STABILISE."

Combined Arms Center, *Military Review*, January-February 2001, Vol. 81, Issue 1, pp. 2-9.

Dickens, David. "Can East Timor Be a Blueprint for Burden Sharing?" *The Washington Quarterly*, Summer 2002, Vol. 25, Issue 3, pp. 29-40.

Dickens, David. "The United Nations in East Timor: Intervention at the Military Operational Level." *Contemporary Southeast Asia: A Journal of International and Strategic Affairs*, August 2001, Vol. 23, Issue 2, pp. 213-232.

Hilkert, David. USARPAC Command Historian Interview with COL Jose Uson, 27 July 2006, transcribed by MAJ Eric Nager.

McPherson, Bill. "The East Timor Tapes: An Interview with COL Randolph P. Strong, Commander, US Forces East Timor (October-December 1999)." *Pacific Voice*, Special Edition, Spring 2000, pp. 6-16.

Nager, Eric. USARPAC Deputy Historian Interview with Wing Commander Steve Kennedy, Intelligence Officer, Royal Australian Air Force, 29 May 2008.

Operation Stabilize Civil-Military Operations Center (CMOC) After Action Review (AAR).

Ryan, Alan. "From Desert Storm to East Timor: Australia, The Asia-Pacific, and 'New Age' Coalition Operations." Canberra, Land Warfare Studies Centre, 2000, Study Paper 302.

US Department of State Website, Bureau of East Asian and Pacific Affairs, http://www.state.gov/r/pa/ei/bgn/35878.htm, April 2008, accessed 19 May 2008.

UNITED NATIONS

S

Security Council

Distr.
GENERAL

S/RES/1246 (1999)
11 June 1999

RESOLUTION 1246 (1999)

Adopted by the Security Council at its 4013th meeting,
on 11 June 1999

The Security Council,

Recalling its previous resolutions on the situation in East Timor, in particular resolution 1236 (1999) of 7 May 1999,

Recalling the Agreement between Indonesia and Portugal on the question of East Timor of 5 May 1999 (the General Agreement) and the Agreements between the United Nations and the Governments of Indonesia and Portugal of the same date regarding the modalities for the popular consultation of the East Timorese through a direct ballot and regarding security arrangements (the Security Agreement) (S/1999/513, annexes I-III),

Welcoming the report of the Secretary-General on the Question of East Timor of 22 May 1999 (S/1999/595),

Noting with concern the assessment of the Secretary-General contained in that report that the security situation in East Timor remains "extremely tense and volatile",

Taking note of the pressing need for reconciliation between the various competing factions within East Timor,

Welcoming the fruitful cooperation of the Government of Indonesia and the local authorities in East Timor with the United Nations,

Taking note of the letter from the Permanent Representative of Portugal to the United Nations to the President of the Security Council of 7 June 1999 (S/1999/652),

Welcoming the conclusion of consultations between the Government of Indonesia and the United Nations on the deployment of military liaison officers within the mission established by this resolution,

99-17413 (E)

/...

S/RES/1246 (1999)
Page 2

Bearing in mind the sustained efforts of the Governments of Indonesia and Portugal since July 1983, through the good offices of the Secretary-General, to find a just, comprehensive and internationally acceptable solution to the question of East Timor,

Welcoming the appointment of the Special Representative of the Secretary-General for the East Timor Popular Consultation, and *reiterating* its support for the efforts of the Personal Representative of the Secretary-General for East Timor,

1.   *Decides* to establish until 31 August 1999 the United Nations Mission in East Timor (UNAMET) to organize and conduct a popular consultation, scheduled for 8 August 1999, on the basis of a direct, secret and universal ballot, in order to ascertain whether the East Timorese people accept the proposed constitutional framework providing for a special autonomy for East Timor within the unitary Republic of Indonesia or reject the proposed special autonomy for East Timor, leading to East Timor's separation from Indonesia, in accordance with the General Agreement and to enable the Secretary-General to discharge his responsibility under paragraph 3 of the Security Agreement;

2.   *Authorizes* until 31 August 1999 the deployment within UNAMET of up to 280 civilian police officers to act as advisers to the Indonesian Police in the discharge of their duties and, at the time of the consultation, to supervise the escort of ballot papers and boxes to and from the polling sites;

3.   *Authorizes* until 31 August 1999 the deployment within UNAMET of 50 military liaison officers to maintain contact with the Indonesian Armed Forces in order to allow the Secretary-General to discharge his responsibilities under the General Agreement and the Security Agreement;

4.   *Endorses* the Secretary-General's proposal that UNAMET should also incorporate the following components:

   (a) a political component responsible for monitoring the fairness of the political environment, for ensuring the freedom of all political and other non-governmental organizations to carry out their activities freely and for monitoring and advising the Special Representative on all matters with political implications,

   (b) an electoral component responsible for all activities related to registration and voting,

   (c) an information component responsible for explaining to the East Timorese people, in an objective and impartial manner without prejudice to any position or outcome, the terms of the General Agreement and the proposed autonomy framework, for providing information on the process and procedure of the vote and for explaining the implications of a vote in favour or against the proposal;

5.   *Notes* the intention of the Governments of Indonesia and Portugal to send an equal number of representatives to observe all the operational phases of the consultation process both inside and outside East Timor;

/...

6.   Welcomes the intention of the Secretary-General to conclude with the Government of Indonesia, as soon as possible, a status-of-mission agreement and urges the early conclusion of negotiations with a view to the full and timely deployment of UNAMET;

7.   Calls upon all parties to cooperate with UNAMET in the implementation of its mandate, and to ensure the security and freedom of movement of its staff in carrying out that mandate in all areas of East Timor;

8.   Approves the modalities for the implementation of the popular consultation process scheduled for 8 August 1999 as set out in paragraphs 15 to 18 of the report of the Secretary-General of 22 May 1999;

9.   Stresses once again the responsibility of the Government of Indonesia to maintain peace and security in East Timor, in particular in the present security situation referred to in the report of the Secretary-General, in order to ensure that the popular consultation is carried out in a fair and peaceful way and in an atmosphere free of intimidation, violence or interference from any side and to ensure the safety and security of United Nations and other international staff and observers in East Timor;

10.   Welcomes in this regard the decision taken by the Government of Indonesia to establish a ministerial team to monitor and ensure the security of the popular consultation in accordance with Article 3 of the General Agreement and paragraph 1 of the Security Agreement;

11.   Condemns all acts of violence from whatever quarter and calls for an end to such acts and the laying down of arms by all armed groups in East Timor, for the necessary steps to achieve disarmament and for further steps in order to ensure a secure environment devoid of violence or other forms of intimidation, which is a prerequisite for the holding of a free and fair ballot in East Timor;

12.   Requests all parties to ensure that conditions exist for the comprehensive implementation of the popular consultation, with the full participation of the East Timorese people;

13.   Urges that every effort be made to make the Commission on Peace and Stability operative, and in particular stresses the need for the Indonesian authorities to provide security and personal protection for members of the Commission in cooperation with UNAMET;

14.   Reiterates its request to the Secretary-General to keep the Security Council closely informed of the situation, and to continue to report to it every fourteen days on the implementation of its resolutions and of the Tripartite Agreements and on the security situation in East Timor;

15.   Decides to remain seized of the matter.

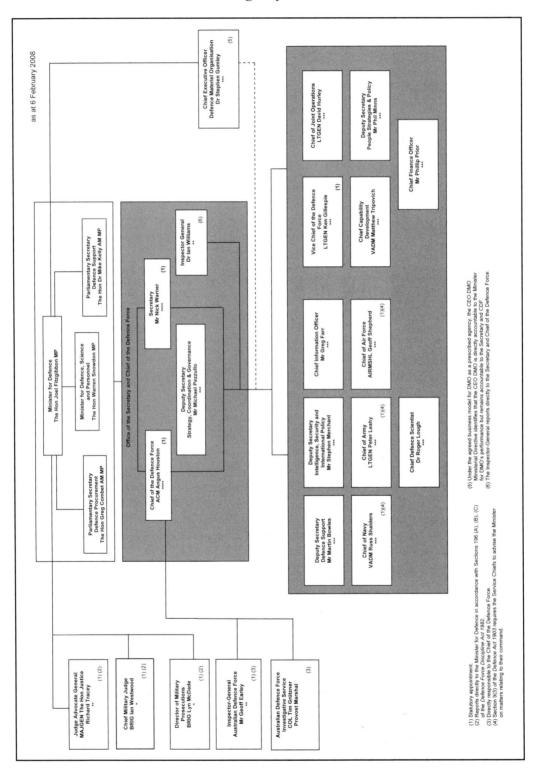

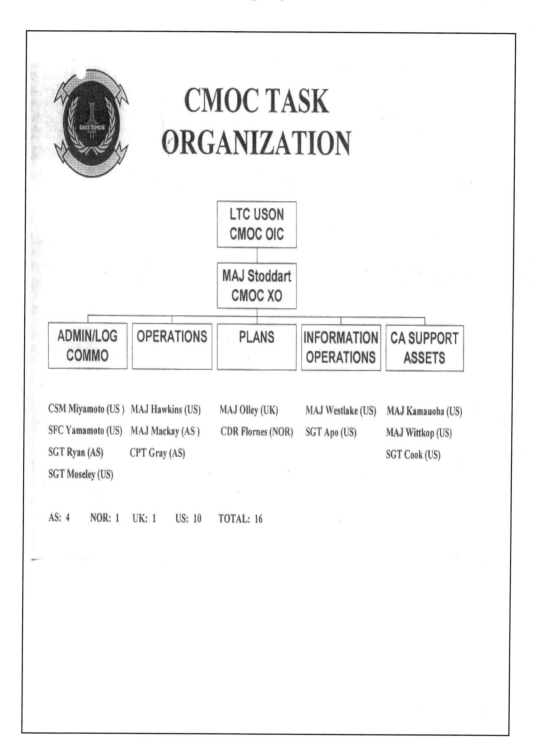

CMOC TASK ORGANIZATION

LTC USON
CMOC OIC

MAJ Stoddart
CMOC XO

ADMIN/LOG COMMO	OPERATIONS	PLANS	INFORMATION OPERATIONS	CA SUPPORT ASSETS

CSM Miyamoto (US)   MAJ Hawkins (US)   MAJ Olley (UK)   MAJ Westlake (US)   MAJ Kamauoha (US)

SFC Yamamoto (US)   MAJ Mackay (AS)   CDR Flornes (NOR)   SGT Apo (US)   MAJ Wittkop (US)

SGT Ryan (AS)   CPT Gray (AS)   SGT Cook (US)

SGT Moseley (US)

AS: 4   NOR: 1   UK: 1   US: 10   TOTAL: 16

EAST TIMOR

MEDECINS SANS FRONTIERES
DOCTORS WITHOUT BORDERS

Dili, 19 October 1999

Officer in Charge
CMOC
Interfet
Dili

Dear Sir,

I wish to confirm to you MSF plans for re-starting medical services in Maliana and raise a concern about the military operations in the area.

MSF initially assessed the Maliana town and surrounding area on 13-14 October when the area became accessible. Based on that assessment and in conjunction with the health co-ordination forum in Dili, MSF planned to re-establish the out-patient and in-patient services in Maliana Civilian Hospital. This decision was also discussed with colleagues in the CMOC.

Based on the immediate needs of the population in the area, including those returning from the surrounding hills and in anticipation of an influx of refugees from West Timor, MSF positioned medical and logistical materials in Maliana on 18 October. Tomorrow, 20 October, the medical team will return to Maliana to begin its programme there. Medical services will begin immediately, becoming fully operational in 7-10 days.

At this time, however, the 3rd Combat Engineer Regiment has moved into the hospital and expressed its intention to remain for 2-3 months, until their task in the area is completed. Like our concerns raised in Baucau, MSF feels that the neutrality of a civilian hospital must be strictly guarded and that the presence of a military agency inside compromises this neutrality. Further it leaves the population of Bobonaro district without a hospital for emergency medical needs.

While MSF greatly appreciates the security provided by Interfet and particularly the significant humanitarian component being offered the 3rd Combat Engineer Regiment both in water treatment and rehabilitation of the hospital, we feel that priority must be given the ensuring that the hospital remains available to treat the sick.

For these reasons therefore, I kindly request that Interfet assist the 3rd Combat Engineer Regiment in re-locating a suitable location in order that both of us can successfully complete our tasks.

With thanks to your attention to this matter.

Yours sincerely,

James Patterson

cc: Mr Ross Mountain, OCHA
       Dr Ahmad Samhari, Unicef

Joint Military-Civilian Civil Affairs

by

Nicholas J. Cull, Ph.D.
University of Southern California

Thanks. As a foreigner living in this country I have a strange relationship with the flag. This is the first time I have actually wondered how much warmer I would be if I wrapped myself up in it. Barry Zorthian would like to have been able to come out to make this presentation himself but was unable to. He was one of the witnesses, not for my national security review piece, but for the book that I have just completed.

The title of my presentation today, "The Armenian Word for Chaos" is one of the jokes from Vietnam. It comes from a journalist who said what does this word JUSPAO mean and was told it was the Armenian word for chaos, an in-joke because Zorthian was from an Armenian-American family and he was in charge of JUSPAO (Joint United States Public Affairs Office). The presentation I am giving you today is derived from research that went into my book which came out a couple of months ago from Cambridge University Press so if you are looking for footnotes this is where you will find them. I will first of all talk about my core terms—public diplomacy and how I see that fitting into the world of counterinsurgency. Then I will look at the way in which public diplomacy worked in Vietnam and something about its interagency context and interagency cases. I will go through each of the presidential phases of the Vietnam War and pick out one case of interagency public diplomacy work as an example to work through. I will finish up by talking about some of the lessons of this period for the present. My theme is one that you have already heard something about, the perennial difficulties in interagency work, especially in the field of information. When we get to the question period I hope that we will be able to get into how we work this problem today because it is a big problem and the system such as it is today clearly still is not working as it should.

So first of all, what is public diplomacy? My own definition is that public diplomacy is quite simply foreign policy through engagement with a foreign public rather than through engagement with a foreign government. The term public diplomacy was created in the Vietnam era because the United States Information Agency wanted to be able to talk about what it did. It wanted a word other than propaganda so that it could say, "We Americans do public diplomacy. Those Communists do propaganda," and filled the term public diplomacy with benign meaning. But it is an old term. There are many ways in which states have always engaged with foreign publics. The most important way of engaging with a foreign public, and the one that everyone forgets, is to listen to that foreign public. The one everybody thinks of first is advocacy—influencing a foreign public through speeches, through all the ways in which we are used to governments addressing foreign publics. But also we have cultural diplomacy, we have exchange work and international broadcasting. Each of these is distinct. We can talk about the distinctions later on if you wish, but they come together in a range of practices which I think can be characterized as all properly civilian, overt, and overseas. Whenever we are doing one of these things in a way that is not civilian, is not overt, and is not overseas, we have a different term for it. But the boundaries, one of the problems with diplomacy in Vietnam is that the boundaries blur. I am feeling like in *The Longest Day* when everyone is trying to get their clickers to work. Successful public diplomacy, I would argue, needs first of all, the right bureaucratic structure. Each element within public diplomacy needs to be preserved as distinct; it needs to have its own credibility. It needs a voice in policy making. It is not enough just to have the public diplomacy element hauled in once you have decided on your policy. You have to be in on the takeoffs of policy, not just the crash landing. Historically public diplomacy is seen as an optional extra and is very much an underfunded element of statecraft. And it needs the right leadership. We know public diplomacy is at its most effective when the person in charge of it has some bureaucratic clout, not so that they can win all the time in the interagency discussion, but at least so they can hold their own. And you need the right policies. The best public diplomacy in the world cannot sell a bad policy, cannot make a bad policy good. The only way in which public diplomacy, I think, can help with a bad policy is by considering that policy early enough that the error of the policy can be pointed out with relationship to its likely impact on international public opinion. This all too seldom happens.

So counterinsurgency, I think, takes place in sort of a parallel space. It is an ancient phenomenon as you all know well. It is hard to do well and very easy to do badly. I like to tell my students that there are no ballads of Prince John. The imagination always runs the other way with the insurgent. The counterinsurgency requires a minimum of force and the subordination of military objectives to the political. It requires a splitting of the enemy from its host population and therefore is often played out in terms of reputation, image, and frames, and a quest to capture the frame, to capture the political future of the area in which the insurgency is taking place. This means that public diplomacy is tied into and parallel to this whole process. It means that counterinsurgency overlaps with public diplomacy and public diplomacy overlaps with counterinsurgency. They come to be about the same thing, about who can engage most effectively with a foreign public.

How was this done in Vietnam? I will now give you a little recap of the foundations of the Vietnam era public diplomacy. The American tradition of public diplomacy is basically this: when we have a crisis we have to do something about it in terms of public diplomacy, but once that crisis is over, quick, let's close it down because we do not like to think of governments having a role in the media. This was the case with the American Revolution, Civil War, First World War—as soon as the crisis was over the public diplomacy element was demobilized. That was not the case in the Cold War. Because of the Soviet threat, the public diplomacy apparatus was maintained and financed under the 1948 Smith-Mundt Act. But the floor within the structure of the Cold War is that public diplomacy has a limited role in the formation of policy. It is a junior partner in the interagency Psychological Strategy Board structure, and there is a very, very confused patent of agencies in Washington, DC. When Eisenhower comes in in 1953 he tries to tidy up the structure creating what he would like to have been a one-stop shop for all American information work. Overt information work had ceased. The United States Information Agency (USIA) winds up not being as tidy as Eisenhower would have liked, thanks to the intervention of Senator Fulbright. The floor in American public diplomacy at its creation is that there is no distinct public diplomacy role in the NSC (National Security Council), no mandated seat. The Director of USIA gets to sit in on NSC meetings at the invitation of the President and when presidents have other priorities, as was the case with Kennedy, Johnson, and thereafter, then increasingly the public diplomacy factor gets marginalized, gets pushed out. The interagency structure, maybe you spoke about this yesterday, is coordination by the Operations Coordinating Board. Was this spoken about yesterday? No, oh well. I talk about it a bit in my book and there are others that have written about it. But the major change that happens in this period is, I think, the Cold War comes to be seen as a zero-sum propaganda war with the Soviet Union in which the image of the United States becomes something that has to be protected. The credibility of the United States becomes something that has to be protected with military force, and of course, once you have made that decision, then force has to be placed on top of force in order to maintain that investment and to protect that investment credibility. The point I am making here is that image was an important part of America's Cold War strategy but without being an important part of Cold War policy making. So there is a bit of a mismatch going on.

When we get to Vietnam the issues play out like this: first phase of course, Eisenhower's mission characterized most famously with Ed Lansdale's CIA mission to help Ngo Dinh Diem in the aftermath of the Geneva Agreements. Now, there is an interagency group. Lansdale tends to be ad hoc with USIA and American military advisors in Vietnam helping him in his work. His big campaign is to encourage North Vietnamese people, especially Catholics, to migrate to the south and he called this campaign "The Virgin Mary Has Gone South." This was part of a project to build an identity for South Vietnam. The problem is that the more American aid is seen as being responsible for helping Diem the less nationalist he seems. America actually begins to undercut the only person who can bring order to South Vietnam. But undeterred by this, USIA expands its network in Vietnam. During the course of the 1950s, we see the setting up of regional offices and more and more investment going into the country. Ironically, even though USIA is putting in all this investment, it cannot use its usual materials in Vietnam. Usually USIA would talk about the benefits of democracy and celebrate freedom and so forth, but they find that in Vietnam they cannot do that. So what they are worried about is that the people will draw a contrast between the virtues of America and the realities of America's client in South Vietnam. So they have to go very lightly in the content in South Vietnam.

Here is a case study from that period. The USIA team wanted to undermine the credibility of the enemy and thought one way of doing this was by thinking up a derogatory title that would undermine their credibility in front of their own population. So some bright spark in USIA came up with the name Viet Cong for the NVA (North Vietnamese Army). Army public affairs used it. Everybody got together to give this term currency. The problem is, of course, that the name emphasizes a national agenda, the Viet component. They would have been better off, as the British did in Malaya, dismissing the enemy as bandits or even calling them the Chi-Cong, undermining them as being Chinese. But every time the term Viet Cong was used, it played into the self-conception of the enemy as being nationalists. In 1962, USIA offered a $50 prize for coming up with a better name; this is for locals, to come up with a better name than Viet Cong. That prize remains unclaimed so if anybody wants to have a go. . . .

The second phase—Kennedy. Kennedy's goal was very much to present a credible image of the United States counterbalancing the successes of communism in Cuba. Vietnam looks like a better place to demonstrate the potency of the American system, America's ability to win a revolution than Laos. Because of this credibility goal of Kennedy's presence in Vietnam, he needs press coverage and this puts the public affairs element in Vietnam into the front line of the campaign. The point man at this time is the public affairs officer in Saigon, former *Time* magazine journalist John Mecklin. The operating doctrine is one of giving guidance to the American press and aiding the government of Vietnam in its psychological warfare effort. The problem is that Diem is keen to protect his sovereignty and resists management and resents the American press reporting of the situation in Vietnam. The United States feels that it ought to be able to take command of the situation, but cannot and you know the rest of the story, support for the coup in 1963, gambling on finding a substitute Vietnamese leader.

But here is a case of how this situation played out, the case of the Invisible Ship. February 1962, the State Department issued a cable, 1006. This was the instruction to provide maximum feasible cooperation, guidance, and appeal to the good faith of correspondents, particularly the American correspondents in Vietnam, so to try and draw the correspondents into the effort. The problem is that both USIA and the military press officers implemented this as a limit on what they could tell journalists and saw this as a mandate to guide in a particular direction. This led to the absurd situation in 1963 of the US mission refusing to acknowledge the presence of the aviation transport vessel, USS *Core* in the Saigon River. Everybody could see it there, but it was officially denied, and it made the mission simply seem incredible. Both the press officers of MACV (Military Assistance Command, Vietnam) and the USIA press teams became part of the oppositional culture. The government people trying to spin the story of Vietnam and the journalists presented themselves as heroically connecting to the young officers in the American Army to get the real story out.

Next phase—Johnson. Now, Johnson has a clearer public diplomacy strategy and the information element in Vietnam settles down to this two-pronged attack—selling the government of Vietnam to its own people and trying to sell the war to the outside world. Now, Johnson decides that in order to give coherence to the information effort in Vietnam, you need a single information czar, and so in April of 1965 he creates the Joint United States Public Affairs Office (JUSPAO) in Vietnam. Here, Army Public Affairs is placed under civilian leadership. The vision of JUSPAO is that there will be one person in command. The differences in institutional culture, which were beginning to emerge between what journalists like Mecklin felt was proper and what the Army Public Affairs people thought was appropriate, could be eliminated just by having one voice in charge. Now, there were non-USIA personnel within JUSPAO. There were Army people filling JUSPAO positions, and also people coming in from the CIA (Central Intelligence Agency), but the whole organization is given a civilian flavor. USIA is eager to take command of JUSPAO to build on its success leading psychological operations in the Dominican intervention in April of 1965. The real reason USIA is so eager to assert itself and take command in JUSPAO is not necessarily to do with the sense that the agency was best fitted to the task. It is a bureaucracy. What do bureaucracies want? They want more—they want more resources, they want more significance. The Director of USIA, Carl Rowan, was very keen to be relevant to President Johnson, and I think, was over-eager to commit USIA to the Vietnam War. Your access to the President at this time is really based on, "What are you doing in Vietnam?" The problem of USIA's role in JUSPAO is that it compromises the civilian element of USIA work and you get USIA, a civilian agency, leading off on psychological warfare operations. The key figure in all this is Barry Zorthian, and I will give you a little bit of background on Barry Zorthian. He is an Armenian-American. He was a graduate of Yale and a Marine. He remained in the Marine Reserve. But his career had been at Voice of America where he was a campaigning program director who resisted McCarthy-ite purges and this gave him great credibility in his work with journalists in Vietnam. They really saw him as one of their own and respected him. He was able to negotiate a voluntary code with the journalists, a sort of self-censorship and then to sell this voluntary code to the Army. So Zorthian was sort of a go-between. He is able to open America's effort in Vietnam to the press and work to seek out good news, particular campaigns he is involved in without playing up Viet Cong atrocity and promoting the State Department white papers that document the levels of North Vietnamese infiltration. The problem is that as Zorthian becomes so efficient he quite swiftly outstrips the ability of the Vietnamese to talk about their own war. And we run into this problem of American ventriloquism when the Vietnamese people themselves start to hear about their own war in American accents.

Here is an example of where Zorthian broke down. While he is technically in control of the messages coming out of Saigon, there were still units in the field that felt that they knew how to appeal to the enemy, and this is a typical theme in a propaganda leaflet, what they call the nostalgia theme, trying to get the soldier to think of his girl friend at home. The only problem is this is what an American thinks the girl friend at home looks like. When a Vietnamese saw that picture they saw a prostitute. So this leaflet was saying not, "Wouldn't you rather be at home?" Instead it was saying, "Wouldn't you rather be at war fighting these Americans who are turning our women into prostitutes?" So it was a counterproductive element in the campaign which shows the need to have some sort of authority, to have some sort of management structure with cultural knowledge. The government of Vietnam and some American military briefers resisted Zorthian's authority and he is trying all the time to get them to say more, to get them to trust more, to open out the war as much as he can. And I think he makes considerable progress.

Here is, however, one of the other cases where things go terribly wrong. March 1965, Peter Arnett and the use of nonlethal gas in operations in South Vietnam. The Army declines to comment on the story and the Associated Press runs the story in its Tokyo bureau. The problem is this term: nonlethal gas. Probably if it had been a tear gas story it would have not attracted the attention it did, but the idea that the United States is involved in gas warfare in Vietnam causes outrage in the West. It becomes a major propaganda story in the East, and Zorthian argues that this is the classic example of the damage that can be done from an initial misstep, the initial mishandling of a story. If the Army had commented and had explained, "What on earth is Peter Arnett talking about? This is the sort of tear gas that has been used all over the world for many years. There is nothing sinister here." But they did not get their voice into the story and so had no control over what happened next.

The fourth phase is the next reorganization of the interagency element regarding public affairs and that is the creating of a greater structure. The first time they try it is 1967 when William Porter is brought in. He was the Deputy Chief of Mission, given the rank of Deputy Ambassador and put in charge of an office of civilian operations. This is then reformed again to create the CORDS (Civil Operations Rural Development Support) bureaucracy. The objective here is to get interagency working for revolutionary development. But it disrupts that sort of perfect structure that Zorthian thought he had back in 1965. JUSPAO and the USIA elements are mixed in with USAID (US Agency for International Diplomacy), CIA, and the government of Vietnam. Porter and his successor trump the information czar in their approach to information. But this is the peak of the public diplomacy budget. A lot of the public diplomacy at this time was seeking to publicize what Johnson called "the other war," the fact that the Americans were building hospitals, were doing positive developmental projects for the Vietnamese people. But of course, America's behavior in Vietnam is the real message, and images of the extraordinary level of destruction in the Vietnam War and of the behavior of American troops in Vietnam on things like search and destroy missions started to become the real story and are much more potent than the official newsreel pictures of a new hospital being opened or kids being given typhoid shots. The blow of the Tet Offensive you know about and the counterinsurgency operations following Tet. The implementation of the Phoenix Program worked and yet they negate the original objective of the entire war which was for something to be seen to be done swiftly, easily and we become bogged down in something quite beyond what was originally intended. At the moment of Tet, Zorthian leaves Vietnam, and he leaves with a profound sense of disillusionment in what he has been a part of.

A case from this period would be the interagency effort around the Vietnamese election of 1967. JUSPAO and CORDS made the election a priority really to prove to the world that democracy was possible in Vietnam. They had special publications, special films. The Vietnamese traditional theater troops—Van Tac Vu theater troops—did shows around the election, there were loudspeaker flights trying to get people out to vote, and special tours for journalists and jurists from around the world to come and inspect the quality of the elections. However, the campaign, both domestically and internationally, raised unreasonable expectations. There was corruption and some recriminations. The losing candidate ended up being jailed. I can see in the wake of the coming election the winning candidate might well wish to jail the losing candidate, but this is not the show piece of democracy. The problem is there is no viable government in Vietnam to be chosen by these elections.

By the time Nixon is in power the game is withdrawal and managing the negatives of the war. USIA's leadership by this time is very wary of its connection to Vietnam. It sees its presence in Vietnam as a negative. The staff does not like going there, not unlike the problem we see today in Iraq. And USIA is also being shut out—even inside Vietnam—of important issues. Zorthian's successor, Ed Nichol, is left out of the planning around the Cambodian operation. Senator Church introduces an amendment to restrict American public diplomacy around the Vietnam War, forbidding the United States from actually conducting any propaganda in support of President Thieu. It is supposed to be to avoid assisting that element which seems very strange, in great contrast to what had been done earlier on. And USIA worked to essentially pull out of the Vietnam operation. By this time there were about 100 US Army personnel working within the JUSPAO structure. As their tours of duty came to an end, USIA merely did not notify the Army that those people needed to be replaced. And so they allowed JUSPAO to wind down until it was no more significant than any other USIA post. Outside of Vietnam, USIA merely stopped talking about the war. They found that if they wanted to reach and engage a foreign audience they talked about something else. They talked about blue jeans, John Wayne, recycling, anything other than the Vietnam War. The American troop withdrawal in 1973 closed this military-civil aspect of the Vietnam War and, interestingly from my perspective, ended the American image problems around the Vietnam War. Okay, some people around the world still say, "What about Vietnam? That was not any good for you people, was it?" But if you look at the polling data within just a few weeks of the withdrawal of American troops there is a big up-tick and you do not see a perpetuation of anti-American feeling based around Vietnam. In fact, the rebound happens very, very swiftly, surprisingly swiftly.

A case from this period would be the Easter Offensive, the NVA attack on Easter of 1972, and of course, was successfully repelled. We see all the JUSPAO elements working very well together and connecting to Voice of America which extended its

Vietnamese service. JUSPAO used its network to post regular updates all around Vietnam so people could go to central places and see what was happening, read the results of the battles, and the particularly effective operation was the rapid turn around of the names of the captured POWs so these could be recorded and then broadcast immediately into North Vietnam where people were able to listen on little radios that had been pretuned to American frequencies and had been dropped coincidentally some months previously, but they were all still working. This was a terrifically effective campaign. And of course, no problem here with coordination of story because it is an obvious story, the defeat of the North Vietnamese Army. It is like the old saying "a rising tide lifts all boats." A really good information campaign needs a good story and if the story is there everything works well. The problem is when you do not have the story. My analysis of what happened in Vietnam is that the wrong policies were followed, aid undermining Diem, the presence of the Americans negated the policy objective of shoring up and protecting South Vietnamese independence and the cure of the American intervention became worse than the disease. But of course, we also have the wrong tactics with the United States initially underplaying counterinsurgency tactics and the military becoming such a large part of the problem. And we see the wrong public diplomacy being used. The public diplomats in Vietnam tried to sell a different story to the world, but they were stuck with the message of America's deeds and stifled the Vietnamese voices.

I see that lessons were learned, but I think they were the wrong lessons. I think that the United States comes out of Vietnam fixating on domestic opinion, planning that its next war should be short with tightly managed domestic coverage and ignoring the role of local media images. There is a problem of the Vietnam War on TV, but here I am not talking about American TV, I am talking about the Vietnamese TV that the US Army installed in Vietnam. This is like the problem of that little flyer of the girl in a bikini, but writ large. America paid for TV sets to be put into Vietnamese villages to be watched communally. These were so popular in the villages that were controlled by the Viet Cong, they would come out of the jungle at night and sit around watching *Batman* and reruns of *Combat!* and other shows they really liked, but when it was the variety shows depicting Vietnamese society, what was showcased was the bastardization of Vietnamese culture. It played directly into their interpretation of what was happening in Vietnam. You could not conceal the impact of the Americans on Vietnamese society and cannot match the national claim made by Ho Chi Minh.

My conclusion is that there is great danger at this juncture between images and action and that an interagency process leaves more than just a pooled resource. It needs an internal consistency of word and deed to making sure that what you claim to be the objective of your policy is worked through in the course of your action. I think there is a big problem here of multiple audiences. Not only the multiple audience having to have information campaigns and public affairs work that played within Vietnam, but also the play to the world. There was no single world audience. Johnson became very frustrated that every time he bombed North Vietnam the Europeans would be very upset with him, but the Asians would really be very supportive and think, "Yes, this is the way to go!" Then whenever he talked to the North Vietnamese the Asians would be very worried—the Philippines, South Korea. "What is President Johnson doing?" The Europeans would think, "This is great!" So he was caught in a juggling act with international opinion. You can see the parallels to present problems in Iraq and Afghanistan where we see lessons of counterinsurgency being successfully applied, but have to ask, "The lessons are being applied, but what is the end state? Is the political center viable? Are we winning a war or is there somebody there to win the war for?" I have already mentioned this lesson. Well, the opinion recovered quickly post-1973. However, the tragedy of the damage done in Vietnam and the damage internally within the United States lived on. I think in some ways, both Vietnam and the United States are still living it. That is all I wanted to say. Thank you very much.

Panel 6—The Interagency Process in Asia
Question and Answers
(Transcript of Presentation)

Major Eric M. Nager, US Army
Nicolas J. Cull, Ph.D.

Moderated by Lieutenant Colonel Scott Farquhar, US Army

Audience Member

Sir, you talked about the problems of multiple audiences. Given the error of instantaneous communication, whether it is video or YouTube or the Internet or text, how do we get the message that we need to craft for specific audiences that do not contradict or appear to be contradictory to different audiences?

Dr. Cull

The first thing you have to do is acknowledge that we now live in a different world where a message crafted for Kansas will be heard in Kandahar. I think we have now moved beyond the phase where a president will go out and say let's have a crusade and then be surprised that there are Muslins listening who do not like that terminology. What I think we are seeing now in terms of public diplomacy is messages for Kandahar being crafted on the assumption that Kansas is listening in, and in a number of countries around the world public diplomacy has become a performance for domestic consumption to give your domestic audience the prestige of the admiration of the world or minimize the absence of the prestige. The real reason the Chinese ran the Olympics was not to impress us. It was to impress their own population because it is their own population that might overthrow them. So much of what happened in the Bush administration was . . . take for example, Karen Hughes when she went overseas. She took the press with her on her listening tour, but she did not take the world's press. She took the American press because she wanted to show that she was doing something to fix America's image problem. I think the first thing we need to have is a public diplomacy that is actually based on engaging its target audience and is created with the courage to defy the four-year political cycle. Otherwise we will be stuck in a perpetual echo chamber of domestic politics. There are some countries that are prepared to do that and are able to do that in other places where it is harder to do, but I think you put your finger on a really important issue. It stresses the need for information to be thought about in a new way. We do not have the luxury of the divisions of audience we had in the past.

Audience Member

I have a question for Major Nager. You had mentioned that there is a need for greater allied involvement in humanitarian assistance, and I was wondering about the potential role of Japan for that kind of help? When the US is so stretched what potential is there for Japan to help out?

MAJ Nager

It is included in my paper, but I did not mention it today in the presentation. Japan substantially funded the INTERFET (International Force for East Timor) Exercise in East Timor. So even though they did not send troops, they gave a substantial financial contribution which they also did, of course, in DESERT STORM/SHIELD. So until they get their constitutional issues resolved where they feel more comfortable sending soldiers overseas, to contribute economically through NGOs (nongovernmental organizations) may be the best way to contribute, but they did contribute financially in that case.

Audience Member

Is that the most we can really hope to expect in the near term? Funding is nice, but sometimes it is helpful to have other things.

MAJ Nager

I think we want greater involvement with troops and soldiers and I hope we will work in that direction.

Audience Member

I had a question about the seven components of future warfare. Could you expand on those a little bit?

MAJ Nager

The Australian lessons?

Audience Member

Yes.

MAJ Nager

All right. I could tell you the other ones. I was just drawing upon a couple that I thought were most pertinent to this discussion. Some are them are just applying to Australia. They did not apply as directly to the United States. I can tell you what all seven of them are if you would like. They talked about command and control, intelligence, surveillance and reconnaissance, tailored effects, force projection, force protection, force sustainment, and force generation. I will give those to you again afterwards if you are really interested. I only examined a couple of them as they pertained to extra-agency coordination which is my focus in this paper. But I would be happy to share those with you afterwards.

Audience Member

I was wondering if in your research prior to the Timor Exercise if any of the lessons learned from Bosnia had any impact on how they conducted that operation?

MAJ Nager

Not that I read in my research. I took this from an Australian standpoint first and how the US plugged into that, so in my research I did not come across anything specifically mentioning that.

Audience Member

My question is for Dr. Cull. Is there an Assistant Secretary for Public Diplomacy now in the State Department?

Dr. Cull

There is an Under-Secretary. What happened was in 1999 USIA was closed down and the public diplomacy elements were integrated into the Department of State. And instead of the USIA Director, the lead person in public diplomacy is supposed to be the Under-Secretary of State for Public Diplomacy and Public Affairs. So that on paper is how it is supposed to work, but it has not been an ideal situation. There is also a Deputy Assistant Secretary of Defense for Support for Public Diplomacy who is Michael Durant, a historian of the Middle East which reflects where they expect the public diplomacy might need to be supported.

Audience Member

I have not heard it brought up yet and this may be the appropriate time. I am curious about how language can be a barrier to the interagency process. I know since we are talking about public information, this may be the appropriate place to talk about that because it was a problem.

Dr. Cull

I think it is an excellent point, and you see how language and institutional culture are both problems.

Audience Member

I thought it was interesting in the study that was done in the *Washington Quarterly* about allied nations to the United States that could take the lead in the coalition as of 2002, and the only two listed were Australia and Great Britain where there was no language barrier. They did not talk about language barrier in that article, but it is a factor as well.

Dr. Cull

We even find that within the UK forces that you can have a language barrier. They found at one of the landings in the Falklands that paratroopers would not get out of a landing craft because the Marines were being told "now" and if you want a paratrooper to move you say "go." Nobody said "go" so they just stood there until they heard the right word to move a paratrooper. So it can even happen within a service. I think the way that we begin to address this is by using the same sort of public diplomacy tools to get nations to understand each other, using exchange of persons, particularly, and increasing mutual experience. It is a slow process, but I think you have hit on an important problem.

Audience Member

Having worked in the US Embassy one of the biggest challenges we found was the constraints that were put on the US Embassy personnel to speak on the record to the press corps. Contrary where in the military, of course, we have taken a total change now where the soldier down at the lowest level is authorized to speak to the press corps provided that they have permission. I just wanted to see from a historical context how is that affecting our ability to get the message out there in a consistent fashion when the military can speak on the record, but yet in a US Embassy there is only one person who is authorized to speak on the record and that is the Ambassador.

Dr. Cull

I think that is the least of the asymmetries, frankly, and the big problem is how much bigger the budget for military public diplomacy is than public diplomacy in the State Department. The State Department retains the title of leading public diplomacy and officially all the DOD can do is support public diplomacy, but we see such an asymmetry. Something really has to be done about that, and I have been impressed by what Secretary Gates has said and done. I think he really understands the problem in a nuanced way, but goodness knows what will happen in January. But that is always the way of things. I think you need to have control of the message and I can understand why the State Department has this restriction. One thing that they are trying to do is to make sure that talking points are made available in real time and according to the news cycles of the rest of the world so that American diplomatic personnel overseas are able to respond accurately and swiftly to issues being raised by the foreign media because for too long America has worked on a domestic news cycle and has ignored the fact that by the time America wakes up a story that was a rumor 12 hours ago is being reported as fact. And American denials and American rebuttals come very late in the day. That is now starting to be addressed by the creation of regional hubs up in Doha, a hub in Brussels, and I believe there is talk about doing one even farther east, but I cannot remember where that is going to be.

Panel 7—Interagency Process in the United States
(Submitted Paper)

Cold War Interagency Relations and the Struggle
to Coordinate Psychological Strategy

by

Robert T. Davis II, Ph.D.
US Army Combat Studies Institute

Introduction

At the beginning of April 1951, in the midst of the Korean War, President Harry Truman issued an executive order establishing a Psychological Strategy Board "for the formulation and promulgation, as guidance to the departments and agencies responsible for psychological operations, of over-all national psychological objectives, policies, programs, and for the coordination and evaluation of the national psychological effort." The creation of the Psychological Strategy Board (PSB) was an attempt to solve years of interagency wrangling over the direction and scope of US information operations in the early Cold War.[1] For two years the Psychological Strategy Board functioned as a planning and coordinating body, but ultimately the Eisenhower administration would find the PSB no more satisfactory than previous attempts at planning and coordination and dissolve the board in the summer of 1953. This paper will review the US government's attempts during the Truman administration to find a solution to the problem of a coordinated information strategy in the context of interagency cooperation, or its lack thereof.[2] By way of introduction, a brief review the previous wartime attempts at information operations that developed in the United States, particularly the Creel Committee and the Office of War Information, is provided.

Antecedents:
The Wartime Psychological Warfare Effort in World War I and World War II

During World War I, President Woodrow Wilson established what is regarded as America's first official propaganda agency. This agency was the Committee on Public Information, but better known as the Creel Committee, since it was headed by journalist George Creel. The Creel Committee was nominally headed by a committee made up of Creel, Secretary of State Robert Lansing, Secretary of the Navy Josephus Daniels, and Secretary of War Newton Baker. This was meant to provide a modicum of interagency cooperation, but as a practical matter, Secretary of State Lansing chose to have little to do with the committee, and the secretaries of the two military services largely left Creel to his own devices. The Creel Committee had responsibility for both censorship and propaganda production, but Creel himself tried to stress that he did not perceive his roll as censor-in-chief. Though the Creel Committee did have both domestic and foreign divisions, the vast bulk of its resources and energy during the war were devoted to maintaining domestic support for the war effort.[3] When the war ended in November 1918, Congress moved quickly to dissolve the committee. Revelations in the interwar period about the nature or the wartime propaganda effort, not least by Creel's own book *How We Advertised America* (1920), as well as the activities of the Communist International and the European fascists governments did a good deal to generate hostility in the United States to the notion of government-sponsored propaganda.

As the world moved toward war in the later 1930s, American concern about again being drawn into a conflict, including potentially by allied propaganda, contributed to the limitations under which President Franklin Roosevelt operated. Roosevelt, for all his desire to convince the American populace of the need for preparedness and his own views on supporting Britain and France, proved cautious about replicating the perceived excesses of the government propaganda effort in World War I.[4] And as a former Assistant Secretary of the Navy during the Wilson administration, he was well familiar with the activities of the Creel Committee. President Roosevelt's own approach to presidential leadership further complicated matters. He had a penchant for fostering bureaucratic rivalries to preserve presidential prerogatives.[5] In keeping with this practice, Roosevelt authorized the creation of a number of different agencies tasked with information activities in the two years before American entry into the war. These included the creation of an Office of Government Reports (est. September 1939), a Division of Information of the Office of Emergency Management (est. March 1941), an Office of Coordinator of Information (est. July 1941), a subsidiary Foreign Information Service (est. August 1941), and an Office of Facts and Figures (est. October 1941).[6] The growing need to counter German propaganda aimed at Latin America in the summer of 1940 occasioned the appointment of Nelson Rockefeller as the Coordinator of Inter-American Affairs in August 1940. Rockefeller's brief gave him wide latitude in the direction of US information operations within the region.[7] This confusing profusion of agencies remained in place until the exigencies of wartime prodded Roosevelt to consider a more centralized public information effort.[8]

Once the United States had entered the war, Roosevelt was pressed by advisors to create a more orderly arrangement of US information activities. Nonetheless, President Roosevelt refrained from duplicating the Committee on Public Information, essentially opting to split its functions between two new organizations, an Office of Censorship and an Office of War Information (OWI).[9] In addition, in June 1942 he expanded the writ of William "Wild Bill" Donovan's Office of Coordinator of Information (OCI) into the new Office of Strategic Services (OSS), although the Foreign Information Service, formerly part of OCI, was transferred to the Office of War Information. Of these three, both the Office of War Information and the Office of Strategic Services played a role in the production of wartime propaganda (psychological warfare, or what we would now refer to as strategic communications) for foreign dissemination. Continuing friction over responsibilities led President Roosevelt to issue further clarification on responsibilities in March 1943. These instructions stated that the Office of War Information was responsible to "plan, develop, and execute all phases of the federal program of radio, press, publication, and related foreign propaganda activities involving the dissemination of information." When these programs were to take place in areas of current or projected military operations, they were to be "coordinated with military plans through the planning agencies of the War and Navy Departments, and shall be subject to the approval of the Joint Chiefs of Staff." Reflecting traditional military views of command responsibility in wartime, the directive also stated that, "parts of the foreign propaganda program which are to be executed in a theater of military operations will be subject to the control of the theater commander." Finally, the authority and functions of the Office of War Information were not to compete with Nelson Rockefeller's bailiwick in Latin America.[10] Appraising the actual contribution of OWI and OSS to the war effort is a very difficult proposition.[11] In any case, in the waning months of World War II, many of the psychological warfare and information operations capabilities (what we would now refer to as strategic communications) that the United States government had developed during the war were wound down. In commenting on the need to study psychological warfare at the end of the war, General George S. Patton well reflected the ambiguous position of many regarding psychological warfare when he stated, "Such a study must be made. Psychological warfare had an important place in the European campaign. It can accomplish much good. It can also be extremely harmful."[12]

With the approach of war's end, the Office of Censorship, the Office of War Information, and the Office of Strategic Services would all be shut down. In the case of the Office of Censorship, Director Byron Price had long advocated that censorship be curtailed as soon as the threat to national security decreased. He proposed that the Office of Censorship be closed as soon as fighting ended. Price's views were endorsed by the Censorship Policy Board as early as 20 November 1943. After the defeat of Germany, Price cancelled program restrictions on the radio code. He subsequently won President Truman's endorsement to declare the end of censorship on the same day that victory over Japan was announced. The work of the Office of Censorship formally ended on 15 August 1945. Byron Price's plan for a voluntary censorship code was generally perceived to have worked admirably well. The editors of the trade journal *Editor & Publisher* stated, "We have never heard anyone in the newspaper business contradict the statement that Byron Price conducted the Office of Censorship in a competent, careful and wholly patriotic manner."[13]

While there had probably never been any question of the Office of Censorship continuing its activities in peacetime, a reasonable case could be made for the retention of the Office of War Information. Unlike the Creel Committee, which had focused the great majority of its propaganda output toward domestic support for President Wilson's policies, from 1944 on, 90 percent of OWI's budget was geared toward international propaganda activities.[14] At the end of August, President Truman signed an executive order that transferred the foreign information functions of the Office of War Information and the Office of Inter-American Affairs to an Interim International Information Service established within the Department of State. Under this order, the remaining functions of the Office of War Information were to cease on 15 September 1945. As of 31 December 1945, both the Office of War Information and the Interim International Information Service were abolished.[15] The War Department retained its Public Relations Bureau, the function of which was increasingly geared toward defending Army appropriations during the massive postwar demobilization. The dawning of the Cold War, however, would soon revive the interest of the US government in pursuing a more aggressive approach to public affairs, public diplomacy, and psychological warfare. Difficult questions about where responsibility for these functions should be vested hampered to implement this program for the next several years.

Postwar Psychological Warfare Planning and the SWNCC

In early 1946 a number of events contributed to a hardening of US attitudes regarding the Soviet Union.[16] Soviet activities in Romania, Bulgaria, northern Iran, and pressure on Turkey all suggested little willingness on the part of the Soviets to work within the framework of the United Nations. A toughly worded speech by Joseph Stalin to the Soviet people on 9 February that warned of the impending final crisis of the capitalist world and called for the Soviet people to be prepared for conflict seemed to some in the West as a warning of more aggressive Soviet intentions.[17] On 22 February 1946, George Kennan cabled his famous "long telegram" from the US Embassy in Moscow which helped to crystallize the attitudes of many in the US government about the need for a firm response to the expansionist impulse in Soviet foreign policy. In March 1946, the Truman administration dispatched the USS *Missouri* to Turkey, ostensibly to return the body of the recently deceased Turkish ambassador, but also to send a clear message to the Soviets that the United States could not look with indifference upon Soviet attempts to pressure Turkey. On 5 March, Winston Churchill, former and future Prime Minister of Great Britain, delivered his "Iron Curtain" speech at Fulton, Missouri. One of the chief advocates in the Truman administration of a firmer US position toward the Soviets was Secretary of the Navy James Forrestal (later the first Secretary of Defense). Forrestal had been much impressed by Kennan's telegram. Shortly after Churchill's speech, Forrestal wrote in his diary that the United States needed a program "to inform the country of the facts, of the complete impossibility of gaining access to the minds of the Russian people, of piercing the curtain of censorship which is drawn over every area they occupy." Though Forrestal thought such a program desirable, he also realized that initiating such a program without infringing on the State Department's own responsibilities would be difficult.[18] It would indeed prove difficult in to reach an accommodation between the Department of State and the nation's military authorities on direction of US information operations for years thereafter.

In 1946, members of the US government were working to find solutions to a number of problems in the international arena while the nation was still coming to grips with its new world role. Part and parcel with this shift toward wider responsibility was a wide-ranging discussion over the best means of determining and directing US foreign and security policy. Any discussion of the struggle to coordinate foreign information operations in the early Cold War must keep this broader context in mind.[19] Since late in World War II, the primary means of coordinating policy between the State, War, and Navy Departments was a series of semiregular meetings held by the three secretaries and interdepartmental meetings of their three principal deputies who met at the State-Navy-War Coordinating Committee (SWNCC).[20] In early March 1946, Secretary of War Robert Patterson approached Secretary of the Navy James Forrestal about the need for bringing together qualified military and civilian experts to assess the utility of psychological warfare in World War II.[21] Then, at the end of May 1946, the Assistant Secretary of the Navy for Air, John L. Sullivan, sent a memorandum to the State-War-Navy Coordinating Committee that suggested the formation of an ad hoc committee under the SWNCC "to study and report on the future status of psychological warfare." He recommended that the committee study psychological warfare with an eye toward developing a peacetime organization that would be prepared for rapid mobilization in wartime and suggest a suitable wartime organization. Ideally, these organizational arrangements would prevent the trial-and-error effort which had characterized the implementation of psychological warfare in World War II.[22] The Assistant Secretary's suggestion was approved by the committee shortly thereafter, and an ad hoc committee formed to study the US government's psychological warfare organization. This ad hoc committee reported its findings in early December. Its assessment was that "psychological warfare is an essential factor in the achievement of national aims and military objectives" whether in time of war or threat of war (the latter to be determined at the President's discretion). At the time, the SWNCC confirmed that the State Department had the primary interest in determination of psychological warfare policy because of its impact on the nation's foreign policy. To improve intergovernmental coordination, the ad hoc committee advocated that a standing Subcommittee on Psychological Warfare should be established under the SWNCC. It was to be composed of two members each from the State Department and the Central Intelligence Group, and one each from the War and Navy Departments.[23]

While the State-Navy-War Coordinating Committee continued to discuss the question of planning for psychological warfare during late 1946 and into 1947, developments in the broader Cold War and questions of the establishment of a more formal US national security apparatus occupied center stage. It proved difficult to formalize mechanisms for policy coordination of psychological warfare—or anything else for that matter—while the debate on military unification (and, ultimately, the creation of a national security council) was under way. However, international affairs continued to press government officials to better marshal US information operation capabilities in the service of government policy. On 12 March 1947, Truman addressed Congress to request aid for Greece and Turkey. The address, which has subsequently been known as the Truman Doctrine, signaled a clear shift in US policy toward the Soviet Union. In June, Congress agreed upon the legislation that became the National Security Act.

The National Security Act (NSA) was meant to help formalize the structure of inter-agency cooperation, though in practice problems continued to plague US interagency relations for many years. The NSA authorized the creation of the new Cabinet position of Secretary of Defense to head a nascent National Military Establishment, established an independent Air Force (within the National Military Establishment), established the Central Intelligence Agency (adding a new bureaucratic competitor to the psychological warfare debate), and established a National Security Council to advise the President. The National Security Act was passed at the end of June, and its provisions went into effect in mid-September 1947.

President Truman attended the first meeting of the National Security Council on 26 September, but was only a sporadic attendee of its meetings prior to the outbreak of the Korean War.[24] Truman initially kept the NSC at a distance, at least in part because he did not want his participation in its deliberations to make it seem that he was bound by its decisions. In his stead, the meetings were generally chaired by the Secretary of State (George Marshall until January 1949, and Dean Acheson thereafter). The first meeting of the National Security Council was primarily an organizational meeting to discuss the policies and procedures of the NSC, as well as discussing the initial directive to the CIA. It would be over six weeks before the next meeting of the NSC convened. Given this lack of clear-cut presidential participation in the NSC's deliberations, its initial role in the policy coordination process was somewhat limited. In the interim between the first and second meetings of the NSC, the State-Army-Navy-Air Force Coordinating Committee (SANACC, as the SWNCC had been reconstituted following the establishment of the Air Force as a separate service) decided there was an urgent need to create an ad hoc committee to investigate whether the United States should "at the present time utilize coordinated psychological measures in furtherance of the attainment of its national objectives."[25] Interestingly, a draft memorandum prepared by Forrestal's staff for consideration by the SANACC anticipated the form that the Psychological Strategy Board would subsequently take.[26] Nonetheless, in early November, the newly created War Council (another product of the National Security Act, made up of the Secretary of Defense, the three service secretaries, and the three chiefs of the military services) reached agreement that SANACC 304/10 (the 304 series of documents pertained to psychological warfare) should be revised to delete any reference to a domestic information program; "fix responsibility for the general direction and coordination of both black and white activities in the State Department"; and have the State Department carrying out white activities and the CIA the black activities. The War Council explicitly rejected the proposal of creating any new board or committee to supervise and coordinate the psychological warfare activities.[27]

At the second meeting of the NSC on 14 November, the Council took under consideration SANACC 304/11, the revised paper on Psychological Warfare.[28] The study stated that the Secretary of State would be charged with the general responsibility for the coordination of psychological measures. The SANACC members wrote that they presumed the Secretary of State would delegate these responsibilities to the Assistant

Secretary of State for Public Affairs, who in turn would be assisted by representatives from the Army, Navy, Air Force, and Central Intelligence Agency. Secretary of the Army Kenneth Royall, speaking on behalf of himself and Secretaries Forrestal and Symington, stated that the National Military Establishment "did not believe it should have a part in these activities." In the course of discussion, Secretary of State George C. Marshall expressed his concern about referring to information activities as psychological warfare. The Department of State was concerned that operational control over covert psychological activities ran the risk, should it become public knowledge, of undermining or outright discrediting US foreign policy.[29] As a result of this discussion, two separate streams of directives on information operations were developed, one which delineated State Department responsibilities, the other, Central Intelligence Agency responsibilities. The NSC approved the first two directives (NSC 4 and NSC 4-A) at its meeting on 17 December 1947. NSC 4, "Coordination of Foreign Information Measures," charged the Secretary of State with responsibility for policy formulation and "coordinating the implementation of all information measures designed to influence attitudes in foreign countries in a direction favorable to the attainment of US objectives and to counteract the effects of anti-US propaganda."[30] NSC 4-A, "Psychological Operations," authorized the Director of Central Intelligence "to initiate and conduct, within the limits of available funds, covert psychological operations designed to counteract Soviet and Soviet-inspired activities."[31] The State Department's leading role in overt information programs abroad was further solidified with the passage of Public Law 402, *The United State Informational and Educational Exchange Act of 1948* (better known as the Smith-Mundt Act), signed into law by President Truman on 27 January 1948.[32]

The Communist coup in Czechoslovakia in late February 1948 energized Western officials, leading to a number of important Cold War initiatives.[33] In this vein, Secretary of Defense Forrestal requested in late March that the NSC review the progress that had been made since the NSC 4 documents had been approved because the international situation made it "more important than ever that our foreign information activities be effectively developed."[34] The Joint Chiefs of Staff now supported the creation of a separate Psychological Warfare Agency which would operate directly under the NSC.[35] State Department remained resistant to losing any authority in this regard in peacetime. For the next three years, the State Department successfully deflected attempts to alter the system established by the NSC 4 series.[36] Indeed, in March 1950, President Truman approved NSC 59/1, "The Foreign Information Program and Psychological Warfare Planning," which reiterated and extended State Department's authority in the field of information operations.[37] Superseding the NSC 4 series and NSC 43, NSC 59/1 stated that the Secretary of State was responsible for the "formulation of policies and plans for a national foreign information program" not only in time of peace, as had formerly been the case, but also in times of national emergency and in the initial stages of war.[38] To carry out these planning and coordinating responsibilities, the Secretary of State, in consultation with the other departments and agencies represented in the NSC, was authorized to name a director of a new organization within the State Department.

This office was initially known as the Interdepartmental Foreign Information Organization. The director, along with a board of consultants representing State, Defense, CIA, the JCS, and the National Security Resources Board, and a staff drawn from State, Defense, and CIA, was charged with the policy formulation and planning functions delineated in NSC 59/1.[39] Complementing this latest reorganization, President Truman inaugurated a "great campaign of truth" in an address to the American Society of Newspaper Editors, with the aim of launching a public diplomacy campaign against Soviet propaganda.[40] This campaign was just getting underway when the Korean War began. The stress of coping with a difficult wartime situation would soon call the appropriateness of the NSC 59/1 arrangement into question.

When the Korean War broke out, the State Department initially retained authority for the conduct of the foreign information program. Acting under the provisions of NSC 59/1, the State Department announced the establishment of a National Psychological Strategy Board on 17 August, though in practice this Board was simply a renamed Interdepartmental Foreign Information Organization.[41] This version of the National Psychological Strategy Board remained under the direction of the Secretary of State. Since mid-July, however, a new arrangement had been under discussion. Pursuant of provisions of NSC 59/1, a draft "Plan for National Psychological Warfare" under the imprimatur of Under Secretary of State James Webb had been submitted for the NSC's consideration on 10 July, and subsequently circulated as NSC 74.[42] NSC 74, which was by and large written before the outbreak of the Korean War, recognized that after the initial stages of any war, the US government would move to establish a national psychological warfare organization of independent status, with direct access to the president, and "the authority to issue policy directives to departments and agencies of the government executing psychological warfare measures."[43] As had been the case in all previous incarnations of a coordinating and planning authority for psychological warfare, the draft of NSC 74 called for a directing board with members from the agencies which had been drawn from throughout the period since 1946–47.

In one of the more interesting critiques of the status quo approach to organization presented in NSC 74, a letter to the NSC Secretariat submitted under the signature of Secretary of Defense Louis Johnson (Johnson had replaced Forrestal in March 1949) argued that, "effective psychological warfare requires perspective over and beyond that available to individuals who must consciously or subconsciously reflect the actions and policies, past and present, of the operating organizations which they represent." He went on to argue that, "with rare exceptions, they do not have the broad, almost intuitive knowledge acquired in varied walks of life which go into the make-up of outstanding experts who have become leaders in their fields by practical application of mass civilian psychology techniques." For Johnson, the only way to avoid the narrow, departmentally focused mindsets which had characterized previous attempts at psychological warfare coordination was to create a board in which the majority of its members were drawn from outside of government. Finally, Johnson suggested—reflecting a long-standing strain of Department of Defense opinion—that rather than simply having "official access" to the President, the Board should be lodged directly within the Executive Office of the President.[44] Interesting though these ideas were, by the time they had been submitted to the NSC, they carried little weight. Secretary Johnson had—for a variety of other reasons—already been forced to resign by President Truman.[45] In his stead, President Truman would recall George C. Marshall to duty.

After the Inchon landings (15 September) and prior to Chinese intervention in the Korean War (late November), when General Douglas MacArthur's pledge of "Home by Christmas" pledge still seemed realistic, the perception that the fighting in Korea would be relatively brief contributed to a sense that there was little urgency to develop the psychological warfare program.[46] Following Chinese intervention, a much gloomier view of the war emerged, which soon prompted Washington to become more interested in the problem of information management in wartime. In order to assert closer control over pronouncements by "officials in the field as well as those in Washington," President Truman issued a memorandum on 5 December requiring all speeches, public statements, and press releases on foreign and military policy to be cleared by the State Department and Defense Department, respectively.[47] However, this was largely a negative policy that did little to address how the US could better craft an information campaign in the context of a Cold War turned hot.

On 16 December 1950, President Truman signed a declaration of limited national emergency. As he had previously explained to key legislative leaders, he intended that this declaration "would have very great psychological effects on the American people."[48] It also helped signal to government officials "the intensity of the current crisis demanded sweeping new initiatives."[49] The State Department and Defense Department had, however, made little progress toward closing the gap between their respective views regarding the appropriate wartime organization of a national psychological organization.[50] Even after the Psychological Strategy Board was established, this difficulty was never fully resolved.[51]

When the NSC took up discussion of the national psychological effort at its meeting on 4 January 1951, President Truman, probably frustrated with the lack of compromise between the respective agencies, directed Sidney Souers, the Executive Secretary of the NSC, and the Bureau of Budget to study the matter and make a recommendation for the organization to him.[52] By 18 January, Souers had produced a "Draft Directive on the National Psychological Effort" which he proceeded to discuss with State, Defense, and CIA officials. Not surprisingly, Souers compromised between the State and Defense Departments by placing his proposed Psychological Strategy Board under none other than the National Security Council (rather than the "independent status" which the State Department had envisioned or directly in the Executive Office of the President, which was favored by the Defense Department). Despite initially strenuous State Department objections, the Psychological Strategy Board would be established broadly within the lines advocated by Souers' compromise.[53]

President Harry Truman issued a directive establishing a Psychological Strategy Board (PSB) on 4 April 1951. The PSB was to serve as a planning, coordinating, and evaluating agency, but it was not to carry out any operations itself. It would take several months to secure a director, staff, and offices. The PSB was comprised of the Under Secretary of State, the Deputy Secretary of Defense, and the Director of Central Intelligence, a military adviser from the JCS, appropriate representatives from other government departments and agencies as determined by the Board, and a Director (designated by the President), who served under the Board and directed the activities of its staff.[54] The PSB's staff was organized into three offices: the Office of Plans and Policy, the Office of Coordination, and the Office of Evaluation and Review. One of the first subjects taken up by the PSB staff was contingency planning for a potential armistice in the Korean War. Should armistice negotiations break down, it was considered important that the fault be attributed to the communist participants. Subsequently, the PSB staff also tried to make an inventory of the US governments "weapons" for conducting psychological operations, and found itself caught up in the long-running wrangle over what should be the national psychological plan for general war.[55] Constraints of time and space prevent a detailed discussion of the efforts made by the Psychological Strategy Board to carry out its mandate, but this author hopes that the interested reader will consult the impressive literature on the subject which has emerged in recent years.[56]

Eisenhower and Psychological Warfare

Once he was elected president, Eisenhower would make the formation of a coordinated national psychological effort one of his key objectives. Back in February 1949, then General Eisenhower foreshadowed his own later views on the Psychological Strategy Board during a NSC discussion of planning for overt psychological warfare operations in peacetime. The summary of the discussion records that Eisenhower "agreed that the Department of State was the proper organization to plan [in peacetime]," but in a wartime situation the NSC itself, "which he saw as a defense cabinet during time of war, could effectively direct psychological warfare."[57] During his campaign for the presidency, Dwight D. Eisenhower frequently invoked the need for psychological warfare as the key component of the Cold War struggle with the Soviets. In a speech in San Francisco in early October 1952, Eisenhower said that in order to avoid the horror of war, it was necessary for the United States government to "use all means short of war to lead men to believe in the values that will preserve peace and freedom." For Eisenhower the end goal was to convince the people of the world that "Americans want a world at peace, a world in which all people shall have opportunity for maximum individual development." One of the key means of spreading this truth, for Eisenhower, was through psychological warfare, which Eisenhower defined as "the struggle for the minds and wills of men." He charged the Truman administration had "never been able to grasp the full import of a psychological import put forth on a national scale." Eisenhower argued that the United States needed a "united and coherent" Cold War effort in which every department and agency of the government would participate. In order to help coordinate this effort, Eisenhower championed the appointment of a "man of exceptional qualifications to handle the national psychological effort" who would have direct access to the President. For Eisenhower, it was critical

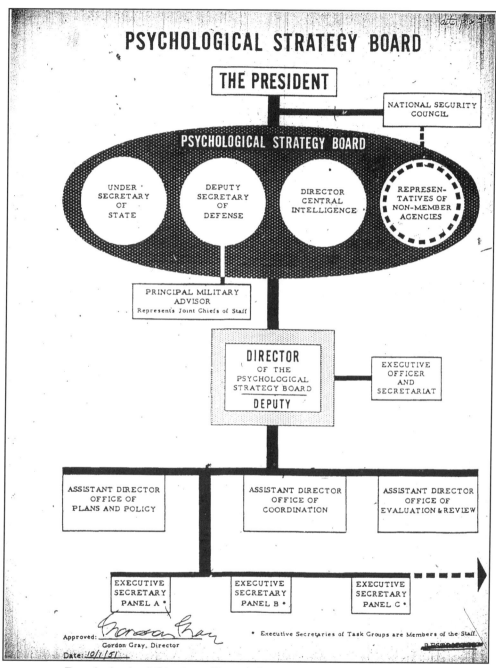

Psychological Strategy Board Organizational Chart (1951)

to the success of such an effort would be the realization that, as a nation, "everything we say, everything we do, and everything we fail to do will have its impact in other lands"[58] What role the Psychological Strategy Board would play was uncertain, given President-elect Eisenhower's interest in reinvigorating the planning and coordinating role of the National Security Council.

The future of the PSB was further clouded when, after his election, Eisenhower appointed C.D. Jackson, a wartime associate at Supreme Headquarters Allied Expeditionary Force (SHAEF) who had experience in psychological warfare planning, to the post of Special Assistant to the President for Cold War Planning.[59] In addition, Eisenhower formed a committee to investigate the foreign information activities of the US government with an eye toward improving their utility and coordination. President Eisenhower announced the formation of the President's Committee on International Information Activities (better known as the Jackson Committee) on 26 January 1953.[60] The Jackson Committee's report was submitted to the President by the end of June, and discussed by the NSC during July and August.[61] The committee members concluded that despite the numerous attempts at improving the policy coordination process in previous years that "a serious gap [existed] between the formulation of general objectives and the detailed actions required to give effect to them." According to the Committee, the PSB had "neither sufficient power to exercise effective coordination nor the techniques adequate to produce meaningful evaluations." Furthermore, the committee considered it inappropriate to attempt to hive off "psychological" planning from wider national policy, in which every diplomatic, military, or economic action had "psychological" implications.[62] Having identified these failings, the Jackson Committee recommended the PSB be abolished, and it be replaced by a new body "capable of assuring the coordinated execution of national security policies." This resulted in yet another organizational shuffle in the ensuing months.

On 2 September 1953, President Eisenhower issued Executive Order 10483 establishing the Operations Coordinating Board (OCB) in order to coordinate and ensure the carrying out of national security policy. As a result of this directive, the Psychological Strategy Board was to be abolished within 60 days, and its functions transferred to the OCB. The Operations Coordinating Board reported to the National Security Council. The OCB was composed of the Under Secretary of State, the Deputy Secretary of Defense, the Director of the Foreign Operations Administration (abolished in August 1954), the Director of Central Intelligence, and a designated representative of the President. Thus, its compositions closely mirrored that of the earlier Psychological Strategy Board, though its writ was intentionally broader. Whether the OCB was any more successful than its predecessor is a matter of debate.

Conclusion

The Psychological Strategy Board represented an attempt to find a bureaucratic/interagency solution to the coordination of America's efforts to more successfully wage Cold War. Mirroring the conclusions of the Jackson Committee, recent historical scholarship has taken a rather critical view of the PSB. According to Walter Hixson, "the PSB produced reams of studies, but failed to marshal the national security bureaucracy behind a coordinated effort."[63] In her recent study, Sarah-Jane Corke has argued that, "The bureaucratic machinery put in place between 1947 and 1952 had the unintended consequence of further removing the planning, development and implementation of covert operations from broader questions of national policy." She regarded this as particularly problematic because it "ensured that American psychological warriors had an unprecedented degree of autonomy" which meant that they did not always act in accord with the wider American national interests.[64] Though it is beyond the scope of this brief overview to assess the PSB's failure to affect broad-ranging interagency cooperation, this author presents the following modest observation. The PSB was established in a period when the United States government had just begun to address the question of what is national security policy, and was doing so after 1950 in the context of a challenging war in East Asia. The emergence of national security policy, as opposed to foreign policy, posed a challenge to the traditional purview of the Department of State. That bureaucratic infighting remained unresolved even after numerous attempts had been made to improve the planning, coordination, and execution of policy may be unfortunate, but it is hardly surprising. The failure of the PSB might not lie in its bureaucratic structure, but rather in the limits of what can be attained by any attempt to better coordinate information policies or psychological strategy. There is also a legitimate question as to what can be construed under the umbrella of "all means short of war." Finally, later day attempts at information management, whether in the Eisenhower administration, during the Vietnam War, or the modern day campaigns in Southwest Asia, have all demonstrated that the PSB's limitations were certainly not unique.[65]

Notes

1. Some of the most extensive early research on the history of the US psychological warfare effort in World War II and the early Cold War was done by Dr. Edward Lilly, a historian, who served as the Office of War Information (OWI) first Official Historian (late June 1944 to late 1945), as a Special Consultant to the Joint Chiefs of Staff from 1946–1952, and as a long-serving member of the National Security Staff thereafter. While working for OWI, Lilly began a history of the agencies activities, which eventually ran to 800 pages. While working for the Joint Chiefs of Staff, Lilly produced two classified reports: a 1,400-page opus on the US psychological warfare effort, generally, and a shorter 104-page report titled "Psychological Operations 1945–1951. None have been published, but drafts and extensive material Lilly collected for research are available at the Dwight D. Eisenhower Library. See Papers of Edward P. Lilly, especially Boxes 15–53, Eisenhower Library. Lilly was finally able to publish a distillation of his voluminous research in an article "The Psychological Strategy Board and Its Predecessors: Foreign Policy Coordination 1938–1953," in *Studies in Modern History*, Gaetano L. Vincitoria, ed. (New York: St. John's University Press, 1968), pp. 337–82. Other early literature on the

subject includes Edward W. Barrett, *Truth Is Our Weapon* (New York: Funk and Wagnalls, 1953); Murray Dyer, *The Weapon on the Wall: Rethinking Psychological Warfare* (Baltimore: Johns Hopkins Press, 1959); Robert Holt and Robert van de Velde, *Strategic Psychological Operations and American Foreign Policy* (Chicago: University of Chicago Press, 1960); Thomas C. Sorensen, *The Word War: The Story of American Propaganda* (New York: Harper & Row, 1968); and Robert Edward Summers, *America's Weapons of Psychological Warfare* (New York: Wilson, 1951); and John Boardman Whitton, ed., *Propaganda and the Cold War* [originally Public Affairs Press, 1963], reprint edition (Westport, CT: Greenwood Press, 1984).

2. A brief note on terminology: psychological warfare was used extensively from World War II on to refer to what an earlier generation would have called propaganda. Due to concern over that "psychological warfare" sounded out of place in peacetime, the phrase "foreign information operations" was applied by some as a substitute. The difficulty in delineating the meaning of these terms (propaganda/psychological warfare/foreign information operations) was recognized by participants at the time. There is fairly extensive literature on this subject, of which the following is certainly not exhaustive. See Nancy E. Bernhard, "Clearer than Truth: Public Affairs Television and the State Department's Domestic Information Campaigns, 1947–1952," *Diplomatic History*, v.21, n.4 (Fall 1997): 545–67; Nancy E. Bernhard, *US Television News and Cold War Propaganda, 1947–1960* (New York: Cambridge University Press, 1999); Sarah-Jane Corke, *US Covert Operations and Cold War Strategy: Truman, Secret Warfare, and the CIA, 1945–53* (London: Routledge, 2008); Peter Grose, *Operation Rollback: America's Secret War Behind the Iron Curtain* (Boston: Houghton Mifflin, 2000); Walter Hixson, *Parting the Curtain: Propaganda, Culture and the Cold War, 1945–1961* (New York: St. Martin's Press, 1997); David F. Krugler, *The Voice of America and the Domestic Propaganda Battles, 1945–1953* (Columbia: University of Missouri Press, 2000); Scott Lucas, "The Campaigns of Truth: The Psychological Strategy Board and American Ideology, 1951–1953," *The International History Review* 18:2 (May 1996), pp. 279–302; Scott Lucas, *Freedom's War: The American Crusade Against the Soviet Union* (New York: New York University Press, 1999); Gregory Mitrovich, *Undermining the Kremlin: America's Strategy to Subvert the Soviet Bloc, 1947–1956* (Ithaca, NY: Cornell University Press, 2000); Michael Nelson, *War of the Black Heavens: The Battles of Western Broadcasting in the Cold War* (Syracuse, NY: Syracuse University Press, 1997); Frank Ninkovich, *US Information Policy and Cultural Diplomacy* (Washington, DC: Foreign Policy Association, 1996); Kenneth Osgood, *Total Cold War: Eisenhower's Secret Propaganda Battle at Home and Abroad* (Lawrence: University Press of Kansas, 2006); Alfred H. Paddock Jr., *US Army Special Warfare: Its Origins*, rev. edn. (Lawrence: University Press of Kansas, 2002); and Shawn J. Parry-Giles, *The Rhetorical Presidency, Propaganda and the Cold War, 1945–1955* (Westport, CT: Praeger, 2002).

3. On the history of the Creel Committee, see Stewart Halsey Ross, *Propaganda for War: How the United States Was Conditioned to Fight the Great War of 1914–1918* (Jefferson, NC: McFarland & Company, 1996), and Stephen Vaughn, *Holding Fast the Inner Lines: Democracy, Nationalism, and the Committee on Public Information* (Chapel Hill: University of North Carolina Press, 1980).

4. For an assessment of the dilemma Roosevelt faced, see Steven Casey, *Cautious Crusade: Franklin D. Roosevelt, American Public Opinion, and the War Against Nazi Germany* (Oxford: Oxford University Press, 2001).

5. William Leuchtenburg has written, "in flat defiance of the cardinal rule of public administration textbooks—that every administrator appear on a chart with a clearly stated assignment—[Franklin Roosevelt] not only deliberately disarranged spheres of authority but also appointed men of clashing attitudes and temperaments." William E. Leuchtenburg, "Franklin D. Roosevelt: The First Modern President," *Leadership in the Modern Presidency*, ed. by Fred I. Greenstein (Cambridge, MA: Harvard University Press, 1988), p. 28.

6. Allan M. Winkler, *The Politics of Propaganda: The Office of War Information 1942–1945* (New Haven: Yale University Press, 1978), pp. 21–28.

7. Ibid., p. 25. On Roosevelt's concern regarding Nazi activities in Latin America, see Robert Dallek, *Franklin D. Roosevelt and American Foreign Policy, 1932–1945* (New York: Oxford University Press, 1979), pp. 233–36.

8. Winkler, *The Politics of Propaganda*, pp. 28–29.

9. On the Office of Censorship, see Michael S. Sweeney, *Secrets of Victory: The Office of Censorship and the American Press and Radio in World War II* (Chapel Hill: University of North Carolina Press, 2001).

10. Reproduced in Appendix A, p. 2 of SNWCC 304/4, 3 September 1947. State-War-Navy Coordinating Committee Records Policy Files, 1944–1949 [Microfilm] (Wilmington, DE: Scholarly Resources, 1977), Reel 26 [hereafter SWNCC].

11. Clayton D. Laurie, *The Propaganda Warriors: America's Crusade Against Nazi Germany* (Lawrence: University Press of Kansas, 1996), pp. 233–40.

12. The General Board, United States Forces, European Theater, *Psychological Warfare in the European Theater of Operation*, Study No. 131. p. 1 https://cgsc.leavenworth.army.mil/carl/eto/eto-131.pdf

13. Michael S. Sweeney, *Secrets of Victory: The Office of Censorship and the American Press and Radio in World War II* (Chapel Hill: University of North Carolina Press, 2001), pp. 209–10.

14. Parry-Giles, *The Rhetorical Presidency*, p. 4.

15. Executive Order 9608, 31 August 1945. John T. Woolley and Gerhard Peters, The American Presidency Project [online]. Santa Barbara: University of California (hosted), Gerhard Peters (database), http://www.presidency.ucsb.edu/ws/?pid=60671.

16. Nonetheless, Truman scholar and biographer Alonzo Hamby argues that "as late as September 1946, [Truman's] personal attitudes toward the USSR remained mixed and allowed room for hope about the future. He did not believe that his move in the direction of a hard line was irrevocable." Alonzo Hamby, *Man of the People: A Life of Harry S. Truman* (New York: Oxford University Press, 1995), p. 360.

17. Hamby, *Man of the People*, p. 346.

18. Walter Millis, ed. *The Forrestal Diaries* (New York: Viking Press, 1951), p. 143. In addition, during 1946 a considerable amount of Forrestal's attention was engaged in the question of unification of the armed services into a single department, which would itself have important implications for the coordination of US national security policy.

19. On the broader context of the emerging national security state, see Michael J. Hogan, *A Cross of Iron: Harry S. Truman and the Origins of the National Security State, 1945–1954* (Cambridge: Cambridge University Press, 1998); and Melvyn P. Leffler, *A Preponderance of Power: National Security, the Truman Administration, and the Cold War* (Stanford: Stanford University Press, 1992).

20. The State-War-Navy Coordinating Committee had been formed in December 1944. It was made up of the Under Secretary of State, the Under Secretary of War, and the Under Secretary of the Navy. Harold W. Moseley, Charles W. McCarthy, and Alvin F. Richardson, "The State-War-Navy Coordinating Committee," *The Department of State Bulletin*, v.13, n.333 (11 November 1945), pp. 745–47.

21. Corke, *US Covert Operations and Cold War Strategy*, p. 28. The General Board, United States Forces, European Theater did commission a study on psychological warfare in the European Theater. It was prepared by Major Ray K. Craft with consultation from BG Robert A. McClure and COL Clifford R. Powell. It is unclear, however, when this report was completed. The General Board, United States Forces, European Theater, *Psychological Warfare in the European Theater of Operation*, Study No. 131. p. 3, https://cgsc.leavenworth.army.mil/carl/eto/eto-131.pdf.

22. Memorandum for the SWNCC Secretariat, 31 May 1946, Ser. No. 00395P35, SWNCC [Microfilm], Reel 26.

23. Report of Ad Hoc Committee—Psychological Warfare, 10 December 1946, SWNCC [Microfilm], Reel 26.

24. President Truman attended only 10 of the first 55 meetings of the NSC. On the functioning of the NSC in the Truman Presidency, see http://www.whitehouse.gov/nsc/history.html.

25. SANA–5781, 24 October 1947, SWNCC Records [Microfilm], Reel 26.

26. Memorandum for SANACC with a draft Memorandum to the President, 25 October 1947, SWNCC [Microfilm], Reel 26.

27. Summary of Decisions of the War Council at its Meeting on 4 November [1947], Declassified Documents Reference System [Microfiche], 2000, Fiche 244, doc. no. 3058. See also *Foreign Relations of the United States, 1945–1950: Emergence of the Intelligence Establishment* [hereafter *FRUS, 1945–1950*] (Washington, DC: US Government Printing Office, 1996), Doc.248, pp. 633–34, http://www.state.gov/www/about_state/history/intel/index.html.

28. Reproduced in *FRUS, 1945–1950*, Doc.249, pp. 635–37.

29. Paddock, *US Army Special Warfare*, p. 50.

30. *FRUS 1945–1950*, Doc.252, pp. 639–42.

31. *FRUS 1945–1950*, Doc.253, pp. 643–45.

32. Krugler, *The Voice of America*, pp. 57–70. The Smith-Mundt Act stated "the Secretary [of State] is authorized, when he finds it appropriate, to provide for the preparation, and dissemination abroad, of information about the United States, its people, and its policies, through press, publications, radio, motion pictures, and other information media, and through information centers and instructors abroad."

33. The communist coup in Czechoslovakia had important repercussions in Western Europe, as well as in the United States. Anglo-French-Benelux discussions on a security treaty got underway in Brussels, Belgium from 4–15 March, which resulted in the signing of the Treaty of Brussels on 17 March 1948. John Baylis, *The Diplomacy of Pragmatism: Britain and the Formation of NATO: 1942–1949* (London: Macmillan, 1993); Bruna Bagnato, "France and the Origins of the Atlantic Pact," and Cees Wiebes and Bert Zeeman, "The Origins of Western Defense. Belgian and Dutch Perspectives 1940–1949," both in Enni DiNolfo, ed., *The Atlantic Pact Forty Years Later: A Historical Reappraisal* (Berlin: Walter de Gruyter, 1991). The British, Canadians, and Americans also undertook preliminary security consultations in March with

an aim toward coordinated their respective emergency war plans. Cees Wiebes and Bert Zee-man, "The Pentagon Negotiations March 1948: The Launching of the North Atlantic Treaty," *International Affairs*, v.59, n.3 (Summer 1983), pp. 353–54; and Steven L. Rearden, *History of the Office of the Secretary of Defense*, v.1: *The Formative Years 1947–1950* (Washington, DC: Historical Office, Office of the Secretary of Defense, 1984), pp. 459–61.

34. All that came of Forrestal's request was a study undertaken by a board of NSC Consultants which essentially verified the status quo. *FRUS 1945–1950*, Docs.262, 266–67, pp. 655, 664–65.

35. Years before, Major Ray Craft stated in his report to the General Board, European Theater of Operations that "psychological warfare operations embrace politics which can only be fixed at the highest government level." *Psychological Warfare in the European Theater of Operation*, Study No. 131. p. 3.

36. The conduct of covert psychological activities abroad remained within the purview of the CIA. NSC 10/2, "National Security Council Directive on Office of Special Projects," approved on 18 June 1948, established an Office of Special Projects within the CIA. The Director of the Office of Special Projects was responsible, through coordination with designated representatives of the Secretaries of State and Defense that covert operations were planned and conducted consistent with the foreign and military policies of the US government. The JCS were also given the right to assist in the drawing up plans for covert operations in wartime and approving such plans as being consonant with approved wartime plans. *FRUS 1945–1950*, Doc.292, pp. 713–15. The State Department position on this matter was consistent under both Secretary Marshall and his successor Dean Acheson. Acheson reiterated the State Department's position that they should retain responsibility for planning at the 34th NSC Meeting on 17 February 1949, Papers of Harry S. Truman, PSF Subject File, NSC Meetings File, Box 177, Harry S. Truman Library [hereafter HSTL].

37. NSC 59/1 was approved by President Truman on 10 March 1950. It had been circulated to the NSC, but does not appear to have been discussed at any formal NSC meeting prior to its approval based on the authors review of NSC Meetings 51–54, dating from 5 January through 6 April 1950. Papers of Harry S. Truman, PSF Subject File, NSC Meetings File, Box 179, HSTL.

38. NSC 43, "Planning for Wartime Conduct of Overt Psychological Warfare" had been approved on 9 March 1949, following discussion at the 34th NSC Meeting of 17 February.

39. *Foreign Relations, 1950–1955: The Intelligence Community* [hereafter FRUS 1950–1955], Doc.2, pp. 2–4; and Lilly, "The Psychological Strategy Board and Its Predecessors," p. 360.

40. "Truman Address to Editors Stresses Facts Urgent in Global Field," *Christian Science Monitor*, 20 April 1950, p. 3; and "Text of President's Plea to Editors to Help Crush Red 'Lies,'" *New York Times*, 21 April 1950, p. 4. Edward Barrett, a former director of the Office of War Information's Overseas Branch and editor of *Newsweek*, was recruited to serve at the Assistant Secretary of State for Public Affairs in order to help launch this campaign. Barrett, *The Truth Is Our Weapon*, pp. 31ff.

41. *FRUS 1950–1955*, Doc.17, pp. 20–21.

42. NSC 74 was never formally approved, but remained the subject of discussion throughout the fall of 1950. The complete report, running to 51 pages, is not reproduced in *FRUS, 1950–1955*. See note 17, pp.20–21. There is a copy of the first nine pages of the report and

comments thereon by Secretary of Defense Louis Johnson and the JCS in the folder for the 77th NSC Meeting, 4 January 1951, Papers of Harry S. Truman, PSF, NSF Files, Box 181, HSTL.

43. NSC 74, p.4, located in the folder for the 77th NSC Meeting, 4 January 1951, Papers of Harry S. Truman, PSF, NSF Files, Box 181, HSTL.

44. Johnson suggested these civilian members might include a public relations counsel, publisher, internationally knowledgeable labor and church leaders, and/or professionals in the field of mass psychology. Memorandum from the Secretary of Defense for the Executive Secretary, NSC re: NSC 74, 13 September 1950. Before the matter was taken up by the NSC at its 77th Meeting, the Joint Chiefs of Staff also submitted a memorandum which supported the suggestion that the director of any new board be independent of any existing agency and responsible directly to the President. Memorandum for the Secretary of Defense from the JCS, 3 January 1951. Copies of both are located in the folder for the 77th NSC Meeting, 4 January 1951, Papers of Harry S. Truman, PSF, NSF Files, Box 181, HSTL.

45. Truman informed Johnson that he wanted him to resign on 11 September, and had Johnson sign his letter of resignation the next day. Thus, it is unclear whether the letter submitted to the NSC on the 13th was actually signed by Johnson, or represented the views of one his office staff. On Truman's decision to remove Johnson, see Hamby, *Man of the People*, pp.546–47.

46. For the best recent study of the US governments public diplomacy/information engagements in the context of the Korean War, see Steven Casey, *Selling the Korean War: Propaganda, Politics, and Public Opinion in the United States, 1950–1953* (Oxford: Oxford University Press, 2008). See also the "Second Progress Report on NSC 59/1," *FRUS 19501–955*, Doc.28, pp.45–46.

47. DDRS, 1992, F127, 1873.

48. Quoted in Casey, *Selling the Korean War*, p.175.

49. Casey, *Selling the Korean War*, p.176.

50. Memorandum from Asst. Sec. State for Public Affairs to Sec. State Acheson, 13 February 1951, *FRUS 1950–1955*, Doc.49, p.92.

51. Osgood, *Total Cold War*, p.45.

52. The National Psychological Effort, min.4, 77th NSC Meeting, 4 January 1951, Papers of Harry S. Truman, PSF, NSF Files, Box 181, HSTL.

53. Memorandum from Asst. Sec. State for Public Affairs to Sec. State Acheson, 13 February 1951, *FRUS 1950–1955*, Doc.49, p.93.

54. *FRUS 1951*, v.1, pp.58–60. The first director of the PSB was Gordon Gray, a friend of Truman's who had previously served in the administration as Assistant Secretary and Secretary of the Army before leaving government to become President of the University of North Carolina. Gray served as Director of the PSB from June 1951 until May 1952. Raymond H. Allen then served as director from May-September 1952, and Admiral Alan Kirk served as director for the remainder of the Truman presidency.

55. "One Year's Development of the PSB," DDRS, 1999, F99, 1155.

56. See note 2.

57. Memorandum of Discussion, 34th NSC Meeting, 17 February 1949, Papers of Harry S. Truman, PSF, NSC Meetings File, Box 186, HSTL.

58. "'Ike' Sees Victory in Cold War," *Christian Science Monitor*, 9 October 1952, p.9.

59. Osgood's *Total Cold War* has the best recent discussion on C.D. Jackson's role in the

Eisenhower administration. Two earlier studies of note are Blanche Wiesen Cook, *The Declassified Eisenhower: A Divided Legacy* (Garden City, NY: Doubleday & Co., 1981), and H. W. Brands, *Cold Warriors: Eisenhower's Generation and American Foreign Policy* (New York: Columbia University Press, 1988).

60. The Committee was chaired by William H. Jackson. Its members included Robert Cutler (Eisenhower's Special Assistant for National Security Affairs), Gordon Gray (the first director of the PSB), Barklie McKee Henry, John C. Hughes, C. D. Jackson, Roger M. Kyes, and Sigurd Larmon. *The President's Committee on International Information Activities: Report to the President*, 30 June 1953, DDRS, 2002, F100-101, 1697.

61. The longest discussion of the report took place at the 152nd meeting of the NSC on 2 July. Subsequent progress reports on the drafting of an executive order to establish the Operations Coordinating Board took place at the 157th, 159th, and 160th meetings. Copies of these meetings are located in Papers of Dwight D. Eisenhower (Ann Whitman File), National Security File, Box 4, Dwight D. Eisenhower Library.

62. *The President's Committee on International Information Activities: Report to the President*, pp. 89–90.

63. Quoted in Osgood, *Total Cold War*, p. 45.

64. Corke, *US Covert Operations and Cold War Strategy*, p. 5.

65. For instance, in 1964, senior officials in the Johnson administration deemed it necessary to establish a "communications czar" in Saigon to improve US information policies in Vietnam; in 1969, President Nixon established an Office of White House Communications to better coordinate its own information policies; and in 2003 President George H. Bush established an Office of Global Communications to improve US information policies in the global war on terror.

"Our Most Reliable Friends"—Army Officers and Tribal Leaders in Western Indian Territory, 1875-1889

by

Mr. William A. Dobak
US Army Center of Military History

Scholars and members of the general public alike often display a warped view of the United States Army's relation to American Indian tribes. This view emphasizes violent conflict rather than the implementation of national policy. The problem has existed for more than a century, at least since Frederic Remington picked the most dramatic incidents in his own and other authors' articles to serve as subjects for illustrations. A generation ago, Robert M. Utley's two classic histories of the Army in the American West appeared, each with the phrase, *The United States Army and the Indian* in the title. Today, the University of Nebraska Press paperback editions of *Frontiersmen in Blue* and *Frontier Regulars* contain only the briefest hints that both books first appeared as part of Macmillan's *Wars of the United States* series, commissioned in the early 1960s—a commission that defined the scope of the author's investigation and necessarily excluded a mountain of documentation about "the Army and the Indian" that would take several historians' lifetimes to work through.[1]

This focus on conflict has led to some very odd assertions finding their way into print. One historian has observed casually: "Not surprisingly, many soldiers were devoted Indian-haters; relatively few were concerned with Indians' rights or their government's guarantees of protection to Indians under federal treaties." Another relates that after General William T. Sherman's narrow escape from an ambush in 1871, he "abandoned the [Grant administration's] Peace Policy and Sheridan, as commander of the Division of the Missouri, was responsible for all military matters in the million-square-mile expanse of the Indian insurrection." An architectural historian writing about 19th century Army posts in Wyoming seems to separate the Army from the Federal government entirely: "The nature of the army's conflict with the Indians changed as the army's goal shifted from protecting emigration to facilitating the process of settlement. US government relations with the Indians changed as well. . . ." The Army has a "conflict"—the US government maintains "relations."[2]

The viewpoint that these and many other authors share might be called the Fallacy of Army Autonomy. One would like to see all of them enrolled in an introductory course in American politics, to learn about the three branches of government and, subordinate to the executive, the cabinet departments, for it was within this framework that the 19th century Army operated. Like any other agency of the Federal government, the Army followed policies decided at a higher level and depended for its budget on funds appropriated from year to year by the legislative branch.

The Army itself is not altogether blameless in this matter. The 1956 edition of *American Military History* (ROTCM 145-20), the military history text used in Army ROTC courses, did not get around to mentioning the administration of Indian affairs until the chapter that discussed post-1865 events. Then, it did so in tones that suggest the screenplay for one of John Ford's Cavalry Trilogy films:

> In 1849 the administration of Indian affairs had been taken away from the War Department and entrusted to the Department of the Interior, whose agents either were inexperienced in the administration of Indian affairs or deliberately cheated their charges. Even under conditions of hostility, these agents furnished Indians with the latest models of repeating rifles and plenty of ammunition, either because it was financially profitable to do so or because they naively believed that the Indians wanted their weapons only to kill buffalo.

Since this textbook concerned itself only with the Army, it did not mention that before 1849, the Commissioner of Indian Affairs reported to the Secretary of War; or, indeed, that administrators of Indian affairs had reported to the Secretary of War since Washington's first term as President. Having noted the 1849 transfer, the authors imply that all Indians are plains buffalo hunters.[3]

The text of the 1969 edition underwent some revision. Gone were the malevolent or naive civilian agents. Instead, shoved even further back in the text, in a brief discussion of conflict with the Apaches, was:

> Throughout the Indian wars there was constant friction between the War and the Interior Departments over the conduct of Indian affairs. A committee of the Continental Congress had first exercised this responsibility. In 1789 it was transferred to the Secretary of War, and in 1824 a Bureau of Indian Affairs was created in the War Department. When the Department of the Interior was established in 1849, the Indian bureau was transferred to that agency. Thus administration of Indian affairs was handled by one department while enforcement lay with another.

This is an improvement—the wording remains unchanged in the most recent edition of the text—but the balance of the chapter concerns itself with a series of unconnected armed conflicts across the West. Granted, the Army between 1874 and 1898 was smaller than today's New York Police Department, or the total number of police (state, county, and municipal) in New Jersey. Granted, too, that the political structure of North American Indian societies—which operated by consensus rather than majority rule, with extremely decentralized authority, and individual membership in any particular band apt to shift at a moment's notice—makes comparisons to official dealings with other peoples difficult. Still, unpublished military and civilian documents in the National Archives reveal a side of Army life largely unknown to readers and writers of the chase-and-shootout school of Western military history.[4]

322

The more usual pattern of relations between the Army, the Indian Service (made up of the agents, teachers, physicians, blacksmiths, and other employees hired by the Office of Indian Affairs and assigned to tribal agencies throughout the country), and the American Indians themselves, grew largely out of the country's need to replenish its treasury after the Civil War. The war had put the United States government deeply in debt, and a secure central corridor became necessary to transport the gold produced in California, Colorado, and Montana mines. In order to do this, the government extinguished the Indian title to lands in Nebraska and Kansas, moving tribes in those north to Dakota Territory, or south to the Indian Territory (present-day Oklahoma). Some of these tribes had moved west only a generation or two before from their ancestral territory east of the Mississippi River, during the era of Indian Removal.[5]

Agents and other employees of the Office of Indian Affairs established offices (called agencies) at convenient locations to administer the different tribes, furnish them with health and educational services, and distribute rations and "annuity goods" such as clothing, blankets, and hardware under the terms of various treaties. Such arrangements were common on Indian reservations throughout the United States. What occurred more frequently in the plains and the Southwest than elsewhere was the establishment of military posts near the agencies. These posts were necessary to restrain the young men of some of the newly confined plains tribes, to avert conflict between tribal factions (some of which stemmed from divided allegiances that dated back to the Civil War), and to prevent the rustling of Indian horses and cattle or the cutting of reservation timber by neighboring white settlers. As Brigadier General Alfred H. Terry wrote when the Crow Indians of Montana asked that white intruders be removed from their reservation: "Every consideration of good faith, humanity and public policy requires that . . . the vagabonds and roughs of the frontier should be taught that they . . . must obey the law . . . an obligation which they now deride." During the last third of the nineteenth century, therefore, soldiers had many opportunities for first-hand observation of American Indians and the white civilians who were appointed to govern them.[6]

Since Richard Ellis published his article, "The Humanitarian Soldiers," in 1969, it has become a commonplace that many Army officers were keen observers of western Indians and devoted a good deal of thought to the condition of those who were restricted to reservations. More recently, Sherry Smith has devoted a large part of her book, *The View From Officers' Row*, to this matter. Yet Ellis's article and Smith's book rely largely on officers' private letters and diaries and on published sources. What does Army officers' official correspondence tell us about how they handled everyday relations with tribal leaders on the reservations? Let's look at some of the letters they wrote about the plains tribes on their reservations in the western part of the Indian Territory during the 15 years after the Army's last campaign on the southern plains, the Red River War of 1874–75.[7]

Land for these reservations became available in 1866 after the United States government imposed a series of punitive treaties on Southeastern tribes that the government had forced west a generation earlier. The punishment was for those tribes having backed the losing side in the Civil War. The 1866 treaties stripped the Five Tribes of their land holdings west of the 98th Meridian. Large tracts of this western land were then set aside as reservations for tribes of the Southern Plains.[8]

This western half of the Indian Territory was the site of three principal Army posts. The oldest of these, Camp Supply, on the North Fork of the Canadian River in the northwest corner of the territory, dated from General Sheridan's autumn campaign against the Southern Cheyennes in 1868. Fort Reno was also on the North Fork of the Canadian River, far downstream from Fort Supply, at the eastern edge of the Cheyenne and Arapaho reservation. Fort Sill stood near the Comanche and Kiowa Agency in the southwest corner of the territory.[9]

Just east of Fort Reno and the Cheyenne-Arapaho reservation, on the other side of the 98th Meridian, lay the Unassigned Lands, popularly known as the Oklahoma District, about 2,950 square miles—which works out to nearly 12,000 quarter sections of farm land.[10] This large tract in the geographical center of the Indian Territory soon attracted the attention of promoters—called Boomers, from the land boom they hoped to create—who urged throughout the early 1880s that the Unassigned Lands be opened to settlement. From 1879 through 1885, the Boomers mounted annual, even semiannual, "invasions" from towns in southern Kansas. Just as often, soldiers from Fort Reno and other Army posts would round them up and deliver them to a Federal court in Fort Smith, Arkansas, or Wichita, Kansas, where the judge would levy a fine—the only legal penalty for trespass on Federal lands—the invaders would protest their inability to pay, and the court would turn them loose to repeat the process. But the Boomers had before them the example of the Black Hills gold rush of 1875 and 1876, and they knew that a popular invasion of Indian lands was likely to make the Federal government disavow its treaty obligations.[11]

They were right. Through the early months of 1889, a lame duck session of the 50th Congress wrestled with legislation to open the Unassigned Lands to settlement.[12] On 12 February that year, a delegation of about 20 Cheyennes and Arapahos called on Fort Reno's commanding officer. "There is considerable uneasiness among these Indians," the colonel reported, "they get all the wild reports about Oklahoma and the Territory from the Kansas papers, and are very anxious to get their views on the matter before the President and Congress."[13]

A Cheyenne named Whirlwind spoke for the delegation. He deplored the projected land grab and asked that the tribes be allowed to keep a contiguous tract "and let us have land enough so we can . . . support ourselves by stock raising and not have to rely on the uncertainties of farming in a droughty country." In his prefatory remarks, he explained the delegation's visit by saying that "in time of peace we have always found the military our most reliable friends."[14]

A little preliminary oil never hurts when asking a favor, but how much factual basis did Whirlwind's assessment of the Army and its officers have? Five years earlier, Whirlwind had been to Washington and met with the Secretary of the Interior and had spoken out against grazing leases for white ranchers on the Cheyenne-Arapaho reservation.[15] He and other tribal leaders in many parts of the west had observed the rudimentary Federal bureaucracy of the 19th century and had some idea, at least, of how it worked. What factors might have moved Whirlwind to call Army officers "our most reliable friends"?[16]

First, there was food. Federal law—the same act of 1834 that created the Office of Indian Affairs—obliged the Army to feed any Indians who visited a military post.[17] Congress wrote this provision into the bill in response to reports of starvation among the tribes that were being moved forcibly from the southeast to the Indian Territory.[18] A generation later, the Southern Cheyennes and Arapahos must have learned from visits with their northern kin about the Army's feeding those of them who visited Forts Fetterman and Laramie, in Wyoming. In 1871, the Army distributed more than $150,000 worth of rations to Arapahos, Cheyennes, and Sioux at those two posts on the North Platte—nearly three-quarters of the money it spent on feeding Indians that year.[19]

More immediate in shaping the Southern Plains tribes' attitude toward the Army must have been the events of 1875, just as the Red River War was petering out. Early that May, under the head "Hungry Indians," the *Army and Navy Journal* printed a dispatch written a month earlier. The Comanche-Kiowa agency had run out of flour, Fort Sill's commanding officer wrote, and he had ordered his own commissary to issue rations to 294 Indians. Moreover, the civilian agency was issuing beef cattle that were too weak to be driven to the Indian camps for slaughter; instead, they had to be butchered where they fell and the meat carried to the camps. A second report from General John Pope, whose command included the Indian Territory, confirmed that the Wichita and Cheyenne-Arapaho agencies lacked provisions too and had since the previous summer. "It is idle to expect that these Indians will remain peacefully upon their reservations with the prospect of starvation in doing so," Pope added.[20]

One way to help feed the people was to allow them to hunt buffalo off the reservations. "Much of the sickness during the past summer among these tribes is attributed to the insufficiency of their rations," Fort Sill's commanding officer wrote in the fall of 1875.[21] To allow hunting parties off the reservations, though, posed the problem of keeping the peace between them and any stray or organized white people they might encounter, whether cowboys or Texas Rangers.[22] The solution was to provide military escorts for hunting parties. Escorts usually included one sergeant or corporal and half a dozen privates. Troops from posts in the western half of Indian Territory furnished such escorts as early as October 1875 for fall hunts by the Arapahos, Cheyennes, Comanches, Osages, and Wichitas.[23] While the troopers protected the hunters from belligerent whites, the leaders of the hunting party were responsible for the escort's safety "should they meet any bad Indians," as the colonel put it.[24] Some of the hunting parties stayed out until January.[25]

The question of escorts for off-reservation hunting parties recalls another, much more unusual, instance of escort duty. In December 1884, soldiers furnished an escort for Quanah Parker and other Comanches on their way to the Big Bend country of Texas in search of peyote cactus.[26] Not that the Army promoted Native American religion; its institutional interest was in keeping order, and since "civilizing the Red Man" was the business of the Department of the Interior's Office of Indian Affairs and its civilian agents, Army officers were apt to dismiss agents' nervous appeals for help in suppressing such manifestations of Indian culture as communal dances.[27] In 1889, the same year in which Whirlwind called army officers "our most reliable friends," the superintendent of Indian schools for the Cheyennes and Arapahos requested military aid in removing a band of about 200 "non-progressive, dirty and indolent Arapahoe Indians, who visit the school buildings begging . . . and every evening to a late hour hold their strangely grotesque dances, with drums and hideous yells, which are heard throughout . . . the school buildings. . . ."[28] Asked to shift these Arapahos, the senior Army officer present replied that the civilian agent was the proper person for the school superintendent to report to, and added: "In regard to the dances . . . reported, they are only the Indian method of amusement and take the place of ball room, theater, saloon, etc., of his more civilized brother."[29] Some years earlier, another Army officer had been closer to the mark when he reported, after attending the Sun Dance at the Spotted Tail Agency, Nebraska: "The annual Sun Dance is to the Indian about what the camp meeting used to be to the Methodists."[30] In both cases, though, recognition of the social side of an annual religious observance indicates a more relaxed attitude towards non-Christian practices than a civilian Indian agent with a civilizing mission could afford to entertain.

To revert to the immediate question of feeding reservation residents: throughout the west, Army officers were present to observe the periodic distribution of food and annuity goods—cloth and clothing, hardware, and other items—and reported on the quality and quantity of what reservation residents received. In the summer of 1875, General Sheridan issued an order affecting the entire Military Division of the Missouri, from Montana to the Rio Grande, that warned inspecting officers to watch out for "fraudulent transactions," but natural causes like drought or freezing weather could affect the rations, especially beef on the hoof, as drastically as could contractors' chicanery.[31] Adverse reports from inspecting officers in western Indian Territory began to reach the Adjutant General's office soon after the end of the Red River War, as most members of the plains tribes reported to their agencies and began to receive rations.

Federal law had mandated these inspections since 1868, but during the three years after the Red River War, Sheridan's Chicago headquarters received so many adverse reports from inspecting officers at Indian agencies that in 1878 he devoted part of his own annual report to the subject. He began by acknowledging the Department of the Interior's responsibility for Indian affairs, "and if it were not for the results which so severely involve the military, this would be none of my business." But "now that the game, upon which the Indians depended for their regular supply of food, is gone," Sheridan continued, "we shall require a greater supply of rations, with perfect regularity in its issue, to meet the needs of these people." He predicted that "wretched mis-management" by the Office of Indian Affairs might lead to a new wave of Indian wars, for, he wrote, "almost any race of men will fight rather than starve."[32]

Sheridan's published report bore the date 25 October 1878. Since that summer though, the general had been embroiled in controversy with the Interior Department about a plan to consolidate the agencies for the Comanches, Kiowas, and Wichitas at a site some 30 miles from Fort Sill. Apart from balking at the expense of building a new post, Sheridan observed that "the removal of Indian agencies away from military posts has for its main motive a desire to cheat and defraud the Indians by avoiding the presence of officers who would naturally see and report it."[33] Secretary of the Interior Carl Schurz bridled at "so insulting an insinuation," and when Sheridan's annual report appeared, he demanded a statement of the facts behind the "general arraignment" of the Office of Indian Affairs.[34]

Fortunately for Sheridan, he had abundant material to draw from. An Army officer at Fort Sill the previous December had reported the issue of beef cattle that were "not equal to the requirements of [the] contract," and that letter was the basis of a corre-spondence file that grew to include more than 50 similar reports from many parts of Sheridan's geographical command by the time Secretary Schurz asked for Sheridan's facts.[35] Just a month after the Secretary's request, Sheridan issued a 10-page "Supple-mental Report" with a 20-page appendix of quotations from letters that had been arriv-ing in his office for the previous four years.[36]

This bureaucratic flare-up does not figure prominently in recent biographies of Schurz and Sheridan, but it was one high point in the decades-long controversy be-tween soldiers and civilians as to whether the War Department or the Interior De-partment should administer federal Indian policy.[37] It was the only occasion during Sheridan's 14 years in command of the Military Division of the Missouri when he pub-lished a supplementary report to substantiate charges of mismanagement by the Office of Indian Affairs. And this incident shows why Whirlwind could call Army officers the Indians' "most reliable friends"—they could usually be relied on to question the ability and honesty of the employees of a different Federal agency.

One series of visits by tribal leaders to Fort Supply's commanding officer in 1883 and 1884 may have produced some results. In July 1883, Colonel Joseph H. Potter re-ceived a delegation of Cheyennes led by Stone Calf, Little Robe, and five others whom the colonel named, to protest grazing leases to large cattle interests on the Cheyenne and Arapaho reservation. The delegation complained of certain prominent Cheyennes and Arapahos who had signed the leases and alleged bribery by the Bent Brothers and an interpreter employed by the Federal government, among other interested parties. The delegates who came to see Colonel Potter "are very emphatic in speaking of these leases," he reported to General Pope's headquarters at Fort Leavenworth. "[T]his mat-ter should be attended to as early as possible."[38]

The grazing leases continued in force for the next 18 months. In December 1884, some of the same Cheyennes, Little Robe and Stone Calf among them, called again on Colonel Potter. This time Potter sent a personal letter to Lieutenant Colonel Michael V. Sheridan in Chicago, where Sheridan served as aide to his brother, the general. The letter is worth quoting at length. Potter told Sheridan:

> "I see by the papers, that the Senate is after the Interior Dept. in regard to the leases of Indian lands, and 'tis about time. I believe this lease business is a most infamous fraud. This whole Territory is fenced in—and covered with cattle. More cattle than grass. The Indians are set aside entirely. There are about eighty of them here now complaining to me. . . . They say that not one of them have agreed to or signed these leases, and those that did sign them were a lot of old bummers who hang about the agency and would sign anything for a handful of grub. These people say that they want to make farms and raise corn, . . . but that the cattle run over their fences and destroy all they have. One of them told me that a year since he had [125] head of cattle and that now he can't find twenty-five of them. All stolen by these cattle men. Small herds stand no chance here—whether they belong to Indians or white men. And the small owners have been forced to leave or sell out. . . . I have told these Indians . . . to go to their agent about such matters, but they say that they won't do it any more, as they have often enough without any satisfaction. He tells them to take the lease money—and that is all. They say that they want some honest people to come here and look for themselves, but they don't want them to come to the agency and listen to the agent and those at the agency only, but to come and listen to the Cheyenne people. . . . I for one think that these people are right, and that the Government ought to look into this matter. It may save trouble hereafter, and 'tis no more than justice. . . . They say that they want General Sheridan to know how it stands with them. They believe in him. Will you tell him so.[39]

How Potter came to know Michael Sheridan well enough to entitle him to send a private letter is uncertain. They may have known each other in Virginia, during the last months of the Civil War.[40] Nor is it clear whether Michael Sheridan called the general's attention to Potter's letter. It is certain, though, that by the middle of the following year there was sufficient disturbance to bring General Sheridan to the Cheyenne and Arapaho agency, and that on 18 July 1885, two years to the day after Potter's reported the complaints of Little Robe, Stone Calf, and the others, General Sheridan telegraphed President Grover Cleveland to recommend that the Army take over the Cheyenne and Arapaho reservation until the matter of the grazing leases could be settled. Five days later, Cleveland issued a proclamation that declared the leases void and gave cattlemen 40 days to move their herds off Indian land. General Sheridan reported soon after that he had interviewed many prominent Arapahos and Cheyennes, and that even those who had signed the leasing agreements "had become sick of the bargain."[41] The 40 day deadline was impossible to meet. Not until the first week of November 1885 was the Cheyenne and Arapaho reservation free of outsiders' cattle.[42]

Of course, we know that there was no happy ending to this story, certainly not for the 210,000 dispossessed cattle: the hard winter that followed caused "frightful losses" in the overstocked Texas panhandle, much as the winter after that did in Montana and Wyoming.[43] The reservations in western Indian Territory were allotted to individual tribal members a few years later and disappeared from maps. Still, the interplay of Little Robe and Stone Calf, Colonel Potter and Michael Sheridan in 1883 and 1884 gives a hint of why another Cheyenne, Whirlwind—one of the grazing lease-signers of whom the other two Cheyennes complained—could call Army officers "our most reliable friends." No matter how much rank flattery it involved.

Let me end with a little speculation. Tribal leaders who used to call at Army posts to talk to the commanding officer—and there were many of them throughout the West during the late 19th century—were expert at factional politics, as when Colonel Potter's visitors characterized their opponents as "a lot of old bummers who hang about the agency and would sign anything for a handful of grub." Little Robe and Stone Calf lived closer to Camp Supply and were slandering Whirlwind—the same Whirlwind who a few years later called Army officers "our most reliable friends"—who lived closer to the agency, as did some other chiefs. Although these leaders had traveled to Washington, DC, some of them more than once, they probably did not have a firm grasp of the intricacies of the Federal government's executive branch. But they could certainly tell that the employees at their agency dressed differently from the soldiers at the nearby fort. In taking their complaints to the commanding officers at Army posts, were they practicing the same factional skills they used in intratribal squabbles? To make that case would require a lot more research and a much longer paper than we have time for this morning, but the evidence suggests as much.[44]

Finally, is there a benefit to be gained from the study of these episodes of quiet, if sometimes tense, diplomacy and long-winded negotiation? A more extensive look at the Army's activities during the last third of the 19th century reveals a different picture from that shown in most books or in countless western films. Knowledge of this seldom-mentioned chapter of military history might be of some help to people starting out in today's Army, who will learn early in their careers that they are expected to accomplish not only tasks that may be distasteful or dangerous, but some for which they have received no training at all. It may help them to recall that an earlier generation of soldiers found itself dealing, all at the same time, with contentious civilian colleagues, angry and suspicious groups of people with unfamiliar customs and beliefs, and whose command of English varied widely, and clamorous, self-interested ordinary American citizens.

Notes

1. Both Robert M. Utley, *Frontiersmen in Blue: The United States Army and the Indian, 1848–1865* (Lincoln: University of Nebraska Press, 1981), p. xiv, and Utley, *Frontier Regulars: The United States Army and the Indian, 1866–1891* (Lincoln: University of Nebraska Press, 1984), p. xiii, mention "the series in which this volume appears," but in neither instance gives the name of the series. Utley himself describes Macmillan's representative approaching him in *Custer and Me: A Historian's Memoir* (Norman: University of Oklahoma Press, 2004), pp. 93–94, 109–10.

2. Barton H. Barbour, *Fort Union and the Upper Missouri Fur Trade* (Norman: University of Oklahoma Press, 2001), p. 182; David D. Smits, "The Frontier Army and the Destruction of the Buffalo: 1865–1883," *Western Historical Quarterly* 25:3 (Autumn 1994), 312–38, quotation, p. 326; Allison K. Hoagland, *Army Architecture in the West: Forts Laramie, Bridger, and D.A. Russell, 1849–1912* (Norman: University of Oklahoma Press, 2004), p. 14.

3. *American Military History* (Washington, DC: Government Printing Office, 1956), pp. 278–79.

4. *American Military History*, ed. Maurice Matloff (Washington, DC: Office of the Chief of Military History, US Army, 1969), p. 314; *American Military History*, ed. Richard W. Stewart (Washington, DC: US Army Center of Military History, 2005), p. 334.

5. James McPherson, *Battle Cry of Freedom: The Civil War Era* (New York: Oxford University Press, 1988), pp. 442–43, 452–53, 818; H. Craig Miner and William E. Unrau, *The End of Indian Kansas: A Study of Cultural Revolution* (Lawrence: University Press of Kansas, 1978), pp. 116–41; Francis P. Prucha, *The Great Father: The United States Government and the American Indians* (Lincoln: University of Nebraska Press, 1984), pp. 461, 490, 493–96; Utley, *Frontiersmen in Blue*, pp. 2–3, 7–8.

6. Brigadier General A.H. Terry, endorsement 8 Aug 1878, in file 4930 AGO 1878, National Archives (NA) Microfilm Publication M666, Letters Received by the Office of the Adjutant General, 1871–80, reel 421. That year, another general estimated that Nebraska settlers had stolen some 2,000 horses from the neighboring Sioux reservation since 1876. Brigadier General G. Crook to Lieutenant General P.H. Sheridan, 4 Dec 1878, in file 8104 AGO 1878, Ibid., reel 446. For "woodhawks" who cut reservation trees for sale to steamboats on the Missouri, the region's principal navigable river, see Commissioner of Indian Affairs to Secretary of the Interior, 10 March 1877, in file 1469 AGO 1877, Ibid., reel 323.

7. Richard N. Ellis, "The Humanitarian Soldiers," *Journal of Arizona History* 10 (Summer 1969), 53-66; Ellis, "The Humanitarian Generals," *Western Historical Quarterly* 3 (April 1972), 169–78; Sherry L. Smith, *The View From Officers' Row: Army Perceptions of Western Indians* (Tucson: University of Arizona Press, 1990), especially 92–138.

8. Prucha, *Great Father*, pp. 430–34.

9. Histories of these posts are Robert C. Carriker, *Fort Supply, Indian Territory: Frontier Outpost on the Plains* (Norman: University of Oklahoma Press, 1970); Stan Hoig, *Fort Reno and the Indian Territory Frontier* (Fayetteville: University of Arkansas Press, 2000); Wilbur S. Nye, *Carbine and Lance: the Story of Old Fort Sill* (Norman: University of Oklahoma Press, 1937).

10. *Historical Atlas of Oklahoma*, ed. Charles R. Goins and Danney Goble (4th edn., Norman: University of Oklahoma Press, 2006), 124.

11. Carl C. Rister, *Land Hunger: David L. Payne and the Oklahoma Boomers* (Norman: University of Oklahoma Press, 1942), 41.

12. Roy Gittinger, *The Formation of the State of Oklahoma* (Norman: University of Oklahoma Press, 1939), 175–82.

13. Colonel J.F. Wade to Brigadier General W. Merritt, 14 February 1889 (932 AGO 1889), NA Microfilm Publication M689, Letters Received by the Office of the Adjutant General (Main Series), 1881–1889, roll 670. The Fort Reno interpreter translated Whirlwind's speech, and Carlisle graduates read the written English text back to Whirlwind in Cheyenne.

14. "Speech of Whirlwind," filed with 932 AGO 1889, Ibid.

15. Donald J. Berthrong, *The Cheyenne and Arapaho Ordeal: Reservation and Agency Life in the Indian Territory, 1875–1907* (Norman: University of Oklahoma Press, 1976), 102–110.

16. Loretta Fowler calls such leaders "intermediary leaders." Their job was to represent their people to agents of the Federal government, civilian or military. Fowler, *Tribal Sovereignty and the Historical Imagination* (Lincoln: University of Nebraska Press, 2002), 21–29.

17. US Statutes at Large, 4:738.

18. For instance, the Choctaws in 1831. Ronald N. Satz, *American Indian Policy in the Jacksonian Era* (Lincoln: University of Nebraska Press, 1974), 73–81.

19. 42d Cong., 2d sess, H.Ex.Doc. 1, Secretary of War Annual Report, 1871 (serial 1503), 232.

20. *Army and Navy Journal*, 8 May and 3 July 1875.

21. Col. R.S. Mackenzie to Assistant Adjutant General, Department of the Missouri, 4 November 1875, NA Record Group (RG) 393, Records of US Army Continental Commands, part 5, Military Installations, Entry (E) 432-1, Fort Sill, Letters Sent, 1:31.

22. The Texas legislature established the Frontier Battalion of Texas Rangers, six companies of 75 men, in April 1874, a few months before the outbreak of the Red River War. Troops of the Frontier Battalion had several encounters with Indians, but within a few years turned their attention to law enforcement along the state's frontier of white settlement. Gary C. Anderson, *The Conquest of Texas: Ethnic Cleansing in the Promised Land, 1820–1875* (Norman: University of Oklahoma Press, 2005), 359; Robert M. Utley, *Lone Star Justice: The First Century of the Texas Rangers* (New York: Oxford University Press, 2002), 144–46, 149–52.

23. Colonel R.S. Mackenzie to Capt T.J. Wint, 26 October 1875, NA RG 393, pt. 1, E 432-1, 1:13; Colonel R.S. Mackenzie to AAG, Dept. Missouri, 4 November 1875, Ibid., 1:33; Captain J.H. Bradford to AAG, Dept. Missouri, 3 November 1875, Ibid., E 457-2, Fort Supply, Letters Sent, 5:297.

24. Colonel R.S. Mackenzie to Captain T.J. Wint, 3 November 1875, NA RG 393, pt. 5, E 432-1, 1:35.

25. Colonel R.S. Mackenzie to J. Richards, 31 December 1875, NA RG 393, pt. 5, E 432-1, 1:89.

26. William A. Dobak and Thomas D. Phillips, *The Black Regulars, 1866–1898* (Norman: University of Oklahoma Press, 2001), 40.

27. J.W. Noble to R. Proctor, 14 September 1889, (4801 AGO 1889), NA M689, roll 708.

28. T.J. Morgan to J.W. Noble, 14 September 1889, Ibid.

29. Colonel J.F. Wade to Brigadier General W. Merritt, September 28, 1889, Ibid.

30. 1st Lieutenant J.M. Lee, Report for June 1878, Spotted Tail Agency Letter Books, vol.

3, NA Central Plains Region, Kansas City, Mo. On the social as well as the spiritual side of the Sun Dance, see Morris W. Foster, *Being Comanche: A Social History of an American Indian Community* (Tucson: University of Arizona Press, 1991).

31. US Statutes at Large, 15:222-23; General Order No. 12, Military Division of the Missouri, 14 August 1875, filed with 4297 AGO 1875, NA M666, Letters Received by the Office of the Adjutant General (Main Series), 1871–1880, roll 225.

32. 45th Cong., 3d sess., H.Ex. Doc. 1, Secretary of War Annual Report, 1878 (serial 1843), 34, 37.

33. Endorsement, 14 September 1878, on P.B. Hunt to Brigadier General J. Pope, 27 August 1878 (6631 AGO 1878, filed with 5641 AGO 1878) NA M666, roll 423.

34. C. Schurz to G.W. McCrary, 7 October 1878 (7144 AGO, filed with 5641 AGO 1878), NA M666, roll 423; C. Schurz to G.W. McCrary, 16 November 1878 (8067 AGO 1878), Ibid., roll 445.

35. 1st Lt. C.M. Callahan to Post Adjutant, Fort Sill, 26 December 1877 (125 AGO 1878), NA M666, roll 389.

36. "Supplemental Report" and "Briefs and Extracts," with Lieutenant General P.H. Sheridan to General W.T. Sherman, 28 December 1878 (8977 AGO 1878, filed with 8067 AGO 1878), NA M666, roll 445.

37. Paul A. Hutton, *Phil Sheridan and His Army* (Lincoln: University of Nebraska Press, 1985), 337–40; Hans L. Trefousse, *Carl Schurz: A Biography* (Knoxville: University of Tennessee Press, 1982), 244.

38. Col. J.H. Potter to AAG Department of the Missouri, July 18, 1883 (4455-DMo-1883), NA RG 393, pt. 1, Geographical Divisions and Departments, E 2601, Department of the Missouri, Letters Received.

39. Col. J.A. Potter to Lieutenant Colonel M.V. Sheridan, 26 December 1884, filed with 1244 AGO 1885, NA M689, roll 253.

40. Michael Sheridan served on his brother's staff late in the war; Potter was chief of staff of the XXIV Corps. Hutton, *Phil Sheridan and His Army*, 140; Mark M. Boatner, *The Civil War Dictionary* (New York: David McKay, 1959), 665.

41. "Brief of Papers in Relation to Disturbances Among the Cheyenne and Arapaho Indians," 26, filed with 3140 AGO 1885, NA M689, roll 362.

42. Berthrong, *Cheyenne and Arapaho Ordeal*, 116.

43. Edward E. Dale and Morris L. Wardell, *History of Oklahoma* (New York: Prentice-Hall, 1948), 218. Shipments of cattle from Dodge City, Kansas, dropped from 32,535 in 1885 to 876 in 1886. C. Robert Haywood, *The Merchant Prince of Dodge City: The Life and Times of Robert M. Wright* (Norman: University of Oklahoma Press, 1998), 158.

44. Compare the "old bummers" remark with a string of quotations from Loretta Fowler's field notes, referring to an incident in Cheyenne-Arapaho politics a century later: "She's a White woman—she has no business here." "Do you want to eat beans while the business committee is in Washington eating steaks?" "I got the Oklahoma [City] Times interested . . . . And I called the US Marshall [sic]." "I got an attorney and sued." Fowler, *Tribal Sovereignty and the Historical Imagination*, 141.

Panel 7—Interagency Process in the United States
Question and Answers
(Transcript of Presentation)

Robert T. Davis II, Ph.D.
Mr. William A. Dobak

Moderated by Ricardo A. Herrera, Ph.D.

Audience Member

This is for Dr. Davis. I am wondering about the relationship between the CIA and the Psychological Strategy Board in the 1950s because I know this is a period where the CIA is famous or infamous for like Wisner's Wurlitzer and I wonder if you could comment on that.

Dr. Davis

Well, the CIA was happier with the Psychological Strategy Board than they had been with the earlier incarnations because the Director of the Central Intelligence Agency sat directly on Psychological Strategy Board and CIA seconded several of its officers for planning purposes, but the Psychological Strategy Board gets shut down as a result of the Jackson Committee's reports. It is essentially abolished from about August of 1953 and then after that Eisenhower creates this new policy making apparatus that functions during his administration where it is a direct feed into the National Security Council process, so the CIA was not unhappy with the shift to the Psychological Strategy Board so far as I know between 1950 and 1952. But it is a moot point in the Eisenhower era because they no longer exist.

Audience Member

I have a question for Mr. Dobak. Going back to one of your early observations in your discussion, you noted the mythology or the so called fallacy of Army autonomy. I am wondering to what extent this fallacy was exploited politically perhaps in Washington. And the reason the question occurs to me is that having done comparable research on the Russian Army in the nineteenth century under somewhat comparable circumstances far from home, the Russian government found it quite expedient on a number of occasions, either for foreign policy purposes or for domestic purposes, to observe, "General so and so is off on his own again. Darn that guy; we did not intend to take that village." I am just wondering to what extent there may have been some comparable maneuvering in DC. Thank You.

Mr. Dobak

Well, the commanding general of every geographical region, Department of Dakota, Department of the Platte, Department of the Missouri which is Kansas, Colorado and New Mexico, and Department of Texas, those are the four departments within Sheridan's military division of the Missouri. There were other departments on the west coast and in the east. They all filed an annual report and these were published each year as part of the Secretary of War's annual report in the Congressional Serial Set. So I am not sure whether that answers your question or not. Does that go part way towards your answer? There was, for example . . . now an outright atrocity like Sand Creek which sparked a Congressional investigation which also wound up printed in the Congressional Serial Set. Colonel Baker's attack on a camp of Piegan Indians in 1870, again, wrong Indians, which was also the subject of a Congressional investigation. But by and large, something large scale like the Sioux War of 1876 actually grew out of a recommendation by an inspector of the Office of Indian Affairs in the fall of 1875, saying that the Crow Indians of Montana and several other tribes were tired of being attacked by Sioux who lived off their reservations. It was time to end this intertribal warfare by forcing the Sioux onto their reservations which led to a conference that included President Grant, General Sheridan, the Secretary of War, and the Secretary of the Interior who decided that it was time to go ahead and try to force the off reservation Sioux onto the designated Sioux reservations. This would, coincidentally, settle the matter of ownership of the Black Hills, and also a right of way for the Northern Pacific Railroad which was going to eventually build through the Yellowstone Valley of Montana. But as I said, these matters were settled at the President and cabinet level. Essentially after 1869, the Army did not lift a finger without a go-ahead from the Secretary of War in Washington who had consulted . . . in response to a request from the Secretary of the Interior.

Audience Member

Yes, thanks. That also does establish for me that the governmental context is a bit different and the analogy I was wondering about is not fully there. Thank you.

Audience Member

Over the last seven years, I think a lot of people close to the US Army have gotten used to hearing about effects-based operations, measuring effects of things like strategic communications information operations. Let me make sure I get the right picture here, it sounds like the US government struggled just to define psychological operations. Was anybody interested in measuring how well the Voice of America and these other instruments worked at all in this time frame?

Dr. Davis

Yes, they absolutely were. The Psychological Strategy Board had three subordinate organizations. One was Plans and Operations, pretty standard and it makes an easy military analogy. Another one was a Coordinating Committee to make sure once the plans and operations had been approved that people in various and sundry government departments were actually following up on that. And then there was a third body that was specifically tasked with measuring those effects. I have not looked at any of the records of these three sub-committees so I could not tell you what they determined in 1951 or 1952 which would be very interesting of course. Did they think they could measure these things? They certainly wanted to.

Audience Member

Exactly. It is a struggle right now to figure out how well a message gets out in any particular province in Iraq. It is amazing to think that they were trying to do this on this very broad scale during the Cold War, but also very indicative of a certain mindset.

Dr. Davis

What I would say, when OWI (Office of War Information) was winding down at the end of World War II, when the Creel Committee came to an end at the end of World War I, the only thing you would see, and still historians today fall back on, "Well, I can tell you how many speeches they gave, how many films they distributed, how many leaflets they dropped over enemy lines." And it is always like, "I can give you a quantitative how much did we do." But I have never seen anybody develop a metric and I do not necessarily believe there is such a thing as a metric that will tell you how much you actually influenced. I do not think it exists, but it is certainly something people have struggled with in the past unsuccessfully.

Audience Member

Just to follow up on that point briefly, they did try after the Gulf War, during the interviews of the sixty thousand prisoners of war, to say, "Were you affected by this particular leaflet or that particular leaflet?" So they were trying to get the measure of what percentage of the folks that saw the "Please Desert" leaflet were affected by it. But you are right, they continually try to come up with metrics, they try to measure, and they fall back on how many they did, almost like the Air Force dropping bombs, "Well, since we dropped sixty thousand bombs, therefore, we were doing our job." And never asking if they were dropped on the right target and what effect they actually had, Psy-War is even more unspecific as that. But for Mr. Dobak, a quick question. Obviously Sheridan got out and he visited, he talked to his people and he followed up with some visits, the Bureau of Indian Affairs, not an organization with the greatest reputation in the world, had its inspectors out there also gathering information. Did Carl Schurz get out there and see for himself the truth of any of these allegations or was it simply a reflex, a fallback and defend my organization, that he would fall back on?

Mr. Dobak

I do not think that Carl Schurz got out of Washington much, as far as I know, during his term as Secretary of the Interior. And of course, he had federal lands to supervise, Indian affairs was only one, although well publicized part of his duties, but he had all of the public lands to oversee; he had a lot of things to do besides Indians.

Audience Member

One of the things that struck me and this is for Mr. Dobak, in your research did you find any documentation of cultural guidance or tribal guidance to the soldiers about how to effectively interact with the different tribes or any kind of training or anything that would help better prepare them?

Mr. Dobak

No, not really. For one thing, each tribe spoke a different language. They certainly did not have linguistic language training. They relied on civilian interpreters. And as someone said, the kind of white man who could stay alive long enough to learn a native language was probably not the kind of guy who was interested in concepts like land ownership and not really well suited to interpret for treaty negotiations and stuff like that. Sheridan did send a young officer out to study Indian sign language and published a book about that, but for the most part, what conflicts there were were over by the time they had enough people to send on detail tasks like that, so really, no, is the answer.

Audience Member

Way back in Andy Jackson's administration, before the Civil War when the Indian Removal Act was signed and passed and the reservations were somewhat codified, was the Department of War responsible for Indian affairs and then it shifted to the Department of the Interior? Was there a feeling that the person who goes out and fights the Indian cannot be the one who governs him and the reservations?

Mr. Dobak

The Department of the Interior was established in 1849, that is, after the great expansion of the 1840s—Texas, the Southwest, Oregon country. If you look at the list of engagements of the regular Army during the period before the Mexican War, Indian fights in 1794, and then nothing until Tippecanoe in 1811, then a couple of fights during the War of 1812, then a short war in Florida, . . . I mean as long as American territory stopped at the Continental Divide, there was no conflict with the Plains Indians. A little raiding on the Santa Fe Trail, but by and large, it was an easy decision to turn the Office of Indian Affairs over to the new Department of the Interior, just at the time when the United States had acquired the Pacific coast and increased traffic through the

Great Plains and the mountains brought about an increase in conflict. So I would say that it probably would have been better to have kept the Office of Indian Affairs as a civilian agency within the War Department, separate from the Army, but that is simply because I have seen how the information traveled from the Indian Agency to Washington, to the Commissioner of Indian Affairs, up to the Secretary of the Interior lateral over to the Secretary of War, then down to the Army and to the regional commands within . . . it would have cut out that much, that stop at the Secretary of the Interior. It would have speeded things up and made for less friction, I would say. But it is a little late for that judgment.

Audience Member

Just to follow up, were there some people in the Department of War and the Army that . . . I believe a lot of soldiers were detailed into these Indian Bureaus anyway because they could not be staffed, that there were some arguments about whether or not this agency should be in the Interior or in the Department of War?

Mr. Dobak

That argument . . . well I mentioned Colonel Baker attacking the wrong Indians in 1870, bad timing because there was a fairly good chance of returning the Office of Indian Affairs to the War Department at that time, but after a gross mistake like that, the Office of Indian Affairs stayed with the Interior Department and right through into twentieth century. So that was a bad screw up, bad timing. But there was that debate back and forth. It is a constant subject like expanding the Army or contracting it. It is one of those things they love to talk about all through the last third of the nineteenth century

[1]Army Lawyers and The Interagency: An Examination of Army Lawyers' Experience With Military Commissions and Habeas Corpus

by

Colonel Gary M. Bowman, US Army
US Army Center of Military History

The Army has periodically played a role in the separation of powers within the federal system of government, most often as the means of enforcing federal supremacy over the states. Two obvious examples of this was the use of the Army during the Civil War against the secessionist southern states and President Eisenhower's use of the Regular Army and federalization of the Arkansas National Guard to enforce the federal district court's order enjoining Governor Orval Faubus from using the National Guard to prevent the desegregation of public schools in Little Rock.

The Army has also played a role in the horizontal separation of powers between the branches of the federal government: the Army has imposed and enforced martial law within the United States and its territories, displaced the judicial branch, and even arrested judges who interfered with it.

However, Army lawyers have only rarely stood on their own against other agencies and officials within the executive branch. Since 2001, Army leaders, and Army lawyers, have resisted policy initiatives of the Bush administration on a number of occasions, involving a number of different issues, including the treatment of detainees, electronic surveillance within the United States, and military support to civilian law enforcement and homeland security officials. This paper will examine several episodes in the Army's interaction with other components of the federal government. These episodes involve military commissions and habeas corpus: Andrew Jackson's conflict with Judge Dominick Hall in New Orleans, Winfield Scott's creation of military commissions in the absence of Congressional authorization during the Mexican War, the Army's implementation of martial law in Hawaii during World War II, and the role of Army lawyers during the 1942 Nazi saboteur case. The paper will also discuss the military commission process during the Global War on Terror and evaluate how the Army's recent experience with military commissions and habeas corpus demonstrate a military legal culture, which may affect how the Army will participate in the interagency process when policy that conflicts with that culture is proposed.

I. Martial Law And Military Commissions Prior To *Ex Parte Milligan.*

It has been argued by many scholars that Congress was the dominant branch in the federal government prior to the Civil War, even in military affairs.[2] However, the historical record does not demonstrate that Congress exercised its dominance. The Army was left to find its own way regarding martial law and military commissions until the Supreme Court's decision in *Ex Parte Milligan*, which was not issued until after the Civil War was concluded.

On December 16, 1814, Andrew Jackson, as the Commanding General of the Seventh Military District responsible for the military defense of New Orleans, issued a General Order declaring martial law in the city. The order required any person entering the city to report to the Adjutant General's office, prohibited anyone leaving the city without military permission, and established a curfew.[3] Violations of the General Order were tried by courts-martial.[4]

After the British were defeated and while peace negotiations were being conducted at Ghent, Jackson refused to rescind the order establishing martial law because he was concerned that the British might return.[5] During that time, Jackson ordered several French-speaking residents of the city deported. When Louis Louallier, a member of the Louisiana legislature, published an article in the local French-language newspaper calling for trials by civil courts, Jackson had him arrested for "inciting mutiny and disaffection in the army."[6] In a March 5, 1815 letter to Lieutenant Colonel Matthew Arbuckle, Jackson directed that any person attempting to serve a writ of habeas corpus for Louallier be arrested and confined.[7] Louallier filed a petition for writ of habeas corpus in the United States District Court.[8] Judge Dominick Augustin Hall granted the petition and concluded that martial law was no longer justified. Jackson ordered the arrest and confinement of Judge Hall, who was jailed with Louallier.[9]

Louallier raised several defenses, including the argument Jackson did not have the authority to establish military tribunals to try civilians without Congressional authority.[10] The court proceeded on the merits of the case, and Louallier was acquitted. Rather than having Judge Hall tried, Jackson ordered him to leave the city and not return until an announcement of a peace treaty or until the British left the southern coast.[11] Hall was marched by troops four miles out of the city and left there. The next day, after confirming that the peace treaty was signed, Jackson rescinded martial law and released Louallier, who had previously been acquitted.[12]

When Judge Hall returned to New Orleans, he ordered Jackson to appear in court and show cause why he should not be held in contempt for failing to obey the order to release Louallier and for detaining Hall. Jackson appeared in court with two attorneys, whom he retained himself, and submitted a written statement. His attorneys argued that the court lacked jurisdiction to punish Jackson. Jackson later appeared to answer interrogatories, and the United States attorney[13] acted as prosecutor, but Jackson refused to answer any questions. Eventually, after several hearings, Judge Hall fined Jackson $1,000.00, which Jackson paid.

Later, Jackson explained his justification for martial law:

[When] invaluable [constitutional rights were threatened by invasion, certain basic privileges] may be required to be infringed for their security. At such a crisis, we have only to determine whether we will suspend, for a time, the exercise of the latter, that we may secure the permanent enjoyment of the former. [Is it wise to sacrifice] the spirit of the laws to the letter [and] lose the substance forever, in order that we may, for an instant preserve the shadow?[14]

In 1842, Congress passed legislation which refunded the amount of the fine, with interest, to Jackson.[15] However, despite contention between the Whigs and the Democrats in Congress as to whether the bill should state whether Jackson or Judge Hall was right, the bill that was passed was silent on the issue.

In an unrelated matter, three years later, in 1818, Attorney General William Wirt issued a legal memorandum on the authority needed to order a new trial before a military court. The specific issue presented to Wirt was whether the Judge Advocate General acted properly in refusing to prosecute Captain Nathaniel N. Hall because he had already been tried by a court-martial on the same charge and the sentence had been disapproved by the President.[16]

Wirt wrote that the President "has no powers except those derived from the constitution and laws of the United States; if the power in question, therefore, cannot be fairly deduced from these sources, it does not exist at all.[17] The Constitution made the President Commander in Chief, but

in a government limited like ours, it would not be safe to draw from this provision inferential powers, by a forced analogy to other governments differently constituted. Let us draw from it, therefore, no other inference than that, under the constitution, the President is the national and proper depository of the final appellate power, in all judicial matters touching the police of the army; but let us not claim this power for him, unless it has been communicated to him by some specific grant from Congress, the fountain of all law under the Constitution.[18]

In 1847, Winfield Scott, who was a graduate of the College of William & Mary, and had practiced law before joining the Army and again when he was suspended from the Army after a court-martial, was appointed to command the army that would enter Mexico from the Gulf and march on Mexico City. Scott's command was the first large American force to fight outside the United States and his force would be composed largely of volunteers. He was anxious to have a well-developed plan for administering discipline within his army and he was also concerned about administering justice fairly to Mexicans in his area of operations so that he would not provoke a backlash or insurgency, as the French had experienced in Spain from 1809-1813.[19]

Before Scott left Washington, he drafted an order establishing martial law in Mexico, which applied to American soldiers and to the Mexican populace. He asked both Secretary of War William L. Marcy and Attorney General Nathan Clifford to review the order, but neither official either explicitly approved the order or expressed disapproval. Marcy requested that Congress enact legislation authorizing military tribunals in Mexico, but Congress did not act on Marcy's request.[20]

When Scott arrived in Mexico, he issued the order which he had prepared in Washington, proclaiming a state of martial law, as General Order No. 20. Scott ordered that certain specified acts, including common crimes and violations of the articles of war, be tried and punished by military tribunals in Mexico. Scott apparently believed that Congress had the exclusive authority to authorize military tribunals, because even after he had arrived in Mexico, Scott urged Congress to authorize the military commissions. However, Congress would not act: Secretary of War Marcy advised Scott that he had discussed the matter with the chairman of the relevant Senate committee and that the Senator did not believe that legislation was necessary because the right of the military commander to impose martial law "necessarily resulted from the condition of things when an army is prosecuting hostilities in an enemy's country."[21] Scott's system of tribunals worked unevenly in Mexico, but his system of military government accomplished his goal of creating the appearance of fairness which, Scott believed, prevented the development of a native insurgency against the occupying Americans.[22]

In 1857, Attorney General Caleb Cushing authored an opinion as to the legality of "martial law" imposed in Washington Territory. He opined that the commander of an army occupying foreign territory is authorized to impose "martial law" in the occupied country as "an element of the jus belli,"[23] Cushing concluded that the United States is "without law on the subject" of whether martial law may be imposed by the armed forces of the United States within the United States or its territories.[24] However, he concluded that, under the Constitution, only Congress is empowered to declare "martial law" and to suspend habeas corpus:

> In the Constitution, there is one clause, of more apparent relevancy, namely the declaration that "The privilege of the writ of *habeas corpus* shall not be suspended, unless when in case of rebellion or invasion the public safety may require it." This negation of power follows the enumeration of the powers of Congress; but it is general in its terms; it is in the section of things denied, not only to Congress, but to the Federal Government as a government, and to the States. I think it must be considered as a negation reaching all the functionaries, legislative or executive, civil or military, supreme or subordinate, of the Federal Government: that is to say, there can be no valid suspension of the writ of *habeas corpus* under the jurisdiction of the United States, unless when the public safety may require it, in cases of rebellion or invasion. And the opinion is expressed by the commentators on the Constitution, that the right to suspend the writ of *habeas corpus*, and also that of judging when the exigency has arisen, belongs exclusively to Congress.[25]

However, Cushing recognized that the declaration of "martial law" is an extraordinary occurrence, in which the "law" ceases to have force because of the exigency of military crisis:

> There may undoubtedly be, and have been, emergencies of necessity, capable of themselves to produce, and therefore to justify, such suspension of all law, and involving, for the time, the omnipotence of military power. But such a necessity is not of the range of mere legal questions. When martial law is proclaimed under circumstances of assumed necessity, the proclamation must be regarded as the statement of an existing fact, rather than the legal creation of that fact. In a beleaguered city, for instance, the state of siege lawfully exists, because the city is beleaguered, and the proclamation of martial law, in such a case, is but notice and authentication of a fact—that civil authority has become suspended of itself, by the force of circumstances, and that by the same force of circumstances the military power has been devolved upon it, without having authoritatively assumed, the supreme control of affairs, in the care of the public safety and conservation. Such, it would seem, is the true explanation of the proclamation of martial law at New Orleans by General Jackson.[26]

Four years after Cushing's opinion, President Lincoln suspended habeas corpus in specified areas of the country. His Attorney General, Edward Bates, reached the conclusion that the President's power to suppress insurrection "is political,"[27] echoing the pragmatic language of Cushing's opinion. Louis Fisher has written that "both Lincoln and Bates acknowledged congressional power to pass legislation that defines when and how a President may suspend the writ of habeas corpus during a rebellion."[28] In fact, in 1863 Congress passed legislation which allowed the President to suspend the writ of habeas corpus "whenever, in his judgment, the public safety may require it."[29]

The Army honored the suspension of habeas corpus and defied judicial orders granting a writ. In *Ex Parte Merryman*,[30] Chief Justice Roger Taney, who was sitting as the circuit judge, issued a writ of habeas corpus to release John Merryman, who was being held in military custody for being a vocal opponent of the war and a suspected insurgent. Taney issued the writ, but the Army commander ignored the order to produce Merryman in court. Taney then issued papers for the commander to appear in court to show cause why he should not be held in contempt, but the marshal was not able to enter the fort to serve the papers. Taney had no means to enforce judicial process, but he issued an order to release Merryman, which went unheeded by the Army.

In *Ex Parte Vallandigham*,[31] the Supreme Court considered the habeas corpus petition of Clement Vallandigham, a former member of Congress and a leading Copperhead, who was arrested in May 1863 for speaking out against the war. He was tried by a military commission and sentenced to confinement for the duration of the war. While he was in military custody, he filed a petition for writ of habeas corpus in the Supreme Court. The Court, recognizing that the Army did not comply with the writ in *Merryman*, held that it did not have jurisdiction to review the decisions of a military commission because, it wrote, a military commission was not "judicial"[32] in nature but was an exercise of military authority.

Ex Parte Milligan,[33] which was decided in the immediate aftermath of the Civil War, is the most significant case on the relationship of martial and civil law. Lambdin P. Milligan, a lawyer in Indiana, was arrested, convicted, and sentenced to hang for conspiring to free Confederate prisoners and lead an insurrection in Indiana. He was tried by an Army military commission even though the civilian courts had remained open in Indiana throughout the war. He filed a petition for a writ of habeas corpus in the Circuit Court for Indiana, which had two judges. The judges disagreed as to whether they had jurisdiction to grant the writ and, by agreement of the parties, the issues of whether the writ should be issued, whether Milligan should be released from custody, and whether the military commission had authority to try Milligan[34] were certified to the Supreme Court.

The Supreme Court held that the military commission which tried Milligan was illegal because the civilian courts of Indiana were open during the war and Milligan could have been tried by the civilian courts. The court wrote:

> It is claimed that martial law covers with its broad mantle the proceedings of the military commission. The proposition is this: that in a time of war the commander of an armed force (if in his opinion the exigencies of the country demand it, and of which he is to judge), has the power, within the lines of his military district, to suspend all civil rights and their remedies, and subject citizens as well as soldiers to the rule of his will; and in the exercise of his lawful authority cannot be restrained, except by his superior officer the President of the United States.

> If this position is sound to the extent claimed, then when war exists, foreign or domestic, and the country is subdivided into military departments for mere convenience, the commander of one of them can, if he chooses, within his limits, on the plea of necessity, with the approval of the Executive, substitute military force for and to the exclusion of the laws, and punish all persons, as he thinks right and proper, without fixed or certain rules.

> The statement of this proposition shows its importance, for, if true, republican government is a failure, and there is an end of liberty regulated by law. Martial law, established on such a basis, destroys every guarantee of the Constitution, and effectually renders the "military independent of and superior to civil power"—the attempt to do which by the King of Great Britain was deemed by our fathers such an offense that they assigned it to the world as one of the causes which impelled them to declare their independence. Civil liberty and this kind of martial law cannot endure together; the antagonism is irreconcilable; and, in the conflict, one of the other must perish.[35]

The Court's finding that the civilian courts in Indiana were open and that war was not being waged in Indiana was at least questionable: Confederate Brigadier General John Hunt Morgan raided Indiana in June 1863, causing the closure of the courts, and Indiana remained under the threat of further raids throughout 1864.

The Supreme Court's decision in *Milligan* was, ultimately, ambivalent. The full Court agreed that the executive had gone too far in trying Milligan in Indiana by military commission. A majority of the Court returned, once the war had passed, to traditional republican language and rejected martial law and military commissions in general; the minority recognized that martial law could and would be used again in a time of "invasion and rebellion." In fact, the United States continued to convene military commissions in the southern states throughout the reconstruction period.[36]

II. Martial Law In Hawaii.

The Army also convened numerous military commissions during the Philippine Insurrection. However, the legality of military commissions was not challenged again until World War II.

On December 7, 1941, after the attack on Pearl Harbor, Joseph B. Poindexter, the Governor of the territory of Hawaii, issued a proclamation transferring all governmental functions to the Army general commanding the Hawaiian Department,[37] pursuant to Section 67 of the Hawaii Organic Act, which authorized the Governor to suspend the writ of habeas corpus and impose martial law "in case of rebellion or invasion, or imminent danger thereof, when the public safety requires it . . . ."[38] Lieutenant General Walter C. Short assumed the role of Military Governor[39] and suspended the writ of habeas corpus (General Short was later succeeded by Lieutenant General Delos C. Emmons, who was succeeded by Lieutenant General Robert C. Richardson).[40] Short issued General Order No. 4, drafted by his Judge Advocate, Lieutenant Colonel Thomas H. Green,[41] establishing a military commission and provost courts to replace the civilian courts, which were closed.[42] Nine days after the initial proclamation, on December 16, 1941, the authority of the civil courts to hear civil matters was largely reinstated.

Conflicts between Army authority and the civil courts in Hawaii were redolent of Jackson's conflict with Judge Hall in New Orleans. The Army detained Hans Zimmerman, an American citizen, pursuant to an order which allowed the detention of persons "for the purpose of inquiring into . . . whether or not such activities are subversive to the best interests of the United States."[43] Zimmerman's wife filed a petition for writ of habeas corpus in the United States District Court. At the hearing on the petition, Judge Delbert E. Metzger refused to take evidence or to require the Respondent—"Captain Walker of the United States Army, Assistant Provost Marshall"[44]—to respond. The judge stated that he believed the issuance of a writ of habeas corpus was justified, but ruled that the court was prevented by "duress" from granting a writ.[45]

On appeal, the United States Court of Appeals for the Ninth Circuit held that *Milligan* did not require the District Court to act upon the petition because the petition, on its face, did not allege grounds that would have justified the granting of the petition. The Court found that "in light of the unprecedented conditions under which present day warfare is waged," the Hawaiian Islands were "peculiarly exposed to fifth-column activities."[46] The Court read *Milligan* to mean that, "[i]t is settled that the detention by the military authorities of persons engaged in disloyal conduct or suspected of disloyalty is lawful in areas where conditions warranting martial law prevail."[47] The Court wrote:

> But taken by its four corners the petition discloses that Zimmerman was being sub-jected to detention by the military authorities after an inquiry related in some way to the public safety, in an area where martial law was in force and the privilege of the writ had been lawfully suspended. The futility of further inquiry was apparent on the face of the petition.[48]

Of course, the Ninth Circuit's reasoning was circular: so long as martial law is declared, the government could hold Zimmerman without charging him, and the government itself was in control of when martial law will be terminated. As Judge Bert Emory Haney wrote in his dissenting opinion, the test of whether the conditions for martial law exist must be by reference to conditions outside of the mere proclamation of martial law:

> [M]ilitary government is not established by merely proclaiming it. It comes into being and exists solely by reason of the fact that strife prevents operation of the civil government. *Ex parte Milligan* . . . 4 Wall. 127, 18 L.Ed. 281, where it is said: "As necessity creates the rule, so it limits its duration." In other words, whether military government prevails is a question of fact depending on the existence of facts in the territory where it is supposed to be controlling, and a proclamation of the military that it exists is su-perfluous and ineffective.[49]

On January 27, 1942, General Order No. 57 provided that the civil courts could exercise their full civil jurisdiction "as agents of the Military Governor."[50] However, the courts could not summon a grand jury, conduct a jury trial, or grant a writ of habeas corpus.[51] A proclamation of February 8, 1943 reinstated trial by jury and indictment by grand jury in the civil courts for violations of non-military law, except for proceedings involving military members or prosecutions of civilians for violating military orders. The proclamation specifically stated that "a state of martial law remains in effect and the privilege of the writ of habeas corpus remains suspended."[52]

After the February 1943 proclamation restoring power to the courts, Walter Glock-ner and Erwin R. Seifer, both American citizens who were being detained by the Army, brought petitions for writs of habeas corpus before Judge Metzger. The US Attorney argued that the petitions should be dismissed because martial law was still in effect. Metzger denied the government's motion on the ground that the threshold element of civil habeas corpus jurisdiction under *Milligan* was satisfied because the civil courts were "open" pursuant to the February 8, 1943 order, and Metzger ordered that the Army produce Glockner and Seifer. When the Army refused, Judge Metzger fined General Richardson (the Commanding General of the Hawaiian Department) $5000.00 for con-tempt. Richardson then issued an order that specifically prohibited the District Court from conducting habeas corpus proceedings and ordered Judge Metzger to expunge the contempt citation or face proceedings in the military courts. The Justice Department intervened; General Richardson rescinded his order and Judge Metzger reduced the fine to $100.00, which was later canceled by a pardon.[53]

US District Judge J. Frank McLaughlin also resisted Army rule in Hawaii. Harry E. White was tried by and convicted of embezzlement by a provost court, which did not include the right to a grand jury and from which there was no appeal. His trial was conducted on August 20, 1942, after the Battle of Midway, and after General Order No. 57 which allowed civil courts to conduct criminal trials, but prior to the proclamation reinstating the grand jury and jury trials. White filed a habeas corpus petition in the district court. Judge McLaughlin held that White was denied his Fifth Amendment right to due process of law and his procedural rights under the Sixth Amendment. Judge McLaughlin held that the territorial governor, who was merely an officer of the executive branch, had no right to delegate the judicial functions of the territory to the Army, and that the Commanding General had no authority to impede the functioning of the civil courts because the threat of Japanese invasion and subversion had been virtually eliminated after Midway.[54] Judge McLaughlin also granted the habeas corpus petition of Fred Spurlock, who was convicted by the provost court of assault, writing:

> [I]t is argued . . . [that the court's] inquiry is limited to determining whether or not the power to declare martial law was properly invoked. This is referred to as the theory of absolute martial law and reliance is placed upon several supporting old cases. The doctrine is that martial law can be terminated only by the one given the original power to invoke it—the Executive. In short the proposition is that whether at any given time factual necessity supports the continuation of a state of martial law is a political and not a judicial question. The same idea is currently expressed in modern garb in the statement that in time of war the courts cannot question the judgment of the Executive—whether that judgment is based on facts or not. By such a specious doctrine did Hitler and his ilk rise to dictatorial power. If ever such was the law of this country it long since has been slain by the Supreme Court. Under our form of government the military even in time of war is subordinate to the civil power, not superior to it. During war the military to be sure is allowed a wide range of discretion, but whether it has abused that discretion is a judicial question.[55]

In *Ex Parte Duncan*,[56] Judge Metzger considered the habeas corpus petition of Lloyd Duncan, a civilian shipyard worker, who was tried and sentenced by the provost court for assaulting two Marine guards in violation of military General Order No. 2. The case reached Judge Metzger after the February 8, 1943 proclamation restoring most power to the civil courts, but which specifically continued the prohibition against the writ of habeas corpus. In the Duncan case—unlike the other Hawaii cases discussed here—it was stipulated that a Japanese invasion of Hawaii was unlikely. Judge Metzger held that martial law did not exist in Hawaii in 1943 because there was no necessity for military rule at that time and the actions of the provost court were void.[57]

The Ninth Circuit reversed Judge Metzger's *Duncan* decision[58] and Judge McLaughlin's decisions in *White* and *Spurlock*.[59] The Supreme Court reversed the Ninth Circuit and held that the Hawaii Organic Act did not authorize the territorial Governor to deny Hawaiian citizens their constitutional right to a fair trial in civil court after the immediate threat of invasion had passed. The Court's decision echoed the language of Milligan:

Courts and their procedural safeguards are indispensable to our system of government. They were set up by our founders to protect the liberties they valued. [Citations omitted]. Our system of government clearly is the antithesis of military rule and the founders of this country are not likely to have contemplated complete military dominance within the limits of a territory made part of this country and not recently taken from an enemy. They were opposed to governments that placed in the hands of one man the power to make, interpret and enforce the laws. Their philosophy has been the people's throughout our history. For that reason we have maintained legislatures chosen by citizens or their representatives and courts and juries to try those who violate legislative enactments. We have always been especially concerned about the potential evils of summary criminal trials and have guarded against them by provisions embodied in the Constitution itself. [Citations omitted]. Legislatures and courts are not merely cherished American institutions; they are indispensable to our government.

Military tribunals have no such standing. For as this Court has said before: ". . . the military should always be kept in subjection to the laws of the country to which it belongs, and that he is no friend to the Republic who advocates the contrary. The established principle of every free people is, that the law shall alone govern; and to it the military must always yield." [Citations omitted]. Congress prior to the time of the enactment of the Organic Act had only once authorized the supplanting of the courts by military tribunals. Legislation to that effect was enacted immediately after the South's unsuccessful attempt to secede from the Union. Insofar as the legislation applied to the Southern States after the war was at an end it was challenged by a series of Presidential vetoes as vigorous as any in the country's history. And in order to prevent this Court from passing on the constitutionality of this legislation Congress found it necessary to curtail our appellate jurisdiction. Indeed, prior to the Organic Act, the only time this Court had ever discussed the supplanting of courts by military tribunals in a situation other than that involving the establishment of a military government over recently occupied enemy territory, it had emphatically declared that civil liberty and this kind of martial law cannot endure together; the antagonism is irreconcilable; and, in the conflict, one or the other must perish.[60]

In World War II, as in the Civil War, the Supreme Court did not exercise its jurisdiction to prohibit martial law while war was actually pending. The Court allowed the military to administer martial law, convene military commissions, and dispense justice without intervention by the courts. But once the emergency had passed, the Court opined on the necessity of civilian primacy and preached the virtues of republican government.

III. The Nazi Saboteur Trial

In fact, the Supreme Court's decision during the war in the habeas corpus case brought by the defendants in the Nazi Saboteur case was not consistent with the sublime principles stated by the Court after the war.

In June 1942, German submarines landed eight Germans, wearing German uniforms (they changed into civilian clothes after they came ashore), in two groups—one group on Long Island, and another group in Florida. All the men had lived in the United States; one claimed to be an American citizen. The men had been trained on sabotage and each had a box of explosives. One of the men soon turned himself in to the FBI, which was then able to capture the others.

Justice Department prosecutors determined that it was unlikely that the men could be convicted of sabotage because they did not have specific plans for sabotage, and it was likely that they could only be convicted in federal court of offenses which carried a maximum penalty of three years.[61] The Judge Advocate General of the Army, Myron C. Cramer, advocated trying the men by a military commission because of the light penalties available in civilian court.[62]

Cramer and Attorney General Francis Biddle discussed the appointment of a military tribunal to try the Germans, and Biddle met with Secretary of War Henry L. Stimson on June 29, 1942, to obtain Stimson's consent for the military tribunal. Stimson wrote in his diary that he was surprised that "instead of straining every nerve to retain civil jurisdiction of these saboteurs, [Biddle] was quite ready to turn them over to a military court."[63]

Biddle suggested the appointment of a military commission to Roosevelt. Several factors supported the creation of a military tribunal. The Justice Department wanted closed hearings so that the fact that the FBI caught the Germans only after one of the Germans turned himself in would remain secret. Also, the Attorney General and the President wanted the Germans to be subject to the death penalty, which was not available for any defined crimes which the Germans had committed, and the President did not want the civilian courts to be involved. According to Biddle, Roosevelt told him: "I won't give them up. . . . I won't hand them over to any United States marshal armed with a writ of habeas corpus. Understand?[64]

On July 2, 1942, the President issued a proclamation which declared that:

> all persons who are subjects, citizens or residents of any nation at war with the United States or who give obedience to or act under the direction of any such nation, and who during time of war enter or attempt to enter the United States . . . through coastal or boundary defenses, and are charged with committing or attempting or preparing to commit sabotage, espionage, hostile or warlike acts, or violations of the law of war, shall be subject to the law of war and to the jurisdiction of military tribunals.[65]

Roosevelt also issued an order under Article of War 38 appointing the tribunal of Army officers, the prosecutors, and the defense counsel. Biddle and Cramer themselves acted as prosecutors, and two Army Judge Advocates, Colonel Cassius M. Dowell, a career Army lawyer, and Colonel Kenneth C. Royall, an experienced trial lawyer from North Carolina who had been commissioned as a Colonel at the beginning of the war, to serve as defense counsel for seven of the eight defendants.[66] The commission was empowered to establish its own rules of procedure.[67]

The prosecutors charged the Germans with four offenses: coming behind the lines of the United States to commit sabotage and espionage in violation of the law of war; assisting enemies of the United States in violation of Article of War 81; lurking as a spy in violation of Article of War 82; and conspiracy to commit these offenses.[68]

Prior to the commencement of the trial, Royall objected to the President's order establishing the commission on the grounds that, under *Milligan*, the Germans should be tried in the civilian courts, which were open and functioning. Royall also objected to the charges, because they were not offenses identified by a prior statute, and thus constituted common law crimes or crimes which were identified ex post facto (after the Germans committed the offenses).

Article 38 of the Articles of War provided that the President prescribe the rules of procedure for military tribunals and "nothing contrary or inconsistent with these Articles shall be so prescribed."[69] Under the Articles of War, the death penalty could only be imposed by unanimous vote, but Roosevelt's proclamation allowed a two-thirds majority vote.[70]

Royall and Dowell, as defense counsel, were in a quandary. They doubted that the procedure of the tribunal was constitutional, and they had a duty to their clients to defend them zealously, but as military officers they also owed a duty to the Commander in Chief, who had appointed the tribunal. Prior to the commencement of the trial, they wrote to President Roosevelt, stating their concern that "there is a serious legal doubt" about the constitutionality of the tribunal. They were advised by an aide to Roosevelt, Marvin McIntyre, that they should act upon their best judgment.[71]

By the twelfth day of the trial, Royall had decided to pursue an appeal to the federal courts, but he had been unable to find a civilian lawyer to appear in civilian court. He told the tribunal that he intended to himself file a petition for writ of habeas corpus in federal court in Washington. The defense counsel representing the other defendants announced their opposition to Royall's strategy, and the tribunal refused to address the question of whether Royall had leave to appeal.[72]

Louis Fisher described the unusual procedure which Royall, an experienced politician, followed:

> With time running out, Royall met ex parte with Justice Hugo Black at the Justice's home in Alexandria, Virginia. When Black said he didn't want to have anything to do with the case, Royall responded: "Mr. Justice, you shock me. That's all I can say to you." Turned down by Black, Royall tried unsuccessfully to reach Justice Frankfurter in Massachusetts. The following morning, he learned that Justice Owen Roberts was in Washington, DC, to attend the funeral of Justice George Sutherland. Royall went to Roberts's office and waited for him to return. After listening to Royall outline the case, Roberts said: "I think you've got something that ought to be reviewed" and suggested that they meet the following day, July 23, at Justice Roberts's farm outside

Philadelphia. Dowell, Biddle, and Cramer joined them. After discussing the matter, Roberts phoned Chief Justice Stone and got the go-ahead for holding oral argument on Wednesday, July 29. Roberts was able to reach all the Justices except two (Douglas and Murphy).[73]

A highly unusual aspect of the appeal is that Royall went directly, *ex parte*, to the Supreme Court justices without first filing his petition for writ of habeas corpus in the lower courts. After the Supreme Court had agreed to conduct oral argument, Royall filed a petition for writ of habeas corpus in the United States District Court for the District of Columbia, which was denied, but no appeal of the District Court decision had been filed when the Supreme Court heard oral argument.[74]

The Supreme Court heard oral argument in the case—styled *Ex Parte Quirin*—for nine hours over two days; the briefs had been hastily prepared and the Court had not had an opportunity to carefully read the briefs before argument.[75] At the conclusion of the oral argument, the Supreme Court issued a brief *per curiam* opinion which upheld the jurisdiction of the military tribunal.[76]

On August 1, 1942, the military tribunal convicted the eight Germans and sentenced them to death. Roosevelt reduced the sentence to life imprisonment for the German who had turned himself in and one other defendant. On August 7, the remaining six were electrocuted.[77]

On October 29, 1942, the Supreme Court issued its full opinion in the case. The lynchpin of the Court's decision was the distinction between "lawful" and "unlawful combatants." The Court held that unlawful combatants may be tried and punished by military tribunals:

> By universal agreement and practice, the law of war draws a distinction between the armed forces and the peaceful populations of belligerent nations and also between those who are lawful and unlawful combatants. Lawful combatants are subject to capture and detention as prisoners of war by opposing military forces. Unlawful combatants are likewise subject to capture and detention, but in addition they are subject to trial and punishment by military tribunals for acts which render their belligerency unlawful. The spy who secretly and without uniform passes the military lines of a belligerent in time of war, seeking to gather military information and communicate it to the enemy, or an enemy combatant who without uniform comes secretly through the lines for the purpose of waging war by destruction of life or property, are familiar examples of belligerents who are generally deemed not to be entitled to the status of prisoners of war, but to be offenders against the law of war subject to trial and punishment by military tribunals. [Citations omitted]. Such was the practice of own military authorities before the adoption of the Constitution, and during the Mexican and Civil Wars.[78]

In *Quirin*, the Supreme Court held that citizens of the United States could be unlawful belligerents and, as such, could be tried by military tribunal.[79] The Court also held that *Milligan* did not apply to the Germans because Lamdin Milligan had not been a member of the enemy armed force and was not subject to the law of war;[80] therefore, even though the civilian courts were open and functioning in the states where the Germans landed and were caught, *Milligan* did not require that they be tried in a civilian court, where they would have received the benefit of civilian criminal procedural law. The Court avoided the question of whether the President had the power to allow the death penalty to be imposed by a two-thirds majority vote, when the Articles of War required a unanimous vote. The Justices agreed that a writ of habeas corpus could not be granted, but they disagreed as to why: some Justices believed that the Articles of War did not govern a military tribunal trying "enemy invaders,"[81] and other justices did not believe that the Articles of War "rightfully construed" required procedures different from the procedures the tribunal followed.[82]

The Supreme Court's procedure and decision in *Quirin* has been widely criticized. In an interview in 1962, Justice Douglas said that it "is extremely undesirable to announce a decision on the merits without an opinion accompanying it. Because once the search for the grounds, the examination of the grounds that had been advanced is made, sometimes those grounds crumble."[83] Louis Fisher has pointed out that Justice Frankfurter asked his former student at Harvard Law School, Frederick Bernays Wiener, who was a Colonel in the Army JAG Corps, for his comments on the Court's opinion. In three letters, Weiner provided sharp criticism of the opinion, arguing that the opinion improperly sanctioned ad-hoc military tribunals which, unlike the established court-martial system, were immune from judicial review and which made up their own rules.[84]

When two more Germans were captured in late 1944, Biddle and Cramer again prepared to personally prosecute them by a military tribunal in Washington. However, Secretary of War Stimson, who was himself one of the most distinguished lawyers in the country and was the former United States Attorney in New York, intervened, and urged Roosevelt to appoint a military tribunal in general accordance with the established court-martial procedure. Stimson wrote in his diary of Biddle: "It is a petty thing. That little man is such a small little man and so anxious for publicity that he is trying to make an enormous show out of this performance—the trial of two miserable spies."[85] A military tribunal was convened at Governors Island, New York, and convicted the two men, but President Truman commuted their sentences to life imprisonment after the war, and they were later released.[86]

IV. The Global War On Terror

Through the Vietnam War, Army lawyers had limited conflict with the other agencies of the Executive Branch. The Judge Advocate General's Corps' main function was to administer the Army's military justice system. Although there were challenges to the military justice system, and even to the administration of military affairs by the Army, in civilian courts, the Justice Department defended the Army as a client, and there was no significant conflict between the Army, civilian lawyers in the Department of Defense, and civilian lawyers at the Department of Justice as to military affairs. Colonel Kenneth Royall's challenge to the Roosevelt's administrations actions in the 1942 Nazi Saboteur case was singular.

However, the culture of the Army JAG Corps changed in the last quarter of the twentieth century. The practice of law by Army JAGs became more complex during and after World War II. As Army forces were stationed around the world, Army lawyers became accustomed to the practice of international law, particularly the negotiation and interpretation of status of forces agreements and other treaties which governed the conduct of Army personnel and activities, such as the 1949 Geneva Conventions. By far, the most important change in military law was the adoption of the Uniform Code of Military Justice (UCMJ) in 1950. The UCMJ "civilianized" the courts-martial system, codified military crimes, created new procedural rights for defendants, and explicitly provided for military and civilian review of court-martial convictions. After the passage of the Military Justice Act of 1968, independent military judges presided over courts-martial, and the participation of Army lawyers as prosecutors and defense counsel was expanded, so that military proceedings became very similar to civilian criminal proceedings. In 1980, the Trial Defense Service was established, and Army lawyers assigned as defense counsel were removed from the chain of command which initiates courts-martial proceedings.

Army JAGs also became more involved in advising commanders on legal aspects of military operations. In 1974, the Department of Defense promulgated Department of Defense Directive 5100.77, which required Army lawyers to ensure that "all US military operations strictly complied with the Law of War."[87] The Law of War program was initiated in response to violations of the Law of War in Vietnam, such as the My Lai massacre. The Law of War program required that military commanders be provided advice on the legal implications of their operational decisions. The implementation of the program in the Army resulted in the creation of a new legal field—"operational law"—and the authorization of positions for JAGs at most brigades throughout the Army. The Army JAG Corps now includes over 3400 lawyers.[88]

At the same time that the Army was becoming more lawyered, after 1974, the power of the executive branch was being curtailed by Congress. In fact, the lawyerization of the Army was itself a result of Congress' effort to limit executive discretion in the military justice system and in military operations. The limitation of command discretion in military affairs enhanced the role and authority of Army lawyers, who became watchdogs of a legal culture within the Army.

When the United States launched its invasion of Afghanistan in retaliation for the terrorist attacks of September 11, 2001, the American military detained numerous individuals who were suspected of being lawful and unlawful combatants, as described in *Quirin*, the 1942 Nazi Saboteur case. On November 13, 2001, President Bush issued a military order, drafted by Timothy Flanigan, the Deputy White House Counsel and David Addington, the Vice President's Counsel, intended to govern the "Detention, Treatment, and Trial of Certain Non-Citizens in the War Against Terrorism."[89] The order applied to any noncitizen for whom the President has "reason to believe" either "is or was" a member of al Qaeda or a person who has engaged or participated in terrorist activities aimed at or harmful to the United States.[90]

According to Charlie Savage, the decision to establish military commissions deviated from the normal interagency process. Initially, the question of how to try individuals who had been detained in Afghanistan was considered by an interagency committee, headed by Pierre-Richard Prosper, the Ambassador-at-Large for War Crimes Issues. However, as Savage has described, the interagency process did not complete its work:

> Prosper's group had met for the next month in a windowless conference room on the seventh floor of the State Department. It had brought together experts from around the government, including military lawyers and Justice Department prosecutors. The group had analyzed a range of options, weighing the pros and cons of each. The Justice Department advocated regular trials in civilian federal courts, as the United States had done after the 1993 World Trade Center bombing. But holding terrorist trials in a regular courthouse on US soil presented security risks. The uniformed lawyers had advocated using courts-martial, which could take place anywhere in the world. But courts-martial had well-established rules of evidence and procedure. Setting up a new system of military commissions, the third option on the table, would provide greater flexibility. There were problems, though: Some lawyers believed that the president might need to go to Congress for specific authorization.

> Then, before Prosper's group could complete its work, Flanigan had abruptly short-circuited the interagency process. Without telling Prosper, Flanigan had secretly decreed that the answer was military commissions, and that the president had inherent wartime authority to create them on his own. Flanigan wrote up the draft order himself. In completing it, he worked with just two other government lawyers. One was Berenson, his junior subordinate, chosen because he had been the White House's representative at Prosper's group and so was already steeped in the issue. The other contributor was Addington.[91]

According to Savage, Flanigan and Addington relied, in part, upon a memorandum written by Patrick Philbin, an attorney in the Justice Department's Office of Legal Counsel, entitled "Legality of the Use of Military Commissions to Try Terrorists." Philbin was Deputy Assistant Attorney General in the Office of Legal Counsel, where he worked with his law school classmate John Yoo, who had been the original advocate within the Bush administration for the legal theory that the President has inherent war powers that are not constrained by Congress, treaties, or other international law. Yoo and Philbin had both been law clerks for conservative Judge Laurence Silberman at the United States Court of Appeals for the District of Columbia Circuit and for Justice Clarence Thomas at the Supreme Court. However, Philbin had little experience with national security law and it is likely that Addington and Yoo provided significant input to the memo. The memo argued that President Roosevelt had demonstrated the inherent power of the President to set up military commissions and prescribe the commission procedures during the 1942 Nazi Saboteur case, and the President's power was vindicated by the Supreme Court.[92]

On November 8, 2001, the Wednesday before Veterans Day, Department of Defense General Counsel William James Haynes II informed Major General Thomas J. Romig, the Army Judge Advocate General, that civilian lawyers in the administration had prepared a draft order on military commissions. Haynes told Romig that "he could send one representative to his office to help review the draft order, but he would not allow the officer to take a copy of the order away or even write down notes about it."[93] Romig sent Colonel Lawrence Morris to review the draft order. The next day, Morris met with Romig in Romig's basement. They were both concerned because the order did not incorporate the changes in international law, such as the 1949 Geneva Conventions, or changes in military justice, such as the UCMJ, which had occurred since the 1942 Nazi Saboteur case. Romig, Morris, and Brigadier General Scott Black, the Assistant Judge Advocate General for Law and Military Operations, worked over the Veterans Day weekend, drafting suggested modifications to the draft order. Morris presented the proposed changes to Haynes, but none of the suggestions were adopted.[94]

On the same weekend—on Saturday, November 10, 2001—Vice President Cheney convened a meeting in the Roosevelt Room of the White House to finalize the order. Defense Department General Counsel Haynes, Attorney General Ashcroft, and "several top White House lawyers" were present. Ashcroft expressed his desire to allow Justice Department participation in the trial process, but was overruled by Cheney. On Wednesday, November 13, Cheney brought the order to his weekly private luncheon with President Bush.[95] Jane Mayer has written that:

> On the Tuesday after Ashcroft's Veterans Day showdown, Cheney presented the draft of the military commission order, which had been secretly written by his legal allies, to President Bush during their weekly private White House lunch. It was apparently the first time that Bush had seen it. The draft ran some four pages. The language was arcane and the subject matter more so. But after the lunch, Cheney advised Addington and the other lawyers that the President was on board. An hour or so later, with no further vetting or debate, and without circulating the draft to any of President Bush's other top advisers on national security matters, the finished document was presented back to the President, this time for his final signature. Deputy White House Staff Director Stuart Bowen told the *Washington Post* that he had bypassed all of his usual procedures, which called for more review, because of intense pressure to get it signed quickly from the lawyers who had secretly written the order. As a result, in the span of little more than a luncheon, Addington's text became law.[96]

The President delegated to Secretary of Defense Donald H. Rumsfeld the responsibility for drafting the rules for the military commissions, and the military JAGs worked with lawyers from the DOD General Counsel's office and the Justice Department's Office of Legal Counsel to develop the rules. Savage has written:

> Initially, Romig said, some of the political appointees were interested in a very draconian system, which, among other things, could convict defendants under a low standard of proof; would deny them the right to have outside civilian defense attorneys; and could impose the death penalty without unanimity by the panel of officers judging the case. The JAGs objected strongly to these and other deviations from military law. One of the top JAGs threatened to resign if some of the harshest proposed rules became final, arguing that they would force military lawyers to violate their legal ethical standards and possibly put them at risk of later prosecution for war crimes.
>
> In the end, the political appointees backed down from some of the most extreme proposals they had been floating. When Rumsfeld signed an order fleshing out what the commission trials would look like in March 2002, the system was closer to what the JAGs wanted: Defendants could be convicted only if guilt was proven beyond a reasonable doubt, outside defense counsel was allowed, and a unanimous vote was required for the death penalty. But the order, Romig said, was still not what the JAGs would have designed had they been allowed to create the commission system from scratch on their own. While less draconian than the political appointees' initial plans, the military commissions were still legally objectionable in several respects. The commission rules, for example, allowed secret evidence that would be kept hidden from a defendant and allowed the admission of evidence obtained through coercive interrogations.[97]

The effort of Republican administrations to marginalize military JAGs had been going on since the middle of the Reagan administration. During the Ford administration, in which Rumsfeld and Cheney served as White House chief of staff, Congress enacted intelligence reforms through the Foreign Intelligence and Surveillance Act, budget reforms which restricted the President's power to impound the funding of programs which Congress had authorized but which the President opposed, and attempted to limit the President's power to wage war through the War Powers Resolution.

In 1986, during the second Reagan administration, Attorney General Edwin Meese received a report from the Justice Department's Domestic Policy Committee, a conservative think tank within the department. The report "noted approvingly . . . [that] the strong leadership of President Reagan seems clearly to have ended the congressional resurgence of the 1970s,"[98] and it described a new approach to presidential power—the Unitary Executive Theory. The premise of the theory is that the President is the unitary authority within the executive branch; that the executive branch can be viewed as an organic being with the President as its brain.[99] Under the theory, there can be no meaningful dissent with the executive branch because the President's will is unitary.

The philosophy of a unitary executive branch conflicts with the trend within the Army JAG corps toward providing independent legal advice to commanders and maintaining a fair military justice system. Attempts to reconcile the tension between the independence of military lawyers with the political will of the President's political appointees has been an issue since the debate over the Goldwater-Nichols defense reform act in 1987. At that time, Congress considered placing the service JAGs under the service's politically-appointed civilian general counsels, but instead merely enhanced the general counsels' prestige by explicitly recognizing them in statute and requiring that they be confirmed by the Senate.

Shortly before the Gulf War, William James Haynes II, the political-appointee civilian General Counsel of the Army, clashed with Army Judge Advocate General, Major General John Fugh, over whose office should control operational legal issues that would arise during the war. Haynes, who was a protégé of David Addington, a close adviser to then-Secretary of Defense Cheney, pressed for greater control over the Army's uniformed lawyers; Fugh resisted. In 1991, Cheney forwarded legislation to Congress seeking to place all military attorneys under the control of civilian political appointees. When Congress failed to enact the legislation, the Defense Department issued an administrative order effecting the change.[100] When Addington was later nominated to become the Department of Defense General Counsel, Senate Armed Services Committee chairman Sam Nunn stated that it appeared to him that the Cheney team wanted to empower political lawyers to force JAGs to "reach a particular result on a question of law or a finding of fact," and that they wanted to create a politically appointed filter between the JAGs and top military decision makers. Addington was confirmed "only after promising that the Pentagon would restore the military lawyers' independence."[101]

The issue was revisited when the George W. Bush administration took office, with Cheney as Vice President, Addington as his counsel, and Haynes as Department of Defense General Counsel. In early 2003, after the resistance of the JAGs to the administration's civilian lawyers' plans for military commissions (as well as resistance to the civilian lawyers' interpretation of the Foreign Intelligence Surveillance Act and the standards for treatment of detainees), Army General Counsel Stephen Morello attempted to change the system for selecting the Army Judge Advocate General from the traditional system, in which a board of Army generals select the JAG. Morello proposed a system in which the uniformed Army would compile a short list of finalists, from which the political appointees in the Department of Army would select the JAG. The civilian leadership backed down after the generals resisted the proposal. In May 2003, Morello attempted to civilianize "a thousand of the army's fifteen hundred uniformed lawyers into civilian positions."[102]

In late 2004, in response to an order by the Secretary of the Air Force placing the Air Force Judge Advocate General and his subordinate uniformed JAGs under the direct supervision of the Air Force General Counsel, Congress included language in the 2005 Defense Authorization Act, which was intended to prohibit the civilian leadership of the Defense Department from "interfering with the ability of a military department judge advocate general . . . to give independent legal advice to the head of a military department or chief of a military service or with the ability of judge advocates assigned to military units to give independent legal advice to commanders."[103]

On October 29, 2004, President Bush signed the Defense Authorization Act, but issued a signing statement which specifically stated that the executive branch would construe the statute in light of "the unitary executive branch" paradigm, so that the legal opinions issued by the Defense Department General Counsel would "bind all civilian and military attorneys within the Department of Defense."[104] In early 2005, Defense Department General Counsel Haynes unsuccessfully attempted to implement the civilian selection of the top JAGs in all the services.[105] In late 2007, the Defense Department proposed a regulation which would have required "coordination" with the civilian general counsels of each service prior to the promotion of any JAG officer. The regulation was widely perceived as an attempt to give the civilian attorneys a veto over the promotion of military JAG officers, and was ultimately withdrawn.[106]

In the three major cases relating to military commissions—*Hamdi v. Rumsfeld*, *Hamdan v. Rumsfeld*, and *Boumediene v. Bush*—the Supreme Court vindicated the concerns which the JAGs had originally raised regarding the procedure which should be applied in military commissions.

Yaser Esam Hamdi was born in Louisiana and raised in Saudi Arabia. In 2001, he was captured by the Northern Alliance in Afghanistan and turned over to the United States military. He was detained at Guantanamo Bay until it was determined that he was an American citizen; he was then transferred to custody in the United States.[107] Hamdi's father filed a petition for writ of habeas corpus in the United States District Court for the Eastern District of Virginia on his son's behalf, which alleged that Hamdi had been in Afghanistan doing relief work for only a short period prior to September 2001, that he had not received any military training in Afghanistan, and that he was being held without charges by the military.[108] The District Court appointed the federal public defender as Hamdi's counsel and ordered that counsel be provided access to Hamdi. The United States Court of Appeals for the Fourth Circuit reversed, and directed the District Court to consider "the most cautious procedures first" and to conduct a deferential inquiry into whether Hamdi was an "enemy combatant."[109]

On remand, the government presented the affidavit of Michael Mobbs ("the Mobbs Declaration") which stated that Hamdi had been captured in Afghanistan as part of the Taliban force and that he was carrying a Kalishnikov rifle at the time of his capture. Mobbs stated that Hamdi had been declared an enemy combatant.[110] The District Court found that the Mobbs Declaration was "little more than the government's 'say-so'" and ordered the government to turn over for the judge's review numerous documents supporting the claims of the Mobbs Declaration. The government appealed, and the Fourth Circuit held that the Mobbs Declaration was sufficient, that Congress' resolution authorizing the use of military force was sufficient authority for the government to detain Hamdi, and that (based upon *Quirin*) the fact that Hamdi was an American citizen did not make any difference as to whether he could be detained without charges or trial.[111]

The Supreme Court applied the balancing test of *Mathews v. Eldridge*,[112] a case involving entitlement to welfare benefits, and ruled that an American citizen who is detained by the government "must receive notice of the factual basis for his classification and a fair opportunity to rebut the Government's factual assertions before a neutral decisionmaker."[113]

Salim Ahmed Hamdan was captured in Afghanistan and detained at Guantanamo Bay. Hamdan's case has been the vanguard of military commission cases since 9/11. On July 3, 2003, the President determined that Hamdan was triable by military commission. In December 2003, Navy Lieutenant Commander Charles D. Swift was appointed as Hamdan's defense counsel; Swift filed a demand for charges and for a speedy trial pursuant to Article 10 of the UCMJ, which was denied by the legal adviser to the Appointing Authority, who ruled that Hamdan was not entitled to the rights provided by the UCMJ.[114] On July 13, 2004, after Hamdan's civilian attorney, Neal Katyal (who had been recruited by Lieutenant Commander Swift to assist Hamdan) had filed a petition for writ of habeas corpus in the United States District Court for the Western District of Washington, Hamdan was charged with "conspiracy" as a member of al Qaeda and as Osama bin Laden's driver and bodyguard from February 1996 until November 2001.[115] Hamdan's case was reviewed by a Combatant Status Review Tribunal ("CSRT"), which determined that Hamdan's continued detention was justified because he was an "enemy combatant" which was defined in the military order establishing the CSRT as "an individual who was part of or supporting Taliban or al Qaeda forces, or associated forces that are engaged in hostilities against the United States or its coalition partners."[116]

The US District Court for the District of Columbia, which received Hamdan's case on transfer from the Western District of Washington, granted a writ of habeas corpus and stayed the commission's proceedings because the commission procedure because it had the power to convict based on evidence the accused would never hear, in violation of Common Article 3 of the Third Geneva Convention. The Court of Appeals for the District of Columbia Circuit reversed the District Court, holding that the Geneva Conventions were not "judicially enforceable."[117]

After the Supreme Court granted Hamdan's appeal but prior to oral argument, Congress enacted the Detainee Treatment Act of 2005 ("DTA"), which deprived the courts of jurisdiction to hear "an application for a writ of habeas corpus filed by or behalf of an alien detained by the Department of Defense at Guantanamo Bay, Cuba," and vested in the Court of Appeals for the District of Columbia Circuit the sole jurisdiction to review the CSRTs' decision to detain an alien at Guantanamo Bay and any sentence imposed on such aliens by a military commission.[118]

The Supreme Court ruled that the DTA did not deprive the Court of jurisdiction in Hamdan's case because the case was already pending at the time of enactment of the statute.[119] On the merits of the case, the Court found that a military commission could not try the charges against Hamdan because none of his alleged acts occurred on a specified date after September 11, 2001, were not alleged to have occurred during the United States' war, and therefore could not have violated the law of war.[120] The Court also held that "conspiracy" was not a recognizable violation of the law of war and was not an offense for which Hamdan could be tried by a military commission.[121] The Court also held that a "uniformity" principle required that Hamdan be entitled to the same procedural protections as defendants in courts-martial unless the President determines that such protections would be impracticable, and such protections were not impracticable at Guantanamo Bay.[122] The Court held that the procedural rules of the military commissions, which did not require that the accused be present or hear the evidence against him, violated Common Article 3 of the Third Geneva Convention which prohibits "the passing of sentences and the carrying out of executions without previous judgment pronounced by a regularly constituted court affording all the judicial guarantees which are recognized as indispensable by civilized peoples."[123]

After the Supreme Court's decision in *Hamdan*, the administration re-examined the procedures followed by the military commissions. On Friday, July 28, 2006, a group of JAG officers met with Justice Department attorneys on the fifth floor of the Justice Department headquarters, just down the hall from the room where the 1942 Nazi Saboteur trial was held. A fundamental concern among the JAGs was that secret evidence could be introduced during the military commission trials outside the presence of the defendant. Shortly after the meeting began, the Justice Department lawyers refused to discuss the secret evidence issue, thus limiting the agenda to less significant issues—not the core concerns which the Supreme Court raised in *Hamdan*. The administration's civilian attorneys submitted new legislation—the Military Commissions Act—to Congress, with virtually no input from the JAG lawyers[124]

Congress decided to invite the top JAG from each service to testify about what they believed should be in the Military Commissions Act. Given an opportunity to bypass the filter of the Bush-Cheney legal team, the JAGs told lawmakers that to be fair and legal, the trials must give defendants the right to see any evidence used against them. The administration continued to argue against such a plan, but Congress ultimately decided that the uniformed lawyers were right; the final bill outlawed the use of secret evidence.

The Military Commissions Act of 2006 (MCA) provided specific Congressional authorization for the President to establish military commissions. The procedure to be followed in military commissions trial is required to "be based upon the procedures for trial by court martial,"[125] and Congress declared that a "military commission . . . is a regularly constituted court, affording all necessary 'judicial guarantees as recognized as indispensable by civilized peoples' for purposes of common Article 3 of the Geneva Conventions."[126] The Act also restated the language of the DTA, which attempted to strip the courts of habeas corpus jurisdiction over enemy combatants,[127] and made the determinations of the Combatant Status Review Tribunals "dispositive" as to whether a person is an enemy combatant.[128]

The Supreme Court's latest decision in this area, *Boumediene v. Bush*,[129] addressed detainees' right to habeas corpus review of their detention after the enactment of the DTA and MCA.

The detainees involved in *Boumediene* commenced their habeas corpus petitions in February 2002, but the cases were dismissed for lack of jurisdiction because the naval station is outside the sovereign territory of United States.[130] The Court of Appeals for the DC Circuit affirmed. In *Rasul v. Bush*,[131] the Supreme Court granted certiorari and reversed, holding that habeas corpus jurisdiction extended to Guantanamo.[132] On remand to the District Court, the cases were consolidated into two separate proceedings. The judge in one case held that the detainees had no rights which could be asserted in a habeas corpus proceeding; the judge in the other case held that the detainees were entitled to due process of law.[133] While appeals were pending, Congress enacted the DTA, which stated that no court could hear a habeas corpus action on behalf of an alien detained at Guantanamo. After the Supreme Court's decision in *Hamdan*, which held that the DTA did not apply to cases which were pending before its enactment, Congress passed the MCA, which specifically stated that the suspension of habeas corpus applied to "all cases, without exception, pending or after the date of the enactment of this Act . . . ."[134] The Court of Appeals held that the new act unequivocally stripped it of jurisdiction to consider the detainees' habeas corpus petitions.[135]

When the detainees appealed to the Supreme Court, five military lawyers—from the Army, Navy, and the Marine Corps—representing detainees at Guantanamo, filed an amicus curiae brief with the Court. The military lawyers wrote:

> The President here asserts the power to create a legal black hole, where a simulacrum of Article III justice is dispensed but justice in fact depends on the mercy of the Executive. Under this monarchical regime, those who fall into the black hole may not contest the jurisdiction, competency, or even the constitutionality of the military tribunals, despite the guarantee of habeas corpus, *see* US Const., Art. I, § 9, cl. 2, and the right to such determination by a "competent tribunal" under the 1949 Geneva Convention. The President's assertion of such absolute supremacy contravenes the bedrock principle that it is "the province and duty of the judicial department to say what the law is," and the similarly "settled and invariable principle . . . that every right, when withheld, must have a remedy, and every injury its proper redress." *Marbury v. Madison*, 5 US (1 Cranch) 137, 163, 177 (1803) (quoting Blackstone). This Court has never given the President the ability to proclaim himself the superior of sole expositor of the Constitution in matters of justice.[136]

In *Boumediene*, the Supreme Court relied extensively upon the amicus curiae brief filed by a large group of "legal historians." The court recited at length the history of habeas corpus, and held that the writ could be suspended only if Congress had provided "adequate substitute procedures for habeas corpus."[137] The Court held that the DTA did not provide an adequate substitute remedy because the only venue for review of a detention decision was the Court of Appeals for the DC Circuit, but the DC Circuit did not have authority to entertain a challenge to the President's authority to detain a particular detainee as an "enemy combatant," to order the release of a detainee, or to heat "relevant exculpatory evidence that was not made part of the record" at Guantanamo.[138] Accordingly, the Court held that the provision of the DTA and the MCA which stripped the court of jurisdiction was unconstitutional and remanded the case to the District Court for consideration of the detainees' habeas corpus petition.[139]

The permanence of the *Boumediene* decision is tenuous, however. The Supreme Court was split 5-4. Three of the justices who voted in the minority (Chief Justice Roberts, Justice Thomas, and Justice Alito) are the youngest justices on the Court, and it is likely that any new appointee by President Bush would vote to strip the courts of jurisdiction to review military commission decisions, shifting the majority on the Court.

In the meantime, the military commission process at Guantanamo Bay continues. On August 7, 2008, a commission acquitted Salim Ahmed Hamdan of conspiracy, and sentenced him to five-and-a-half years in prison, although he was given credit for the 61 months that he had already served. Hamdan was the first detainee to be tried by a military commission (a prior case, involving an Australian citizen, ended in a plea agreement). However, the Justice Department continued to assert that the President has the inherent authority to hold detainees until the end of the war on terror, so it is uncertain whether Hamdan will be released even after his sentence is completed.[140]

The Army's relationship to the judiciary's power to intervene in military affairs has turned about since Andrew Jackson's expulsion of Judge Hall from New Orleans. The Judge Advocate Generals' testimony against the Military Commissions Act and the military lawyers' amicus brief are an unprecedented divergence between military lawyers and the policy of the executive branch, which undermined the ability of the President to act as a unitary executive.

In his dissent to the Court's decision in *Boumediene*, Justice Thomas wrote that:

The Founders intended that the President have primary responsibility—along with the necessary power—to protect the national security and to conduct the Nation's foreign relations. They did so principally because the structural advantages of a unitary Executive are essential in these domains. "Energy in the executive is a leading character in the definition of good government. It is essential to the protection of the community against foreign attacks." *Id.*, [The Federalist] No. 70 at 471 (A. Hamilton). The principle "ingredient" for "energy in the executive" is "unity." *Id.* at 472. This is because "[d]ecision, activity, secrecy, and dispatch will generally characterize the proceedings of one man, in a much more eminent degree, than the proceedings of any greater number."[141]

In *Youngstown Sheet & Tube Co. v. Sawyer*,[142] the case in which the Supreme Court invalidated President Truman's seizure of steel mills during the Korean War, Justice Jackson articulated a test for reviewing the President's wartime decisions. He wrote that the President's executive power varies, depending upon whether he acts on his own or with the approval of Congress. When the President acts entirely on his own, in a manner which is contrary to the "express or implied policy of Congress,"[143] his power "is at its lowest ebb, for then he can rely only upon his own constitutional powers minus any constitutional powers of Congress over the matter."[144] When the President acts in the absence of Congressional action, "he can only rely upon his own independent powers, but there is a zone of twilight in which he and Congress may have concurrent authority, or in which its distribution is uncertain."[145] However, when the President "acts pursuant to an express or implied authorization from Congress,"[146] "his authority is at its maximum, for it includes all that he possesses in his own right plus all that Congress can delegate."[147]

The controversy over the use of military commissions—like the dispute over the treatment of detainees and electronic surveillance within the United States by military agencies—is a dispute over Presidential power. The Bush administration, motivated primarily by Vice President Cheney, who has sought to redress the diminution of Presidential power that he witnessed as Chief of Staff during the Ford Administration, has acted on the conviction that the President has unitary power to act without dissent or dissonance within the executive branch, and has the inherent authority to wage war without interference or limit by Congress, the judiciary, or international agreements.

When the Supreme Court considered *Hamdi*, who was an American citizen, Congress had not authorized military commissions or specified the procedures for treatment of unlawful and enemy combatants. Therefore, the Judge Advocates General, the detainees' military defense lawyers, Congress, and the courts had a role in determining what policy should be, because the President was acting when his power was at its "lowest ebb." However, after Congress enacted the MCA, the President's power was at "its maximum," and, although it turned out that the Supreme Court held that Congress did not properly suspend habeas corpus, the President should have been able to expect that the executive branch would act as a unitary whole in effecting his policy.

However, the JAGs who resisted the administration policy on military commissions have advanced an institutional agenda which they consider to be independent of, and to impose a duty higher than, administration policy. As Rear Admiral Donald Gutner, the Navy Judge Advocate General, told Jane Mayer: "We were marginalized. We were warning them that we had this long tradition of military justice, and we didn't want to tarnish it. The treatment of detainees was a big issue. They didn't want to hear it."[148] Since World War II, the Judge Advocate General's corps—and the Army Judge Advocate Generals' Corps, in particular—had developed a robust institutional identity and jurisdiction as the military's conscience. This identity was bolstered by the overhauls of military justice in 1950 and 1968, which made the military justice system perhaps fairer than its civilian counterpart. The systemic implementation of the Department of Defense law of war program, the development of operational law as a field of substantive law, and the widespread internalization of international law and limits to warfighting, all overseen by the JAGs, caused the JAGs to believe that they were the keepers of a higher standard of warfighting which matched American values and that their input into policy-making that involved the military was indispensable.

This attitude is a departure from Army tradition. The Army has been a reliable agent of executive power throughout its history, and has defied officials outside the executive branch to do so. The basis of Army action has been its understanding of its mission from the president—its organizational jurisdiction—even in the absence of specific instructions. Andrew Jackson ignored Judge Hall, and even removed him from office because he believed that it was necessary to accomplish his mission of defending New Orleans, and that his power to act flowed from the President who had given him his mission. Winfield Scott imposed martial law in Mexico, even without explicit approval, because he believed it necessary to accomplish his military mission. Army lawyers during the Civil War prosecuted civilian citizens before military commissions even after courts issued writs of habeas corpus requiring the Army to produce their prisoners. The Army commanders and lawyers in Hawaii enforced martial law pursuant to the President's proclamation in the face of continued resistance and orders from the federal judges in the territory. Colonel Kenneth Royall, who aggressively defended the Nazi saboteurs, only challenged the President's military commission proclamation in the courts after the President's representative authorized him to act within his own discretion.

The JAGs resistance to the current administration's military commission process may pass into history as a footnote—the Bush administration may terminate without the appointment of a new Supreme Court justice who would vote to overturn *Boumediene*, and the new administration will probably not adopt the aggressive position on the unitary and inherent power of the executive which was sponsored by Vice President Cheney. Nevertheless, this episode raises important questions about the institutional jurisdiction and independence of the Army as a service and Army lawyers as advocates, since the JAGs continued to resist the President's policy on military commissions and treatment of detainees even after Congress enacted the MCA. The military commission policy was supported by both of the political branches of the government. The aggressive resistance to the military commissions may be a sign that the Army (and its JAGs) are committed, or are becoming committed, to a philosophy of military justice and treatment of enemies that, as an institution, it considers superior to the policy of the President himself, or even of the President and Congress acting together. At this point—in September 2008—after the *Hamdan* conviction, the military is implementing the military commissions policy, but this should not obscure the fact that the substance of the policy was changed because of the military's resistance to it and that the policy which the military is now implementing is fundamentally different from the policy which the administration originally promulgated.

According to Jane Mayer, Major General Romig claims that Addington was "overheard" to have said: "Don't bring the TJAGs into the process. They aren't reliable."[149]

The wariness of the leading administration's civilian lawyers of military lawyers, who have now demonstrated an unprecedented independence, is a fundamental issue of civil-military relations which may affect the role which military lawyers are allowed to play in the interagency process, especially when policy which conflicts with military legal culture is proposed.

Notes

1. See Louis Fisher, *Military Tribunals & Presidential Power: American Revolution to the War on Terror*, (Lawrence: University Press of Kansas, 2005).

2. 3 Papers of Andrew Jackson, 206–7 (Harold D. Moser ed. 1991).

3. Fisher, 25.

4. 2 Correspondence of Andrew Jackson, 179 (J. Bassett ed. 1927).

5. Ibid.

6. 2 Correspondence of Andrew Jackson, 183.

7. Habeas corpus is the process by which a prisoner may seek review of his detention. Article I § 9 of the US Constitution provides that the writ of habeas corpus may not be suspended except in times of rebellion or invasion. A prisoner may file a petition for a writ of habeas corpus with a court and, if the court grants the writ, the person responsible for the prisoner's detention must produce the prisoner in court and justify the legal basis of the detention. If the detention is found to be illegal, the court may order the prisoner to be released.

8. Robert Remini, *Andrew Jackson and the Course of American Empire, 1767–1821*, (Baltimore, MD: The Johns Hopkins University Press, 1977), 310.

9. Jonathan Lurie, "Andrew Jackson, Martial Law, Civilian Control of the Military, and American Politics: An Intriguing Amalgam, *Military Law Review* 126, 133.

10. 2 Correspondence of Andrew Jackson, 189 (letter to Captain Peter Ogden and order to Judge Hall, both March 11, 1815).

11. 3 Papers of Andrew Jackson, 310.

12. Lurie, 140 (relying upon Dart, Andrew Jackson and Judge D.A. Hall, 5 *Louisiana Historical Quarterly*, 551–53 (1922)).

13. Remini, 313.

14. Fisher, 27 (referring to Jonathan Lurie).

15. Ibid., 37.

16. Ibid.

17. 1 Op. Att'y Gen. 233 (1818).

18. Fisher, 32.

19. Ibid., (citing 2 Justin H. Smith, *The War with Mexico*, 220 (1919)).

20. Ibid., 33.

21. John Pinheiro, *Manifest Ambition: James K. Polk and Civil-Military Relations During the Mexican War*, (Westport, CT: Greenwood Publishing, 2007), 91–95.

22. 8 Op. Atty Gen. 365.

23. Ibid.

24. Ibid.

25. Ibid. The argument that martial law is imposed in the absence of law recalls Wellington's rationale for the imposition of military rule in Portugal. He wrote in his dispatches from Portugal that: "Let us call the thing by its right name; it is not martial law, but martial rule. And when we speak of it, let us speak of it as abolishing all law, and substituting the will of the military commander, and we shall give a true idea of the thing, and be able to reason about it with a clear sense of what we are doing." (cited by David Dudley Field in his argument before the Supreme Court in *Ex Parte Milligan*, 71 US 2, 35 (1866).

26. 10 Op. Atty Gen. 74, 85.

27. Fisher, 44.

28. 2 Stat. 755 (1863).

29. 17 Fed.Cas. 144 (Cir.Ct. Md. 1861).

30. Wall. (68 US) 243 (1864).

31. Ibid., 253–54.

32. 71 US 2 (1866).

33. Ibid., 108–109.

34. Ibid., 124–125.

35. *Ex Parte McCardle*, 74 US 506 (Wall.) (1868).

36. *Ex Parte White*, 66 F. Supp. 982, 984 (D. Hawaii 1944).

37. *Duncan v. Kahanamoku*, 327 US 304 (1946) (*citing* Hawaii Organic Act, 31 Stat. 141, 153).

38. *The Army Lawyer: A History of the Judge Advocate General's Corps, 1775–1975*, (William S Hein & Co, 1975), 178.

39. Fisher, 131.

40. *The Army Lawyer*, 178.

41. *Ex Parte Zimmerman*. 132 F.2d 442, 444 (9th Cir. 1942).

42. 132 F. 2d at 443.

43. Ibid.

44. Ibid, 444.

45. Ibid.

46. Ibid.

47. Ibid., 446.

48. Ibid., 450.

49. 66 F. Supp. 982, 984.

50. Ibid.

51. Fisher, 134–35.

52. Ibid.

53. *Ex Parte White*, 66 F. Supp. 982 (D. Hawaii 1944).

54. *Ex Parte Spurlock*, 66 F. Supp. 997, 1004 (D. Hawaii 1944).

55. *Ex Parte Duncan*, 55 F. Supp. 976 (D. Hawaii 1944).

56. Ibid., 981–82.

57. *Ex Parte Duncan*, 146 F. 2d 576 (9th Cir. 1944).

58. *Steer v. Spurlock*, 146 F. 2d 652 (9th Cir. 1944).

59. 327 US at 323–24.

60. Fisher, 95.

61. Ibid., 96.

62. Fisher, 96 (quoting Diary of Henry L. Stimson, June 28, 1942, Roll 7, at 128–29, Manuscript Room, Library of Congress, Washington DC).

63. Fisher, 99 (quoting Biddle, In Brief Authority at 331 (1962).

64. 7 Fed. Reg. 5101.

65. Fisher, 99–100

66. 7 Fed. Reg. 5103 (1942).

67. Fisher, 103–106.

68. "Petition for Writ of Certiorari to the Court of Appeals for the District of Columbia," reprinted in 39 Landmark Briefs and Arguments of the Supreme Court of the United States 296, 550 (1975).

69. Fisher, 112.

70. Ibid., 106.

71. Ibid. (Fisher relied upon "The Reminiscences of Kenneth Claiborne Royall," *Oral History Project*, Columbia University, New York, 1965, 35).

72. Ibid., 106–107.

73. *Ex Parte Quirin*, 317 US 1, 18 (1942).

74. Fisher, 107.

75. 317 US at 48.

76. Fisher, 114.

77. 317 US at 30–31.

78. Ibid., 37.

79. Ibid., 45–46.

80. Ibid., 47.

81. Ibid.

82. Conversation between Justice William O. Douglas and Professor Walter F. Murphy, June 9, 1962, at 204–5, Seeley G. Mudd Manuscript Library, Princeton University, quoted in Fisher at 124.

83. Fisher, 121–24.

84. Ibid., 127, quoting Stimson Diary, January 5, 1945 at 18–19.

85. Ibid., 128.

86. Frederic L. Borch, *Judge Advocates in Combat: Army Lawyers in Military Operations from Vietnam to Haiti*, (Honolulu, HI: University Press of the Pacific, 2001), 319.

87. http://www.goarmy.com/jag/history.jsp

88. 66 Fed. Reg. 57833.

89. Ibid., 57834.

90. Charlie Savage, *Takeover: The Return of the Imperial Presidency and the Subversion of American Democracy*, (Boston: Little, Brown and Company), 135.

91. Ibid., 136–37.

92. Ibid., 137.

93. Ibid., 138.

94. Ibid.

95. Jane Mayer, *The Dark Side: The Inside Story of How the War on Terror Turned into a War on American Ideals*, (New York: Doubleday Publishing, 2008), 86.

96. Savage, 139.

97. Ibid., 47.

98. Ibid., 48.

99. Ibid., 62.

100. Ibid., 63.

101. Ibid., 287.

102. H.R. 4200 "Ronald W. Reagan National Defense Authorization Act for Fiscal Year 2005" at § 574.

103. President's Statement on the Ronald Reagan National Defense Authorization Act, 2005, at http://www.whitehouse.gov/news/releases/2004/10/print/20041029-6.html

104. Ibid.

105. Charlie Savage, "Control Sought on Military Lawyers," *Boston Globe*, December 15, 2007, http://www.boston.com/news/nation/washington/articles/2007/12/15/control_sought_on_military_lawyers/

106. *Hamdi v. Rumsfeld*, 542 US 507, 510 (2004).

107. Ibid., 512.

108. Ibid.

109. Ibid., 513.

110. Ibid., 516.

111. 424 US 319 (1976).

112. 542 US at 533.

113. Swift was passed over for promotion in 2006 and left the Navy. See Shukovsky, "Gitmo Win Likely Cost Navy Lawyer His Career," *Seattle Post-Intelligencer*, July 1, 2006 at http://seattlepi.nwsource.com/national/276109_swift01.html

114. 548 US at 569.

115. Ibid., 570.

116. Ibid., 571.

117. Ibid., 572.

118. Ibid., 584.

119. Ibid., 599.

120. Ibid., 612.

121. Ibid., 623.

122. Ibid., 630, quoting Geneva Convention (III) Relative to the Treatment of Prisoners of War, August 12, 1949, [1955] 6 UST. 3316, T.I.A.S. No. 3364 (Third Geneva Convention).

123. Ibid., 279–81.

124. 10 USC. § 948b(b).

125. 10 USC. §948b(f).

126. 28 USC. §2241(e).

127. 10 USC. §948d(c).

128. 128 S. Ct. 2229, 171 L.Ed.2d 41, 2008 US LEXIS 4887, 76 USL.W. 4406 (2008).

129. 128 S.Ct. at 2241.

130. 542 US 466 (2004).

131. 128 S.Ct. at 2241.

132. See *Khalid v. Bush*, 355 F.Supp.2d 311 (DC 2005); *In re Guantanamo Detainee Cases*, 355 F.Supp.2d 443 (DC 2005).

133. 10 USC.A. § 948(a) *et seq.* (Supp. 2007).

134. 128 S.Ct. at 2242.

135. Brief of the Military Attorneys Assigned to the Defense in the Office of Military Commissions as Amicus Curiae in Support of Neither Party, *Al-Odah v. United States*, Supreme Court of the United States, No. 03-343 (February 14, 2004) at 5–6.

136. Ibid., 2270.

137. Ibid., 2272.

140. Ibid., 2277.

141. William Glaberson, "Bin Laden Driver Sentenced to a Short Term," The *New York Times*, August 8, 2008, http://www.nytimes.com/2008/08/08/washington/08gitmo.html?

142. 542 US at 580.

143. 343 US 579 (1952).

144. Fisher, 233.

145. 343 US at 637–38.

146. Ibid.

147. Fisher, 232.

148. 343 US at 637.

149. Mayer, 88–89. "TJAGs" is the acronym for The Judge Advocate Generals.

Panel 8—Military Governments and Courts

(Transcript of Presentation)

Provincial Reconstruction Team (PRT): A Grass Roots Interagency Counterinsurgency Methodology in Afghanistan

by

Lieutenant Colonel Lynda M. Granfield, US Army
Former PRT Commander for Jalalabad

Thank you for that humbling welcome. While you are focusing on my title and symbology in the photo here (Slide 1), I am going to provide some thoughts, so if I do not move through the slides as quickly as you think I should be, I am not missing something. I have some things that I want to just put out there and it is not a paper because I am not an academic. I am a practitioner and maybe that is a good thing or maybe it is a bad thing; I am not sure. This paper represents my personal views on what the interagency and whole of government process currently looks like through the lens at the field-level for over fifteen years in a variety of assignments on active duty as a civil affairs officer, and now in my current role in Washington DC at the strategic-level. So with that, I have to caveat that I do not speak for Secretary of Defense nor do I speak for Secretary of State on their policies, views, or doctrines. This is my personal opinion. Working in an interagency environment as a member of the United States Armed Forces from my perspective is more of an art and not necessarily a science. Although there is a plethora of doctrine and books on this subject, much of what needs to take place is based on personal relationships, networking, and more importantly, being seen as value added and not only the USG (United States Government) process, but with a variety of potential partners found in complex operations. But now is the time, and we must get this right, as much depends on creating an environment in which each agency's goals and objectives can be met, thus creating synergy to play off each other's strengths and make up for areas that are known weaknesses. If we do not find balance then the enemy will continue to exploit our very bureaucratic tendencies and agency stovepipes. There is room for the US military to take the lead, but in many cases, there are other agencies in the USG that are more suited based on their expertise. In my position in civil affairs, most recently at the PRT (Provincial Reconstruction Team) in Afghanistan, there are often obstacles to creating that much needed synergy. I plan to briefly set the stage of this presentation on PRTs first from the USG's strategic perspective so that you have an overarching framework to build on (Slide 2). In many cases, the tactical part is where the friction can end up either resulting in the cooperating graduate, i.e. working in concert to achieve results or the downside of herding squirrels, and notice I did not say cats because squirrels can jump from treetop to treetop as going to policy makers and key decision makers. The Jalalabad PRT was a model of successful integration of interagency best practices, necessary to achieve a whole of government approach. I will then conclude with my perceptions at the strategic tactical level of USG integration.

There are three parts to the three Ds—diplomacy, development, and democracy that I would like to cover (Slide 3). Unfortunately, I will not get to each bullet in detail today due to my time limits. That leaves you to do your homework on the USG's strategic publications. This is only the tip of the iceberg and is solely focused on the USG. Not only do we need a USG integrated approach to counterinsurgency and stability operations, but a multi-national framework as well. As currently seen in Afghanistan, NATO, and the United Nations and a variety of interagency bodies are partners as well. We must contend with not only USG policy, but with those as well. There are international agreements and treaties that the US is a signatory to and a couple of examples include the Sphere Guidelines and Oslo Accords. Every military member should be familiar with these two documents. The reason why this is important is in the case of ISAF (International Security Assistance Force) currently in Afghanistan. There is a PRT handbook, but not all participants follow the strategy based on country caveats that hinder operations. The National Security Presidential Directive 44, otherwise known as NSPD–44, was signed on 7 December 2005. Interestingly, it replaced the Clinton Administration Presidential Directive 56. Both documents cover a similar document, and interestingly, that is what we have heard throughout the theme of today. Produced in May of 1967, the National Security Action Memo 362, outlined the responsibility for the US role in pacification. Of interest, NSPD–44 was issued one month after DOD Directive 3000.05. It seems like the cart and the horse, but I am not quite sure. NSPD–44 designates the Department of State (DOS), notice not the Department of Defense, as a lead agency for interagency coordination and planning. DOS is to develop strategies and plans for reconstruction and stabilization as well as develop a civilian response capability through the recently created Office of the Coordinator for Reconstruction Development (S/CRS). The second responsibility for Department of State is to harmonize such efforts with US military plans and operations, to include complex emergencies, failing or failed states, and environments across the spectrum of conflict. This is a fairly large undertaking for DOD, let alone Department of State. It probably has less than twelve hundred Foreign Service Officers worldwide.

The second part of the "D" is development (Slide 4). Since the Foreign Aid Act Reform of 1961, USAID (United States Agency for International Development) has come a long way in developing a comprehensive policy framework for bilateral aid by implementing transformational diplomacy for development. There are some institutional changes within DOS and USAID that you should be familiar with. As the current administrator, Henrietta Fore, equivalent to a Deputy Secretary, has been appointed, designated by the Secretary of State as the Director of Foreign Assistance. This change took place in January 2006 when Secretary of State announced a major change in her transformational diplomacy strategy that not only included DOS but USAID as well. Prior to this policy shift USAID had organized several new offices to better address coordination with the US military during complex operations in humanitarian response. The Office of Military Affairs (OMA) was established in 2005 and was created in response to the national security strategy to collaborate between the two organizations in order to achieve national security. OMA is the day-to-day interface with DOD and helps integrate the USG policy between the two organizations and has recently provided senior development advisors to the five geographic combatant commanders, AFRICOM (United States African Command) and JFCOM (United States Joint Forces Command), respectively, and has senior military personnel represented in the OMA to help with three focused areas—interagency planning and implementation, policy development, and training and education. I would say that one of the most important aspects to OMA is the large piece of training and education. This includes providing trained US representatives at JRTC (Joint Readiness Training Center) in JFCOM exercises. The administrator for USAID takes the interaction between the two groups very seriously and has recently issued the Civilian Military Cooperation Policy designed to facilitate a whole of government approach to draw on the strengths of the two agencies to deal with interactions during complex emergencies. DOD Directive 3000.05 *Military Support for Stability Security Transition and Reconstruction Operations* was released November 2005. It establishes the framework for DOD to plan, train and prepare, and to conduct stability operations (Slide 5). It assigns responsibility across DOD. The policy highlights that stability operations are a core US military mission that DOD should be prepared to conduct and support. Paragraph 4.4 goes further to state, "Integrated civilian and military efforts are key to successful stability operations." Our recent experiences in Afghanistan and Iraq highlight the importance of successful integrated strategies in complex environments. Not only do we need a national framework, but a field-level guide that helps prepare our forces to work in environments that define the populations at the center of our efforts. FM 3-07, due to be released in the upcoming months, builds on FM 3-0. It describes an operational concept where commanders employ offense, defense, and stability or civil support operations simultaneously as part of an interdependent joint force. Both documents build on FM 3-24, counterinsurgency that really was the first real hard look at the Army's doctrine and strategy and played on the painful lessons of Vietnam.

This slide (Slide 6) shows the overlap of S/CRS as an essential task matrix and how it corresponds with USAID programs and tasks found in FM 3-0. This is a methodology in a systematic way that the interagency should conduct mission analysis in whole of government planning efforts. It is a framework, and a critical component to the USG effort, it offers a foundation to build upon during stability and reconstruction activities. Whether one follows the essential task matrix or stability task found in FM 3-0, the application at the field-level is critical to determine which agency is best suited for being the primary or secondary effort across the spectrum of conflict. Another example of the complexity is the Department of State's Humanitarian Information Unit, a small interagency organization created showing the overlap and alignment of the three Ds at work. As you can see from the slides, the lines are often misaligned and as we all well know, conflict has no borders (Slide 7). Often issues become regional in focus and from this map of the world, it shows how our current institutional organizational constructs might not be as flexible and adaptable as it needs to be in complex emergencies or stability operations. So to the meat of my presentation, something that I like to talk about.

Provincial Reconstruction Teams (Slide 8) first found their roots in Coalition Humanitarian Liaison Cells (CHLCs), established in early 2002 in Afghanistan and were designed to begin coordinating the activities of the USG and other international partners to include the United Nations Assistance Mission to Afghanistan (UNAMA). It had established small outposts in Jalalabad, Kandahar, Herat, and Mazar-e Sharif. CHLCs had three functions—assess humanitarian needs, implement small-scale hearts and minds projects, and establish relationships with UNAMA. The scope and concept of CHLCs grew into a new organization called PRTs. They were established in Gardez, Bamian, and Kunduz. These three original PRTs were of strategic importance to the interim government of Afghanistan and President Karzai, as they represented areas of tribal and rival factions that could have disrupted the progress made thus far by the US and UNAMA. Originally PRTs were called Regional Reconstruction Teams but the name was changed based on the historical baggage of strong regions, i.e. warlords versus a strong central government. PRTs were instrumental in creating the political desire of the international community to share the burden of nation building to avoid the perception that the US and the international community were invading Afghanistan. PRTs, due to their provincial focus and civil-military resources, have wide latitude to accomplish their mission of extending the authority of the central government by improving security, supporting good governance, and enabling economic development. This had been, in the past, the strength of the PRT as it provided a flexible approach and resources to support the provincial government, improve security, and gain population support of not only the USG actions, but more importantly, to recognize the efforts of their own government and that the population in Afghanistan recognize that their government was working for them at three levels—central, provincial, and district. To give you a bit of context of our area of operations, Jalalabad is the capital of Nangarhar Province due east of Kabul and borders Pakistan (Slide 9). Nangarhar is considered a strategic gateway due to several factors. History shows us if Jalalabad falls, Kabul falls. The economic importance of the Khyber Pass border crossing, the rich agricultural capacity that has the potential to produce roughly sixty percent of Afghanistan's food requirements with the right development, an illicit trade route that has a long history with the outside world, it has over the last five years produced fifteen to twenty percent of the world's opium depending on the year. Haji Kadir welcomed Bin Laden after he was forced out of Sudan, hosted the infamous Tora Bora battle, and claims the successful launch of the first Stinger missile by the mujahedin against the Soviets and the list goes on and on. Here is the PR (Public Relations) mission statement from 2005 that was agreed upon by the interagency at the CJTF-76 (Combined Joint Task Force)-level in an effort to further refine our focus at the field-level (Slide 10). As previously, there was much dissent in the interagency on the original mission as many in the US did not feel that they were to be involved in a counterinsurgency operation.

PRTs worked to bring the Afghan government to the population. It was not an Afghan-face strategy by the USG, but Afghan led as now was the time to push the provincial and district leadership to make decisions for the population and increase the security situation so that development efforts could begin to take hold. I would also like to highlight, from my perspective, PRTs are not a new concept. If you refer to the current configuration of the Civil Military Operations Center (CMOC), I would argue from a doctrinal perspective, that PRTs are CMOCs on steroids. It is the first time that I have had all the pieces and parts, i.e. the interagency, and a large budget to effect the PRT mission. You can call them whatever you want, CORDS, CMOCs, PRTs, clearly they are an ends to a mean.

As you can see from this original 2003 task organizational chart, not much consideration was given to the interagency (Slide 11). You can see over here on the right hand column that the interagency is on the outside of the PRT and creating the much necessary synergy expertise that they brought to the table. In the early days, DOS and USAID were merely living on the Forward Operating Base and their stated purpose was outside the mission PRT construct. The base of relationship was information sharing only and that was normally accomplished through the Civil Military Operations Center. The relationships were dysfunctional as the team normally worked cross purposes, and the military dealt with issues back with the military chain of command and Department of State and USAID went back to Kabul for guidance and program strategies. It was extremely disconnected. As you can see from this slide (Slide 12), the purpose of the PRTs did not change over time, but the efforts and strategy evolved to capture synergy. DOS, USAID, and the US Department of Agriculture are no longer outside the PRT construct, but an integral partner with an equal voice. This was in part due to the recognition of the CJTF-76 Commander, Major General Kamiya, in mid 2004, prior to US SETAF's (United States Southern European Task Force) deployment to Afghanistan, that more emphasis on the benefits of interagency coordination and the application of the three Ds at the field-level, should follow the lessons from the earlier efforts of the Civilian Operation and Revolutionary Development Support (CORDS) program of the late 1960's in Vietnam. This unified effort of interagency representation was not only implemented at the PRT level, but at RC (Regional Command) East and South, at CJTF-76 in Bagram and at CFC (Combined Forces Command) Alpha in Kabul, providing interlocking efforts at each level. Having the interagency team as part of the decision cycle at each level allowed for unity of effort and created a command group, or as we coined at the Jalalabad PRT, the Interagency Executive Team. The reason why the Jalalabad PRT was so successful is that this executive team created short-, medium-, and long-term objectives that nested with CFC Alpha's lines of operations and more importantly, the Afghan National Development Strategy, developed by President Karzai and supported by the international community through a series of Donors' Conferences. We visualize this concept of unity in our infamous rope slide, progress though interagency synergy (Slide 13). And it symbolizes the need for each agency to work together to achieve successful conditions in the province. Even our provincial governors would say that they were only as weak or as strong as their weakest governor. Creating the optimal conditions requires all entities to come together to maximize resources and capitalize on available assets without compromising the different agencies' missions and goals. After all, at the end of the day, team members to the PRT are stakeholders in the provinces. We are there to create conditions for the central and provincial government to be responsive to the people and self sufficient Afghanistan government. The primary and secondary effort concept was created by a USAID representative on the PRT and illustrates my earlier point, referring back to the three PRT lines of operation—promoting good governance and justice, enabling the Afghan security apparatus, and facilitating reconstruction. Developing economic growth was no small task. Understanding the operational environment and the priorities for the military and the interagency partners became the priority effort and we leveraged the strengths and expertise of each member. So how did the PRT create unity of effort without blurring the lines and crossing into each other's strengths? As you can see in this slide, in the areas where security was an issue (Slide 14), the main effort was on creating a safe and secure environment so that the USAID implementing partners could begin to build capacity and infrastructure and the Department of State could begin working with local district leadership in governance and transparency. Likewise, in areas that were more permissive for development, USAID programs became the main effort and the PRT programs executed through the Commanders Emergency Response Program (CERP) money were focused in other areas such as mentoring and training the Afghan national police and border police at the provincial, and more importantly, the district level.

So I am not going to talk about each one because I am not going to have enough time, but I am going to highlight the first three (Slide 15). In the case of 2006, the US captured best practices in an effort to leave behind a strategy that could continue to be implemented. I have talked about the interagency piece for our PRT, but what did that mean in the day-to-day operation of the PRT? It was as simple as the four of us in the executive team getting together over coffee prioritizing the week's events, reviewing the three to six month strategy, and by the way, figuring out what we were going to do with the thirty visitors coming to the PRT that week. In all, we had three hundred and sixty seven for the year, not our primary focus, but we had to coordinate our efforts. It could also be as complex as building programs that captured all three stakeholders and their resources in a unified effort. One of the biggest successes of the joint interagency work that our PRT managed to pull off was to create the Nangarhar Construction Trade Center. It was a USAID funded program that was 4.5 million dollars. A joint assessment early on from the executive team determined that the original schools and clinics that were built by USAID did not necessarily hold up to the harsh climate. As well, many of the construction firms used local labor that was not trained. We determined if we could create a center that brought the workers from those companies together and trained them in five focus areas, we could then send them back to repair the clinics and schools and then begin to build the capacity of the local labor. The measure in this program was not the money spent, which is what you hear, effects equal money and that is not necessarily the case, but the sustainment of the program and what do I mean by that? It does you no good to create a program of skilled workers that do not have a job when they leave the program, and by the way, if the government cannot sustain the program over the long term, then it is not viable. In the case of this program, it is still there today. The University of Nangarhar has assumed responsibility for the program and the sustainment piece and the school is now part of the engineering department. Lastly, the PRT created a Project Nomination Review Committee that included senior members from the staff as well as the interagency team. Each element could nominate a project in support of the lines of operation that were nested in our system's approach and each member of the PRT had a veto vote. This forced PRT members to do their homework and it ensured that it nested with the provincial lines, director's priorities and allowed the USAID and the Department of State rep to coordinate with the ministries in Kabul, extending the reach of the central government. This allowed all members on the PRT to have visibility of what each agency was doing and in many cases, if the project was denied, a more senior discussion with the executive team would be held to determine if the program was viable or not. Most of the time a compromise was reached or a project was shelved. The only exception was USAID opposition to the PRT efforts with the Director of Hajj and Mosques in the mullah community. However, since the mullah community was part of the PRT's overall key communicator strategy, the PRT ended up funding several programs to support the Director and government controlled mosques in our Temporary Work for Afghan Program.

So are we there yet? What is a presentation without some quotes (Slide 16)? As you read through these next few bullets I think it is extremely important to realize that the CORDS program was an effort, albeit late in implementation, to bring together the interagency. If you have not taken the time to read some of Komer's thoughts, you should. I think his fourth point about creating unified management at each level where multidimensional conflict situations dictate integrated multifaceted responses really points out that, despite the lessons from Vietnam concerning the interagency, not much has changed or improved. We are still, from my point, working at cross purposes and paper thin efforts. John Paul Vann also hits the nail on the head and it does not apply just to Vietnam (Slide 17). We are seeing this in Afghanistan. The Afghans see a new group come in every nine to twelve months and make the same mistakes culturally, religiously and tribally. And they are quite honestly, tired of teaching us. In 2006, we were well on our way to creating momentum with what the PRT's mission and focus were in capturing the strengths of the interagency. However, we have lost progress. It is a complex environment and requires trained experts that command and staff PRTs. No offense to my Navy and Air Force counterparts that are now running PRTs, but they do not have fifteen years of civil affairs experience, and we have not set them up for success. It is not about sending them off to three months at Fort Bragg in a crash course and that being good leaders that they will just figure it out as they learn, often making critical errors or building schools that have no teachers or program students. An example of comparison is that I have forty-five hours in a Cessna 152 so I guess that means the Air Force should send me off to three months of F-16 training and put me in the air. I would argue that we have not applied the lessons from the past and, in many points, have further created friction at senior levels. CORDS, NSPD–44, and S/CRS are good examples. Recently, events in Georgia required a USG unified and coordinated effort. S/CRS set up a team to go to the Embassy and yet the diplomacy piece of our response was secondary and the US military engaged in humanitarian relief despite the Oslo Accords framework, that any military supportive relief should be the last resort. Further, S/CRS has not gotten the traction needed within Department of State and there is still much friction within the regional bureaus. In the CORDS program, both civilian and military were placed under command and control and unity of effort of the Military Assistance Command. And the program was run by a civilian, creating leveraging in the USG unity of effort, again, a little bit too late. So if the strategic-level is not there, are the tactical efforts any better? I would argue that the PRTs should be the main effort in Afghanistan and yet they lack the expertise, civilian and military, the right equipment, the right force composition and adequate funding. The CERP funding differences in Iraq and Afghanistan are so disparate and should be cause for alarm. Perhaps a single source funding authorization stream for the USG in these types of operations would create more favorable conditions for the agencies to work together. Finally, Congress must give the other pieces of the interagency the resources the President has directed them to carry out. Without significant changes in personnel and budget authorizations, I do not have much faith that things will improve to allow S/CRS and Department of State to take the lead for the task they have been given, interagency coordination and planning, and we will all continue to work at cross purposes.

The US Army and the Interagency Process:

"Provincial Reconstruction Team (PRT): A Grass Roots Interagency Counterinsurgency Methodology in Afghanistan"

LTC Lynda M. Granfield
Former Jalalabad PRT Commander June 2005-May 2006

UNCLASS

Slide 1

UNCLAS

Agenda

Part I: United States Government (USG)- Macro

- Over view of key USG and policy changes

- 3'D's at work

Part II: Interagency Coordination- Grass Roots Level

- Understanding the Operational Environment

- PRT-Command and Control (C2) Relationships

- Best Practices

Part III: Are we there yet?

UNCLAS

Slide 2

UNCLAS

U.S National Security Strategy (NSS):
3 D's Diplomacy, Development and Defense

Department of State
•NSPD-44: Designates DoS as the lead agency for interagency coordination (through S/CRS

• 2003-Created the Humanitarian Information Unit

• 2005-Office of the Coordinator for Reconstruction and Stabilization (S/CRS)

•Secretary Rice initiative: "Transformational Diplomacy"

•Creation of "F"-Dual-hat USAID Director as DoS lead for "Foreign Assistance"

•Secretary Rice launches the Civilian Response Corps on July 16, 2008

UNCLAS

Slide 3

UNCLAS

3 D's (cont)

United States Agency for International Development (USAID)

• 2003: Establishes the Office of Conflict Management and Mitigation

• 2005: Establishes the Office of Military Affairs

• 2006: Policy Framework for Bilateral Foreign Aid: Implementing Transformation Diplomacy through Development

• 2008: Civilian-Military Cooperation Policy

UNCLAS

Slide 4

3 D's (cont)

Department of Defense
• DoD Directive 3000.05 – IA coordination for "Military Support for Stability, Security, Transition and Reconstruction Operations"

• Irregular Warfare Roadmap

•JFCOM-Establishment of Joint Interagency Coordination Group (JIACG)

•J.P 3.08-Interagency, Intergovernmental Organizations, and NGO Coordination Volume I and II

•FM 3.0 Operations Feb 2008

•FM 3.7 Stability Operations (pending release)

•Provincial Reconstruction Teams- Afghanistan(2003) and Iraq(2006)

Slide 5

How the 3 D's overlap

S/CRS Essential Task Matrix:	USAID Programs:	FM 3.0 Stability Tasks:
Security	Conflict Prevention	Civil Security
Justice and Reconciliation	Rule of Law	Civil Control
Humanitarian Assistance and Social Well Being	Humanitarian Assistance and Global Health	Restore Essential Services
Governance and Participation	Democracy and Governance	Support Governance
Economic Stabilization and Infrastructure	Economic Growth, Agriculture and Trade	Support Economic and Infrastructure Development

Slide 6

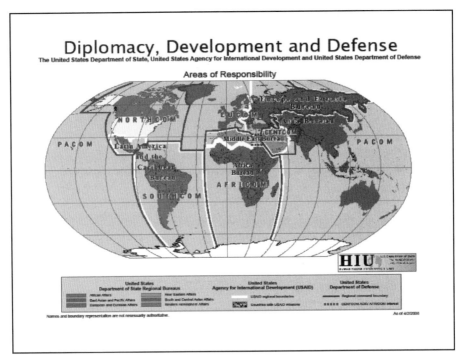

Slide 7

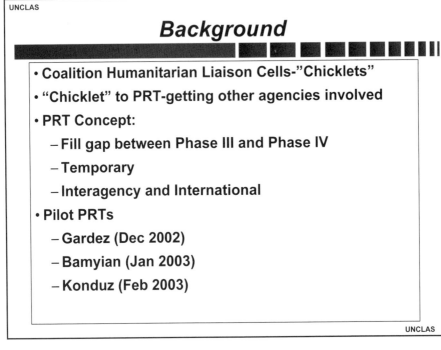

Slide 8

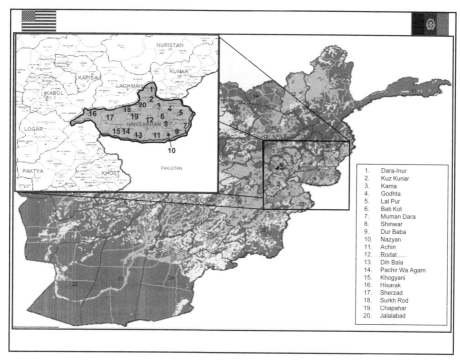

Slide 9

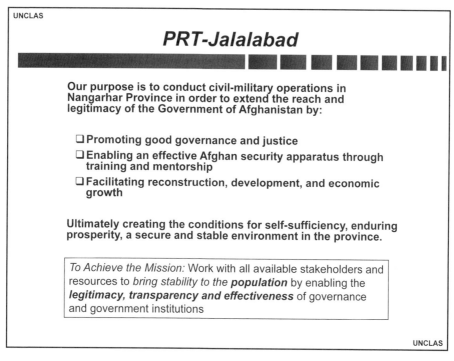

Slide 10

2003 PRT Task Org

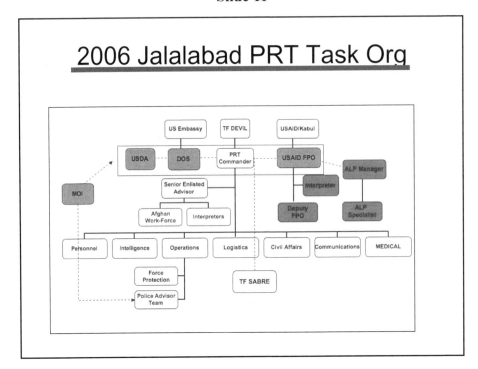

Slide 11

2006 Jalalabad PRT Task Org

Slide 12

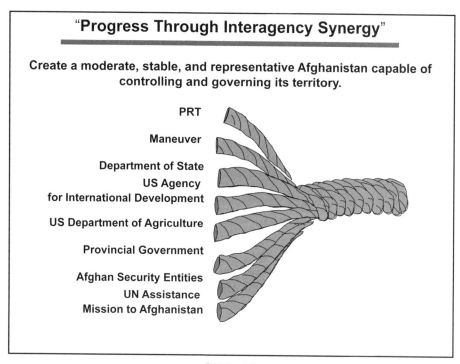

Slide 13

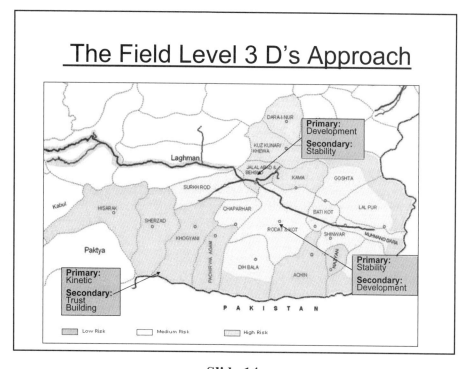

Slide 14

Jalalabad PRT -"Best Practices"

- Interagency "Executive" Committee Meeting

- Internal Weekly Project Nomination and Review Board

- Creation of "Center of Construction Excellence"- Nangarhar Construction and Trades Center

- Weekly Provincial Security Coordination Meeting

- Monthly Provincial Coordination Council

- Hospital Task Force Working Group

- Roads Technical Working Group

- Construction of District government and Judicial Center's

- Governmental Conferences/Symposiums

- Monthly "Key Leader " Engagement Strategy

Slide 15

Learning Lessons?

"But our Vietnam experience also shows how difficult it is to translate such general…and in hindsight obvious…lessons into the requisite performance. This demands a constant, deliberate effort to offset the inevitable tendency of bureaucracies to keep doing the familiar and to adapt only slowly and incrementally. In particular, such efforts requires:
1) specially selecting flexible and imaginative conflict mangers at all levels:
2) revising training and incentive systems in place to place a higher premium on adaptiveness instead of the "school solution";
3) setting up autonomous ad hoc organizations to manage specially tailored programs which are not in conventional organizational repertories:
4) *creating unified management at each level where multidimensional conflict situations dictate integrated multifaceted responses*:
5) assigning adequate staffs to single managers…"

Robert W. Komer, Deputy to USMAC-V, CORDS

Slide 16

Learning Lessons?

""We don't have twelve years' experience in Vietnam. We have one year's experience twelve times over"

John Paul Vann- former Deputy for CORDS in the Third Corps Tactical Zone of Vietnam

"I could probably write a lengthy book about being a PRT commander in Afghanistan. I have told many Army officers that in many respects, it is even a tougher job than battalion (squadron) command. *Interagency partnership at the grassroots level*, integration of security and reconstruction/development, working with Reserve and Active forces... with coalition partners, not to mention the Afghan leadership such as provincial governors and tribal elders...an accumulation of experience that is simply invaluable."

MG Jason Kamiya- former CJTF76 Commanding General

Slide 17

Panel 8—Military Governments and Courts
Question and Answers
(Transcript of Presentation)

Colonel Gary M. Bowman, US Army
Lieutenant Colonel Lynda M. Granfield, US Army

Moderated by Mr. Robert Ramsey III

Audience Member

I looked at your organizational chart and what I saw was three boxes with vertical lines going up to three horizontal boxes with no connection at the top and you mentioned that you think there is a problem with unity of effort. It seems to me the problem is unity of command and that starts at very high levels. We had a former Ambassador speaking to us the other day. He was very jealous of his prerogative to be the President's personal representative in a certain country. Well, who is in charge in Afghanistan? Who makes the call on what the PRTs do? My experience over there was, and you have much better experience than I do obviously, was that somebody from USAID, for example, might completely disagree with the military commander's objectives for the PRT. Where does this need to be solved so that one person is on the blame line and prioritizes the efforts of the United States government as a whole in that region?

LTC Granfield

Absolutely, that is an excellent question and it probably would take me about two or three hours to really cover where I think things were. As I highlighted, in 2006 we had representatives of the interagency at all four levels and so the coordination and the synchronization of the efforts of the USAID reps at the PRT-level to the brigade-level at the combat team up to 76 and then Combined Forces Command, were all nested at those levels and it worked quite efficiently when I was there from 2005 to 2006. There was obviously friction points because again, it is personality based. There is not any doctrine. There is an executive agreement in the interagency MOU (Memorandum of Understanding) that was signed in 2003 that really, quite frankly, has not been adhered to. There is supposed to be an executive steering committee that met monthly with all of the players in Kabul that included the USAID Mission Director, the Ambassador, General Eikenberry, and the ISAF (International Security Assistance Force) because they were a component, to come up with what the priorities and the strategies for the PRT should be. Then that was to trickle down in guidance. Were there disconnects? Obviously, there were. We made it work at our PRT just because it was personality based. How do we fix it? I think at the national-level when we get ready to do a mission, we have to rely on some sort of agreement and it has got to be a quick responsive agreement. If we do not have a quick responsive agreement just like we saw in Georgia, S/CRS (Office of the Coordinator for Reconstruction Development and Stabilization) stood up a team as they outlined in their matrix and their integrated management system (IMS), but they are supposed to stand up a team that is directed by the Principles Committee, and that includes all the senior reps at the Secretary-level and the Executive Branch, come up with a plan of action, who is going to be the priority of effort, whether it is going to be the military or not. And then they launch a team in to help the Ambassador and the Embassy simultaneously. An integrated planning team is supposed to go off to the combatant commander that is supposed to be an interagency team to help—in this case, EUCOM—come up with a military response to that whole of government plan. They do not have the right people, and they do not have the right funding; it is just a very frustrating process. I do not have an answer for it. There are systems in place but we have to get past "Who is the main effort; who is the supporting effort?" Maybe the answer is to create a new agency, outside another executive branch, that has the interagency pieces and it is not in S/CRS. In S/CRS's case, the regional bureaus often trump what S/CRS has been tasked to do by the President. So I do not know if that answers your question but it is not a simple solution. And it becomes personality based.

Audience Member

My question is, you talked about measures of success in terms of what you see. What tools do you see out there that have been useful in determining measures of performance as the people within Afghanistan see it?

LTC Granfield

Well, that is excellent because if the people do not see what our efforts are, I think that we are not making progress, and they do not see that their government is working for them. In our case, in 2005 and 2006, our key communicator strategy was one of the most effective ways that we could measure performance and the reason being is Jalalabad had riots into May of 2005 over the Gitmo-Koran incident. And they literally burned the city down as many provinces in Afghanistan did. My predecessor came up with a program to deal with key communicators who they thought were the instigators of the riots, one of which was the mullah community. So we continued that system and that is why it was so important that we continue the work despite USAID's objections. It is not their objections from a personality base, it is because there are legal objections that they work with religious institutions, which is again, something that Congress probably needs to take a look at. But what we did then was create this very systematic approach to all of the key communicators so when the cartoon incident happened about seven months later, we were the only Province that did not have riots. That is, I think, a direct reflection of not only our key communicator program, but also to what we were doing in the provinces and in the districts that were some of the problem areas. Additionally, we bothered to train the police in Jalalabad in riot control and get State to get them the right gear so they could deal with riots. We had a big program for University of Nangarhar students, and again, they are a part of our key communicator strategy so all of these different lines and extremes fit into our measures of performance. And from our perspective at the interagency executive team level, we could not have done it without each other. So we built off of each other's strengths. What is happening now? Unfortunately, now you are seeing the PRT Civil Affairs Teams going out and building schools and clinics in areas that really is an USAID function. And by the way, it is an international function because of donors. So if you build a school and you do not have teachers that are nested with the ministry at Kabul's education program, you just have an empty building. And so you have to have the right people that understand that it has to be a systems approach. It is not just one area. We are going to work on schools, USAID is going to do this, and State is going to work on governance to bring that framework together. Why did it work for us? I think a lot of it was personality based and that we had a commanding general that had vision, both at the JTF–76 level and also down in Kabul, who worked very closely with the Embassy to build those relationships, and we had a very strong relationship with General Eikenberry and Alonso Fulgham who was the USAID mission director, and Ambassador Neumann. So again, it is based on those personalities. We really have this momentum going in not only creating effects, but its performance as well.

Audience Member

I have a question for Colonel Bowman. Sir, what are the implications of the extension of habeas corpus to detainees? Is it restricted only to those in Gitmo or does it extend to detainees under US military control in theater?

COL Bowman

The implications are extensive and most of my papers are about those implications actually but you cannot understand what the current system is without understanding where it came from. Yes, habeas corpus extends to Gitmo. One of the initial challenges to detention were from a group of detainees, the Rasul group of detainees. The lower courts held that habeas corpus did not extend to Guantanamo; the Supreme Court held ultimately that it did. So those detainees' cases came back through the system and that group, or the core of that group, were the petitioners in the latest Supreme Court case called *Boumediene* which just came out this summer. And in that case, the Supreme Court said that the Military Commissions Act of 2006 which was Congress' attempt to validate the administration's policy for determining whether, and to what extent, the government could detain these people in Guantanamo. It was Congress' attempt to get on board with the administration, having a unified executive policy that was validated by Congress; that was in the Military Commissions Act. The Supreme Court struck down the Military Commissions Act and the administration has said that they are going to abide by that. So at this point, basically the major implication of this is that the military and the Army had stood up to the administration's system for detaining these people at Guantanamo. They slow rolled that for a long time, for seven years throughout the war. The executive made its intention clear, Congress passed two acts—The Detainee Treatment Act and The Military Commissions Act. Military lawyers recruiting civilian lawyers finally got that to the Supreme Court like Royall in the Nazi saboteur case resisting what the President wanted and they were eventually able to break the spear on what the executive's policy was.

Audience Member

My question is for Lieutenant Colonel Granfield. It is about organizational culture. It is very interesting that when you mentioned that USAID could not really get past the mission statement, and it is as if as an agency their cultural predisposition is more in tune with that of maybe the Peace Corps when you compare it to ours which would be more to see ourselves as an extension of policy by other means. So I was just wondering if you could dime out some of the key agencies that we could work with in the future and describe for us some of their cultural predispositions across the agency that make them easy or difficult to cooperate with.

LTC Granfield

Well, I can only speak from my experiences. In this most recent experience it is the first time really that I have worked within USG (United States Government). I can talk about the UN and some of the NGO (Non-Governmental Organizations) agencies that I have worked with over the last fifteen years, but in the case of Afghanistan, I think we have seen a significant improvement in USAID's thought process in that they have created this new Office of Military Affairs. Henrietta Fore has recently issued the civilian-military cooperation policy, not only an external audience piece but also internally. How? That means that the USAID personnel will interface with the OD at that tactical-, operational-, and strategic-level. So I think that is a huge paradigm shift from the mentality of probably 2004 where the environment in Afghanistan was, most places there was obviously kinetic stuff going on, but in many cases it is semi-permissive and there was freedom of movement so they thought they could just run off and go look at projects and have their implementing partners work to where the security situation got a little bit more dicey, especially in areas such as Kandahar and Kunar where they did not have that freedom of movement. And so I think that is a credit to them. I gave up one of my NCOs out of my Civil Affairs Team to work with my USAID rep so that she could map out where we had villages that were sitting on the fence so that we could then use her QUIP (USAID Quick Impact Project) money at the PRT-level instead of her spending it because she was spending 17.5 million for counternarcotics and she really did not have the time to manage three million in QUIP money. To look at information and where we saw targeting of the Marines, the maneuver force, where we saw targeting of the Government of Afghanistan entities, whether it was satellite dishes or their government facilities. And then go out and interview those villages, sit down with the Shura and develop a strategy of projects that may help them come into the fold of the government. And so for her to say, "Hey, I want to know what the root causes of this insurgency are and how do we as the PRT come together and develop a strategy?" Again, maybe I was fortunate because I had a forward thinking USAID rep, but you saw that in other PRTs too and so, I think, you are seeing some of the old gray hairs, no offense to the gray hairs in the audience, leave the agency and retire and you have all these young guys that are coming in that do not necessarily buy into that whole mindset that we are just here to do development. Likewise, I see it in the State Department with a lot of the young Foreign Service Officers coming in, volunteering to go to Iraq and Afghanistan where their counterparts may not necessarily want to. And in the case of USDA (United States Department of Agriculture), we got some great people, but the problem is they were volunteers and their agency had to foot the bill for them being gone and hire somebody temporarily in their place. So I was fortunate in that I had a cooperative extension specialist that came to my PRT which was a perfect fit. But in the case of Panjsher, they had a veterinarian and they did not really need a vet there. So those are just some of those institutional . . . where there is not a mechanism that the USG, those normally non-deploying kinds of people, really want to participate in where we are going with this counterinsurgency framework.

Audience Member

I have a question for Lieutenant Colonel Granfield. Recognizing the differences or the lack of passing the experience on from the military and the interagency perspective, what were you better able to do when you were leaving? So we do not repeat learning and relearning the lessons every year, what were you able to better do for the interagencies so that their replacements were coming with a better experience base than when you came there with your interagency folks?

LTC Granfield

That is a good question. Actually, because I am at Department of State, my boss has given me flexibility to go out and talk to different organizations. So I have been asked to come out to FSI (Foreign Service Institute) when they train up the State Department and workers that are going into Afghanistan. About the PRT experience as well, I have worked with USAID. Additionally, I teach at the Joint Special Operations University, courses on interagency operations, and so typically there are a lot of the State and USAID reps that are going out. I also left behind about . . . well, we had a three week transition which was really good. NESA (Near East South Asia) Center at the NDU, hosted a weeklong course or two week course for all of the PRT commanders, the USAID reps and the DOS reps that were going into theater, and so I also spoke in that. That program has been done away with and now the Air Force and the Navy have a program down at Fort Bragg and they bring in former PRT commanders with more recent experience to talk to not only their folks but also the USAID and the DOS reps that participate, not in the three month program, but they participate in the last week of a combination exercise. So there is some low hanging fruit; the Office of Military

Affairs and S/CRS and JFCOM (United States Joint Forces Command) has tried to capture those folks coming out to try to really get at best practices and continuing some of the great efforts that worked well when we were there.

Audience Member

I have a question for Colonel Bowman. Sir, can you explain to me the logic of how a Canadian or American soldier captures say a Saudi or Pakistani in Afghanistan, explain to me the logic and the point where they are granted the rights of a US citizen? For example, habeas corpus, as opposed to not being held to the standard of what we call the Law of Land Warfare which are the Geneva-Hague agreements.

COL Bowman

That question assumes . . . there are some mistakes in the question, which I really do not have time to correct all those, but I can address it to some extent. American citizens get different rights than non-American citizens, generally. But in one of the early cases which gets to the question before about the Padilla case. The Supreme Court held that American citizens can be unlawful combatants, in fact one of the Nazi saboteurs claimed to be an American citizen. American citizens do get more rights than just regular detainees. The fact that they get some rights does not mean that they get the same rights as American citizens. You would have to compare what American citizens get versus what these foreign unlawful combatants, the array of rights they get and we do not have good comparisons of those. But the point at which habeas corpus attaches though, there is a process that is called the Combatant Status Review Tribunal. When a detainee is detained to the point that they get a review by this Combatant Status Review Tribunal and that tribunal determines that the person is an unlawful enemy combatant and is going to be detained, then at that point there is something to challenge. If you just catch a detainee in Afghanistan and the guy just goes to a detention facility in Afghanistan or even if he is moved around in theater, that person does not have the same array of rights to challenge detention or habeas corpus as when the person is determined to be an unlawful enemy combatant by a tribunal. At that point they have already gotten some process. Now the real question is whether they are entitled to even that because there are lawyers in the administration that say that the President, when he has the power to wage war, can do whatever he wants to the enemy, within some standards. And the question then becomes what those standards are and some lawyers within the administration, which the most powerful lawyers within the administration arguably, would say that there are not any standards. It goes back to what Nixon said in the Nixon-Frost interview. "If the president does it, it isn't illegal." Well, we are at that point where the administration at the highest levels, its lawyers believe that the President has that inherent power. The irony is, that the military, which you would think would sign onto that, the lawyers in the military do not believe that and so one of the unusual aspects of this that is hard for people to understand is that the roles are kind of reversed here. The military stands up for greater rights for these people than the civilian administration does and that is why it is like Kenneth Royall's situation here with the Nazi saboteurs. Even though he was an Army colonel, he was standing up for the rights of these Nazi saboteurs, challenging it in the Supreme Court when the weight of the administration was pushing for the quick execution of these people.

Audience Member

My question is for Lieutenant Colonel Granfield. Do you consider that due to the fact that the security situation in Afghanistan seems to be deteriorating. Do you see a possible evolution in the bureaucratic concept? This was considered six years ago in order to address this new security environment.

LTC Granfield

That is a great point. I think that the PRT should still be the center of gravity because the population is the center of gravity. General Eikenberry will argue that, "No, it is the government." But I truly believe that it is the population. I think that the PRTs should be more integrated and we were fairly well integrated with the maneuver battalions. But now there is talk of sending over more maneuver brigades into Afghanistan, and I do not know if kinetics is necessarily the right answer. And I will give you an example. When we were there the Marines were getting significantly IEDed (Improvised Explosive Device) in Khogyani district, however, the PRT could travel freely through Khogyani and not get IEDed. And the reason for that being that the population saw the PRT as doing work for the provincial and district leadership and benefiting the people. It was actually the first time that it was done in Afghanistan, but if you refer back to the slide, you have Pachir Wa Agam, Khogyani, and Sherzad which are all Khogyani tribes. The Marines are getting IEDed down there, so we actually, myself and the maneuver battalion, went down to meet with the elders of Pachir Wa Agam and did a Nana Vati ceremony. Nana Vati means sanctuary of last resort. And basically we brought enough lamb and rice and stuff to feed the village down there. And we said, "Okay look, we know that things have gone wrong and the Marines are here. We are guests. We are sorry we have done things wrong. We would really hope that you would accept our request for Nana Vati and bless the lambs and sheep and say, "Okay, we are going to try to start this from scratch and not blame each side anymore because there is still blood chits for any member that is killed in the Pashtun areas." We—to include the governor and the deputy governor—went over to Sherzad because our problem area was Khogyani so we were trying to put pressure, and I actually had members from the Wazir tribe which is the tribe from the wild, wild west there, and they are Khogyani but they do their own thing outside the Khogyani. They were in both of those Nana Vati ceremonies. And so both elders in Sherzad and Pachir Wa Agam came up with the document with all the signatures that said, "We are going to divide up our province into safe zones of responsible elders that are going to take control of those elders and we are going to guarantee you safe passage." The intent was to go to Khogyani but unfortunately I ended up rotating out before we could get to Khogyani. But the story with that was we left that system in place for the next guys coming in. We also saw significant reductions in IEDs against the Marines moving in those areas because word got out that, "Hey, the PRT and the Marines actually understand Pashtunwali." I know you still have to kill bad guys. We all understand that and there are bad guys to kill, but if the Afghan government and the police and the ANA (Afghan National Army) are not the lead in going after those guys, and the US is seen as going after that, those are the tactics that have changed in 2007 and 2008, then maybe that is not necessarily the best way. I used to challenge the Special Forces guys because they would go after low value targets and kick in doors. "Hey, guys just tell me who you want and if I cannot get them here through the government system then go kick in the door." And so I challenged them and they eventually gave in and they did and guess what? We got the guy that they were wanting and they did not kick in any doors. There are techniques to all of this and there are strategies. I am not advocating that we not do kinetics but we have to balance the kinetics with the non-kinetics and I think that the PRTs should expand. If I would have been able to have three or four more security platoons so that I could move in the districts and a helicopter to get to places where I needed to get to quickly, it would have made my life much easier. Maybe the brigades become a super PRT in the provinces and that is where you put the emphasis because you have to have balance.

Day 3—Featured Speaker

Interagency at the Washington, DC Level
(Transcript of Presentation)

Mr. Mark T. Kimmitt
Assistant Secretary of State for Political-Military Affairs
United States State Department

Well, what comes around goes around and it is good to see my return to Fort Leavenworth is for the coveted after lunch spot. That could only be attributed to the fact that my classmate and good friend, Bill Caldwell, is the Commander here—a friend of mine for almost 36 years now, somebody who I have had a tremendous amount of respect for over the years and I can just tell you that you are very lucky to be at Fort Leavenworth at a time when Bill Caldwell is out here because I think you are going to find extreme innovation tempered with combat experience so it is not a theoretical exercise. Everything that is being done out here is being done for the right reason and has direct application. I am sorry to have missed the earlier discussions but as I read through the symposium notes and the agenda, it looked like a significant amount of the discussion was about interdepartmental interagency operations at the operational and tactical levels. Most of the discussions of the symposium seemed to be focused on agencies on the ground that have been developed over the last few years—Provincial Reconstruction Teams (PRT) inside both Iraq and Afghanistan, Joint Interagency Task Forces at combatant commands in the field and at their headquarters, and joint agency and other country teams with the United States Embassies. But that is simply not enough. These are important to get the interagency piece right down at that level, but what I would like to talk today is about what happens in Washington, DC at the interagency-level and talk about what is going right and what is going wrong. I would say that if these operations, and the guidance that is getting down to our troops in these interagency constructs on the ground, are not guided as an outcome of DC-based interagency decisions, then no matter how many times you put the teams together on the ground, no matter how much work you do trying to integrate, no matter how much enthusiasm is shown by those teams, if at the end of that meeting, the members of that meeting go back to their offices, call their organizations back in Washington, DC and get separate and sometimes contradicting guidance that reflects more agency equities than it does synchronization of military operations, interagency operations, then it is just not going to work. If we do not get it right at the top, it will never work itself right at the bottom. Now this is a military audience and quite frankly, you should be familiar with this challenge. It is a challenge for any commander that operates in a coalition environment and as Clausewitz said, "Always attack the coalition when given two options." Coalitions have inherent strengths but they also have inherent weaknesses. We see that today in Iraq, and we see that today in Afghanistan and the Balkans. No matter how many orders you give to other national forces in the coalition, one can expect that they will call back to their capitals for permission or refusal. These national red cards diminish the ability for forces to synchronize on a coalition battlefield and quite frankly, those national red cards that are pulled out are very similar to what I would consider the departmental red cards that are raffled off and pulled out by our fellow agencies on the ground. So with that in mind, I would like to spend a few minutes on the impact on operations of the processes and personalities inside of DC reflecting on the role of the departments, the National Security Council staff and the translation of interagency decisions made at the highest level of our command authority into actions on the ground, because as I said, if we cannot get it right in DC, it does not get better on the ground. Now before that, among this group talking about interagency, I would like to offer a brief public service announcement. I hope that as you have discussed interagency discussions, you have not only talked about the horizontal integration of government agencies into operations, but that you have also explored the notion of vertical integration as well. Frankly, our mission sets are becoming vertically integrated as well and they are expanding. Those missions that the military either participates in directly or indirectly are getting bigger and more numerous. When one brings in State, USAID, Justice, Homeland Security, Commerce, and Treasury, that implies that the commanders on the ground are directly or indirectly taking on missions done by these agencies. When doing so it is probably as a result of the demand by a commander on the ground who understands the importance of what these teams can bring to the battlefield. Nonetheless, we might as well call it what it is. We are back in the business of building nations, and nation building will probably remain a core competency for some time to come. Now while this is not the time or place to normatively debate that issue, it does emphasize that as we are doing nation building, we need to get these decisions right in DC. But those decisions also must simply not be idiosyncratic but they ought to be based on sound principles and policies. If we do not get the policy right as I said earlier, no amount of ad hocracy on the ground by an enthusiasm by our officers and our troops will get it right. And this is the impetus behind a policy paper that we have been developing inside of DC between State, USAID, and DOD. This policy paper reads:

> Provide State, DOD, USAID, and other policy makers and practitioners with guidelines for planning and implementing security sector reforms with partner nations, reform efforts directed at the framework, institutions, and forces that provide security and promote the rule of law.

> This expanded view of security, one that not only seeks to achieve a tolerable level of violence inside of society, but actively promotes and facilitates activities to establish and respect the rule of law, supporting the establishment of relevant legal frameworks, planning, oversight, civilian management, and budgeting capacities will be an instrumental part of this policy development and how our troops operate in the field.

That is a pretty ambitious set of objectives but I think if we take a look at what our troops are doing on the ground right now in Iraq and Afghanistan, that is exactly what they are doing and in many ways it is a disappointment that the troops are ahead of the policy. It is our hope that in this process we can get some policy out finally, between

State and DOD, USAID and others to provide that backbone for the decision making process. We came across a quote that was very, to our mind, instrumental to this. This quote says, "It reflects the imperatives for our soldiers to know something about strategy and tactics and logistics, but also economics and politics and diplomacy and history. Our soldiers must know everything they can know about military power and must also understand the limits of military power. They must understand that few of the important problems of our time have in the final analysis been finally solved by military power alone." Anybody know where that quote came from? That quote was delivered by John F. Kennedy to the graduating class of 1961 at the US Naval Academy. That is something we seem to have known for about forty-seven years, at least at the presidential level. And our troops certainly understand it at the ground level. And our view is it ought to be time that we get everything in between right for them.

So with that in mind, I would like to move on to a couple of points regarding the interagency process as it currently exists and has existed in the recent past. There is a cottage industry out there for people that are promoting interagency reform and I am not here to join in that industry but simply to offer an insider's perspective. Over the last two years, I have either participated in or prepared my principle, either the Secretary of Defense or the Secretary of State, for over two hundred PC's (Principles Committee), DC's (Deputies Committee), or PCC's (Policy Coordinating Committee). And as a result, I can tell you that these observations are not drawn from any particular expertise or native intelligence but simply, on-the-ground experience. Of these three observations to this industry of interagency reform, I think two of them are a little bit controversial but I hope I can convince this audience of the thought process behind them.

The first point that I want to make is that the NSC (National Security Council) staff are advisors to the President and should not have a fixed role. There is a suggestion out there that we ought to have a roles and missions definition for the NSC staff, that they ought to have exactly the same roles and missions from administration to administration. People talk about the halcyon days, the NSC staff before the first Gulf War, the experiences of the Berger NSC, and what was going on under Rice, and now Hadley. Each of these are different and equally valid models for the NSC because each of them reflect the President that they serve and attempting to formalize or cookie cutter the staff would not be helpful to the process nor would it serve the President. The second point that I would like to make concerns the notion of the War Czar. The selection of a War Czar to run the operations in both Iraq and Afghanistan was one of the most controversial decisions taken by the President with regards to the NSC during his eight years. While it has been controversial, I think history will bear it out to be probably one of the best decisions that was made and I will explain. And third, my view is that Goldwater-Nichols legislation to impose better behavior and outcomes on the interagency is not the answer. The executive branch did not want an external actor to come in and clean up the interagency and Congress will not allow for a more efficient process.

Let me first start talking about the issue of the NSC staff and the roles of personalities. On this role of the staff and the departments, I suspect you have already talked about the 1947 National Security Act and the statutory responsibilities of the National Security Council. But it is important to recognize that the staff is first and foremost part of the President's personal staff. It is an advisory body to the chief executive. The cabinet members who make up the NSC by contrast are nominated by the President, they are confirmed by the senate, and they are subject to request for testimony in front of the United States Congress. They serve the President, but they also answer to the will of the Congress, particularly when it comes to resources. By contrast, the NSC staff is not confirmed by the senate, cannot be subpoenaed by the committees, and does not testify in front of Congress. As a result, there is a dual nature between those two elements that make up the National Security Council and the supporting staff for it. And then of course, there is the third element, and a very important actor in this entire process, the Vice President who not only has a role inside the executive branch and also has a role inside the United States Congress. Because of the latest controversy about what his authorities and responsibilities are, I am not going to get into that subject. All I would say is that this current office of the Vice President has significant influence and has a significant voice at the table for all of the National Security Council debates and decisions. So what I would tell you is that who a President picks as his Vice President, who he picks as his National Security Advisor and his cabinet Secretaries will, in a large sense, dictate how the National Security Council staff interprets and implements its role. You take a look at the role that was held by the National Security Advisor and the cabinet Secretaries for the first Bush, there you had James Baker as Secretary of State, you had Dick Cheney as Secretary of Defense, strong personalities, and all held the confidence of the President. As a result, you saw Brent Scowcroft as the National Security Advisor, someone who saw his role more as a facilitator, and you saw the National Security staff somewhat like a weak battalion staff working for a strong battalion commander and strong company commanders, in somewhat of a backward role. But just because it was a backward role or a back foot role, it did not see itself as a decision making body, it did not see itself as an authoritative body, but it is often recognized as one of the most effective National Security Councils and National Security Council staffs in interagencies that we have seen since 1947. If anybody would like to read more of that, and I recommend this highly, it is the *Living History Project*, the National Security Council project that was run by Ivo Daalder up at Brookings where he brought a significant number of people in to talk about those days and what went right and what went wrong. And what was remarkable, particularly since I do not think a lot of people recognized this or realized it, nor did I for that matter, is inside this seventy-eight hundred and fifty-page document, there is a little subtext to all of this and it was the role of Brent Scowcroft's Deputy and his role running the DCs, was able to elevate the DCs to Under Secretary-level, and was a tremendous catalyst in getting these meetings done right and getting these meetings done effectively, efficiently, and one that to this day, (this study was done about fifteen years after that), where people continue to point out this individual as a catalyst for making this all happen. Anybody have an idea who Brent Scowcroft's Deputy was? The current Secretary of Defense,

Bob Gates, is roundly seen as the hero of this entire story and the hero of the National Security Council in the interagency process during the first Bush administration. Well worth reading. Clinton was a little bit different. The Clinton era seemed to put more emphasis on the role of the National Security Advisor and when you take a look at the cabinet Secretaries, that might explain why. The cabinet Secretaries at that time did not necessarily have as much . . . you had Les Aspin, Madeleine Albright . . . these were not really dominant figures at the NSC table. As a result, when you needed things to get done, you went to Sandy Berger. Sandy Berger kind of cracked the whip and kept things going and as a result, the National Security staff reflected that as well. I remember as a young Army lieutenant colonel sitting in a PCC being run by Susan Rice and this was about the time of the Burundi situation and I can tell you, you went away from that meeting knowing what your homework was, you knew what you had to turn in, and if you did not get it right you were going to hear from the NSC staff. They were not there to facilitate. They were there to orchestrate, and they were there to make decisions that would then go to the President to underwrite. There was a little bit of resentment in the first Bush administration. In fact, perhaps not resentment, but it was probably more of a difference in tact. I do not know if most people recognize that there was an article about seven years ago where Condi Rice and Steve Hadley laid out their view of how the National Security staff should operate and used the word facilitators. There have been a lot of people that have criticized the current National Security staff as one that just has a lot of meetings and doesn't really demand action and implementation the way the previous NSCs did and the interagency meetings did. I think if one takes a look at the National Security team that the President chose, that is understandable, a very strong Secretary of State in the form of Colin Powell, a very strong and verbal Secretary of Defense, and Condoleezza Rice, while an academic, really was not one that was going to be forcing her opinion at that table. So it really was the staff supporting the subordinate commanders, so to speak, in the forms of the Secretaries and the President and the staff facilitating those decisions, moving them up to the President. I bring that out simply because I think as there are some out there who would be prescriptive about the way the National Security staff should run and the way the interagency should run, it is important to recognize that first and foremost, it remains an advisory staff whose responsibility is to the President and prescribing their behavior or describing their behavior, I think, does not do good service to either the National Security staff nor, more importantly, the President nor the cabinet Secretaries. So my first observation is, I would treat with great suspicion any project that came forward and said, "We are going to improve the interagency by dictating and demanding that the National Security staff operates in this particular way."

The second point is the issue of the War Czar. It is very controversial. There were a number of people that were asked to be the War Czar and chose not to be the War Czar, and Doug Lute finally stepped up to the plate. The role that the War Czar has played has not been necessarily in taking over the war responsibilities from the departments, but following through on the decisions made by the National Security Council. For anybody that has ever worked in the interagency, it is great at coming up with decisions and absolutely lacking when it comes into implementation. It is an organization, there are probably, five to ten DCs and PCs every week and at each one of those decisions are made. The decisions are made at the principles level, deputies agree, the PCCs analyze, but at the end of the day you would think that by the end of the week, you would have a nice set of decisions that would percolate through the chain of command because that is the way it works in the military. It absolutely does not work that way inside the interagency process. Great decisions are made, but in the past very few of those decisions have been implemented or followed through. For anybody that knows Doug Lute you know that first and foremost this is a former J3 at numerous levels and he is a guy that tracks decisions. You go to the meetings with Doug Lute and he is going to ask where your homework is. If you say, "My dog ate it," there is going to be some naming and shaming and blaming at the table. He is a guy that understands that it is not simply enough at the highest level of the government to make decisions, but those decisions not only have to be executed and implemented, but they have to go down the chain of command. Or you end up with the paradigm we talked about earlier, the simple issue is, if you go into the interagency decisions, make those decisions and if the principles walk away back to their departments and stay focused upon their own department equities and hope that decision never comes back to haunt them, you get away with a lot. Doug Lute does not let things ride and when decisions are made by the principles, the decisions are made and you move out smartly. There are many in this community that would suggest that is not the way to go, that is an unbridled granting of power to a uniformed officer, and it does not necessarily have to be a uniformed officer. There are significant numbers of areas that you have to ask, "Who goes to sleep at night worrying about this particular issue? Who wakes up in the morning and says, is this going to get done?" It seems normal to all of us in the room that that would happen. It is very abnormal inside Washington, DC, but I think quite frankly, what you want to do is you want to have more of these czars. They may not have that particular term, but there has to be somebody who holds responsibility for implementing and monitoring the decisions that are made by the National Security Council. It is not being done and it needs to be done. All I would say is for anybody that wants to be a czar, always remember, when it comes to czars, the nobles normally hate them and the peasants normally assassinate them.

The last point I would like to make before taking some questions is the notion of Goldwater-Nichols for the interagency. I would like to throw this thought out and just quite simply say, my personal view is it would not work. We do and we are making incremental improvements inside interdepartmental operations and interagency operations, but Goldwater-Nichols is not the ultimate solution. First of all, for those of us that lived through the Goldwater-Nichols era, in many ways, the results of Goldwater-Nichols are a success, but the very fact that the services had to have an external force, the United States Congress, legislate what needed to be done, you have to consider that to be a failure. The services knew what to do. We all knew what to do. We knew the importance of working in a joint environment, but the incentives systems and the rewards were not set up for either acting joint and certainly not for going joint. There were a couple of wise old heads inside the United States Congress that recognized that they could use the power of the purse to do this. They recognized the necessity to get the services to act joint and I think twenty years hence, we can all look at that as a success. But I would also tell you that trying to take that extension from the United States military into the interagency is bound for failure. The primary reason it would not work is number one, Congress will not allow it to happen because what Congress would be allowing to happen is granting more authority to the President of the United States. In fact, it is more of a comparison to the separation of powers clause in the Constitution and not Goldwater-Nichols. Congress intentionally limits the power of the President and limits the power of the department by separating the structures by which the different departments respond to Congress. That is the reason you have a 600-700 billion dollar defense budget while the State Department has about a 55-60 billion dollar budget. Different committees operate, different committees appropriate, different committees obligate money to those organizations and it does not seem, in my mind, rational that we would expect the United States Congress to give up that power to the President to be able to take money from the Department of Defense, give it to the Department of State or do anything that would make it more efficient. This separation of powers, these checks and balances currently held by the United States Congress which would be forfeited if we had a Goldwater-Nichols for the interagency, just does not seem to pass the common sense test. People say, "Well, what is wrong with ultimate synchronization? Why shouldn't that power be consolidated under the responsibility of the President?" Well, if you really want to do that, the ultimate form of synchronization, of course, is the dictatorship and I do not think America is going to be heading to a dictatorship anytime soon and I do not think that we are going to see a Goldwater-Nichols for the interagency anytime soon. It just does not seem to be the right way to do it and it just quite frankly, does not seem that it would be plausible in the first place to expect it to be achieved. But with that in mind, let me give some summary comments. We clearly have a lot of work to do in DC with regards to the interagency, and I think that there is a lot of internal work that needs to be done because until it is done, I cannot be confident that it is going to take care of the soldiers in the field and the interagency mechanisms that we are setting up on the ground. That is particularly acute given that there is a new administration that will be coming in town and it is going to have a completely new interagency National Security team. It will emerge, it will not look anything like the one we currently have, and I would expect that the National Security team under an Obama administration would look very different from one that would emerge under a McCain administration if he is elected. I think both of them could go from either the very strong managerial model that we saw under Sandy Berger. It could perhaps be the case that under a very strong McCain administration with very strong service Secretaries and department Secretaries, might seek somebody who is less assertive and less demanding upon the departments in the form of the National Security Advisor. We will certainly find that out in the next six months or so. But before I do finish up and turn it over for questions, I think this issue that you have taking on for the last couple of days is extremely important, and what I very much appreciate is sort of the top to bottom approach that you have taken. There is a tremendous amount of success going on at interagency, whether it is down at the PRT level, whether it is brigade-level headquarters inside of Iraq and Afghanistan, the JIATFs (Joint Interagency Task Force) that have been set up at CENTCOM (United States Central Command) and other places and frankly, the interagency which is a joint interagency task force inside of Washington, DC. What I suspect is you have taken a look at which way to go in the future on the ground. All of us recognize there is still a lot of work that needs to be done at the highest levels, and quite frankly, it is not entirely inaccurate to suggest that Washington, DC does not yet know that, does not yet understand it because, quite frankly, Washington, DC is not on a war footing. But I would tell you that there are still small pockets of individuals out there and groups that never fall asleep at night nor wake up in the morning without thinking about these issues because of the consequence it has for our soldiers in the field. And I just want to finish up by again remarking on the importance of our soldiers in the field. They are carrying the work of a nation in their rucksacks and we should be enormously proud of what they are doing. If this conference marginally improves, in any way, the interagency capability of this nation, insofar that it facilitates and supports what our troops on the ground are doing, then I would consider this conference a tremendous success. God bless those troops for what they do. They are the ones that ought to be the first people we think of as we try to improve interagency operations and we should just be enormously grateful for what they are doing every day. So thank you very much and rather than take the entire hour and a half to talk, I thought it might be helpful, either to let you guys get out early or take a number of questions in the interim, so thank you very much.

Day 3—Featured Speaker
Question and Answers
(Transcript of Presentation)

Mr. Mark T. Kimmitt
Assistant Secretary of State for Political-Military Affairs
United States State Department

Audience Member

If you could comment briefly and I know you have only been at State now since the beginning of August it looks like. What do you feel like as you move from Defense? Do you have a sense yet from State? Is there an institutional perspective that you as an individual feel that you are going to be pressured to do things with State that you would not have had to do with Defense or vice versa? How does interagency work for an individual who has moved back and forth between both departments?

A/S Kimmit

That is a good question. First of all, we have to be loyal to our institutions. We have to be loyal to our leaders. As I said at my swearing in, it is remarkable that when you take the oath of office as an Assistant Secretary of State, it is almost word for word the same oath of office you take when you are commissioned as a United States officer or promoted. You swear to support and defend the Constitution of the United States, not the Department of State, not the President of the United States, but the Constitution of the United States. The last time I checked, when I drove over the bridge and got to the other side, it was still the same Constitution that the Department of State and the Department of Defense adheres to. That was point one. Point two, it was remarkable and perhaps *[loss of audio]*. I have often had the Under Secretary from policy call me up and say, "Mark, I want to talk to you about the paper." I said, "What paper? I have not even seen it yet." But that is just the way it works and you have to learn to live with that. But fundamentally, I do not see . . . well, let me put it this way, there are old-time State Department types out there that do not believe in all this interagency stuff. There are old-time senior Army officers that do not believe in this interagency stuff. Well, what is remarkable is this young generation of Foreign Service Officers and Army captains and majors that have served together, worked together, lived together, ate together, and in many cases, been shot at together, that are growing up inside these organizations and it is very refreshing. In fact, you get a bunch of young Foreign Service Officers together with a bunch of military officers and soldiers at a social occasion where they have served together in combat and you would not be able to differentiate one from the other. So what I would say is, those sharp distinctions which used to exist between our departments, war creates strange bedfellows and in this case that is a good thing. So I would just tell you that I was refreshingly . . . well, when I got over to State, I was not brought into a secret cloak room, given the secret handshake and told to say, "Okay, everything you did over in DOD does not apply here." It is very similar. I think that we all work at the same objective, we all serve the same President, but most importantly, it is important to defend the same Constitution. So that was a very refreshing observation to learn that after I had been there about two weeks.

Audience Member (Blogosphere)

Sir, if you could comment about what your feelings are about the National Center for Strategic Communications that would be the premier education and training facility for both DOD public affairs and Department of State public diplomacy. There have been congressional movements to establish this center to unify the strategic message across the USG (United States Government).

A/S Kimmit

In terms of a process or a building, I am indifferent. But in terms of an outcome, I think all of us have recognized, and Bill Caldwell and I in particular who have spent our last combat tour at a podium, understand the importance of information. And it is becoming such an incredible component of what we do on the ground. General Caldwell and I were talking about it over lunch. The first people he introduced me to this afternoon were some of the people that had worked with him in Iraq, supporting him in his job. I see Matt *[inaudible]* who was my right arm while I was in Iraq, who taught me how important it was to get the information out, be first with the truth. There are many that still may think, "This 'five o'clock follies,' press conference stuff is not what real military men are all about, what real military women are about. They are wrong. It is an essential part of what we do when we are fighting somewhere around the world—tell our message, get our message out, and be first with the truth. If for no other reason than the fact that even if you do not believe it, your enemy does. Your enemy is probably reading that same blog right now and trying to figure out how he or she can use a blog to disseminate this information to unwilling and unfriendly neighbors. So if there is a movement afoot to try to formalize this in a building, in a process, in an organization, I think that is great as long as it achieves the outcome which we all seek which is a core of foreign service, military, and civil service professionals that are able to get up in front of audiences, whether they are newspapers, whether they are our own citizens, whether they are our own press, or whether they are foreign nationals and explain what we are doing, why we are doing it, and why our soldiers are putting their lives on the line to do it. That is, in my mind, an absolute essential aspect of leadership for the 21st century. And if people do not get it, they do not get it. But your enemy does, and if for no other reason, you ought to understand that.

Audience Member

You mentioned that Washington is not on a war footing. How would you compare and contrast the level of war footing between Defense and the State Department?

A/S Kimmit

Listen, do not go into the Pentagon hallways at 4:30 on a normal afternoon. You will get run over. I would not necessarily put the cloak of purity on the Department of Defense. If you are working in the policy shops for East Asia, life is not bad. If you are working in the policy shops for the Middle East, you work a little bit harder. I think everybody recognizes that we are in a war. I do not think people recognize that we are at war. I would say that comment applies in every bureau and remarkably, there are some departments that you would never expect to be thinking that they are at war that are so incredibly important these days. If you want to find some of the toughest, meanest people in Washington, DC fighting this war on terrorism, go to the Treasury Department. You run into guys like Stu Levy who spends every day of his life trying to figure out where the money flows are coming from in Iran and where they are going to. These guys are absolutely vicious. So I think as I said in my final comments, there are small pockets out there that really understand, not only that we are in a war, but we are at war. Then there are larger pockets that understand we are in a war, and then there are some other people that just acknowledge, "Hey, that is somebody else's business and I am not worried about it." So short answer is, not enough people understand we are in a war, not enough people understand that we are at war. But I would not make that department specific.

Audience Member

Earlier this academic year, we received a brief from Dr. Thomas Barnett and he spoke to us about his idea of creating a SysAdmin force which I understand to be an extreme form of interagency cooperation, a force structure that is able to deploy anywhere around the world with military, DOD, State Department, and other agencies. How do you feel about that idea? Is that something that is even possible with the current structure?

A/S Kimmit

Maybe I got confused on this notion of a SysAdmin force. I thought you were going to take me down into the cyber space fighting environment, but he is suggesting that we ought to have a stand-by capability to deploy anywhere in the world?

Audience Member

Yes Sir, in a joint interagency force that has . . .

A/S Kimmit

I absolutely agree with you. In fact, that is one of the initiatives that is being pushed right now to the Civilian Reserve Corps at the Department of State. Oddly enough, the European Union figured this out ten years ago. They have a quote, "Rapid Reaction Force" of not only soldiers but lawyers, economists, so on and so forth. That is the effort that is being pushed by Ambassador John Herbst right now in S/CRS (Office of the Coordinator for Reconstruction and Stabilization) inside the State Department. It is probably not getting as much traction as it needs, but it is certainly the case that, depending on a civilian RFF (Request for Forces) process to try to find yourself an economist, a forensics expert, and a foreign policy advisor once the war starts is not the right way to go. I think there is debate in Washington, DC, there is actual money that has been put against that project, but it does not seem to have the traction that I think both of us would recognize that it needs over time.

Audience Member

I want to ask you about dynamics within the State Department. Robert McNamara made the argument once that in the 1950s we gutted the Asian specialists. As a consequence of that when we went to Vietnam, we were not as knowledgeable as we should have been. Another argument has been made as we went into Iraq that we did not have sufficient number of Arabists, people that had a view from the position of Arabs in the Middle East. As a consequence of that, we did not have all the information we needed and all the understanding that we needed when we went into the Middle East. And the argument has been that within the State Department there is a dynamic that these various areas of the world have friction between them and one becomes dominant and influences policy more so than others. Have any comment on that?

A/S Kimmit

You brought up a couple of points there and different points. It is true that inside the State Department and the regional bureaus, there is somewhat of a hierarchy within the regional bureaus and much of that hierarchy is personality dependent. Obviously, right now the Near East Bureau, under David Welch, gets a lot of attention and gets a lot of resources because of what is happening inside the Middle East. By contrast, Secretary Jendayi Frazer down in Africa is always having to make the argument that, "Look, this is all about phase zero operations. Let's prevent the next war rather than spend all our efforts fighting the current war." But I think you are going to find that in all organizations, it was certainly the case in OSD policy, that as the DAS (Deputy Assistant Secretary) for the Middle East, I probably was able to secure more of what I needed whether it was resources, focus, attention, than say, the DAS for Western Hemisphere. So, true. The other point you made, I think, is even more important, which is developing that cultural base of expertise early on because cultural and language skills, to use the Japanese term, are not a "just in time" inventory. You cannot just call up and all of a sudden get a hundred Asian specialists, a hundred Arabic speakers, so on and so forth. And even if you can quickly get the language speakers, the time it takes to develop the cultural expertise well surpasses the amount of time you are going to have in advance warning. This is something that General Caldwell and I were talking about at lunch and I was impressed to hear the initiatives that the Department of the Army in general, but Leavenworth in particular, are pushing in that regard. So I wish I could disagree with you, but I think in both cases, in terms of the cultural requirements, that we are going to need in the future and in terms of the bureaucratic behaviors inside the State Department and other departments, I think you are spot on.

Audience Member

Sir, could you talk a little bit more about the cultural difference between your thirty plus years in DOD and what you are now seeing as far as the culture within the Department of State and specifically about the career incentive programs within the Department of State. Are you rewarded for these types of tours, career progression patterns, etcetera, and is there some cultural resistance within Department of State to not only interagency but just deployments in general?

A/S Kimmit

Interesting question and first of all, it was more of a cultural leap for me to take off my uniform and work as a Deputy Assistant Secretary of Defense than it was to go from wearing a tie in Department of Defense to working in the State Department. It was a tremendous leap between the two and some of the mistakes I made were what caused me to be delayed a little bit in my confirmation process, well deserved. That said, I think you would be remarkably impressed by the similarities in the cultures between the two places. Although the rules are different, the norms are very similar in terms of organizational loyalty, how you can progress, what is the path for success in the military versus the path for success in the State Department. And that is something that the State Department is examining right now. The State Department, in a lot of ways internal to itself, is going through a Goldwater-Nichols examination of itself. What do I mean by that? I also run the POLAD Program (Political Advisor Program). We just got finished selecting a new group of political advisors. As I was talking to a couple of them the other day, it sounded like twenty years ago, before Goldwater-Nichols, "Oh gosh, getting off the regional bureau track is going to hurt me." Or the ones that are there, "This is the best job I have ever had, but it is not going to get me anywhere in the State Department." And you take away the term "POLAD" and you put in the word "joint" and it is very similar to us. What I have been trying to amplify because I certainly did not invent this, but this notion that within the State Department that if they truly want to acknowledge how important interagency tours are, they have to be rewarded. People by and large are going to go where the motivations and incentives lie. Twenty years ago, we would not have gone to a joint assignment; it would be the kiss of death. Nobody gets promoted, nobody gets selected for command. And in many ways in the foreign service, the answer is to be in a bureau, start off as a desk officer, go to an Embassy, work as a political officer, come back to the bureau, become a director, go back to an Embassy, be a DCM (Deputy Chief of Mission) or to a different one, and go back and forth. While that sounds nutty, I bet if you talked to most senior officers in the United States Army, we grew up in an environment where you go to a battalion, you go to a brigade, you go to successive commands. You do something crazy like working joint or take a White House fellowship and you are putting yourself at risk. We got over it inside the military and I think the State Department is trying to get through it now. I would tell you though, that they have not yet taken the final step which is to establish incentives, motivations, and penalties for those that do not, and I would suggest that human nature will say, "Until you actually penalize those who do not and reward those that do, you are not going to get anywhere." They made a couple of steps two years ago when we were coming out of Iraq. Almost every award given that year in State Department was given to somebody who had served in Iraq, so they are getting rewarded. Just as we went through this in the Army, as we transitioned over to joint, they are going through that now. Does that help? I am absolutely impressed with this crowd for one that started at one o'clock in the afternoon right out of lunch. I was very carefully watching to see if anybody nodded off. Anybody in the class of 1976 that ever had a class with me knows that between the hours of one o'clock and three o'clock, I could be found at my desk with my head on the desk, drooling from the side of my cheeks as I was falling asleep, so if for no other reason, thank you very much for the opportunity to speak today, thanks for what you are doing out here, and congratulations on a great symposium.

Day 3—Closing Remarks

Conclusion
(Transcript of Presentation)

Dr. William Glenn Robertson
Director, Combat Studies Institute

In the Army pedagogical technique known as the Staff Ride, there is a brief but critical phase known as the integration phase. It begins the process of integrating all that we have learned through disparate means into a coherent framework. Let me very briefly begin that process of integration here at the close of our symposium, knowing that the integration process will go on individually within your minds as you leave this symposium, and further as you interact with others on this very important topic. What have we heard here this week? The interagency process is, first of all, not a new phenomenon, stretching back centuries. We've also learned that it is difficult to accomplish successfully. We've learned that personality difficulties and differences complicate the interagency process. Time and again, speakers have commented on the role of personality. We've learned that structural differences between and among agencies complicate the process. We've learned that organizational cultural differences complicate the process. But the good news is that we've also learned that people of good will and great skill can overcome all these difficulties; and when those difficulties are overcome, the synergies available in service to our nation are immense. The historical record is clear. Indeed, it's a commonplace. So, why don't we learn? Or, if we do learn why do we forget? How can we institutionalize what works and avoid or discard what doesn't? These are hard but critical questions, and this symposium has made a contribution toward framing an answer. The papers and discussions that have taken place over the last three days have told us what happened in the past. You have heard examples transcending single centuries and single places. You have heard those examples examined for significant insights and the details that each of them brings to the table, but the question of institutionalization remains. How does historical insight translate into policy? This symposium has turned up the rheostat on an already bright light. We may have to leave it to others to direct that light onto the path that is the future. Still, we've made a beginning, and I want to thank all of you—participants and attendees alike—for a most stimulating symposium. Next year, we have tentatively selected for our topic, "The Military and the Media in Wartime: Historical Perspectives." We hope you will find the time to attend. Our proceedings of this symposium we hope to have published by the end of December. I want to add my thanks to those people that Ken Gott thanked, and I want to extend my thanks to Ken as well for managing this symposium so well.

Appendix A
Conference Program

Day 1
Tuesday, 16 September 2008

0800 – 0825 **Registration**

0825 – 0835 **Welcome/Administrative Announcements**

0835 – 0845 **Opening Remarks**

Lieutenant General William B. Caldwell IV
Commanding General,
US Army Combined Arms Center

Keynote Address

0845 – 1000 **Keynote Presentation**

Interagency Coordination
Brigadier General Robert J. Felderman
Deputy Director J5, Plans, Policy & Strategy,
USNORTHCOM

1000 – 1015 **Break**

Panel 1

1015 – 1145 **The Difficulties in Interagency Operations**

The Interagency Process and the Decision To Intervene in Grenada
Edgar Raines, Ph.D.
US Army Center of Military History

The Interagency in Panama 1986-1990
Dr. John T. Fishel
National Defense University

1015 – 1145   Panel 1 (cont'd)

Lebanon and the Interagency
Lawrence Yates, Ph.D.
US Army Center of Military History

Moderator
Richard Stewart, Ph.D.
US Army Center of Military History

1145 – 1300   Lunch

**1200 – 1230   *Optional Virtual Staff Ride Demonstration –
Operation ANACONDA* (Room 2104)**
Mr. Chuck Collins
US Army Combat Studies Institute

Panel 2

1300 – 1430   The Interagency Process: Southeast Asia

*The Interagency Process and Malaya:
The British Experience*
Benjamin Grob-Fitzgibbon, Ph.D.
University of Arkansas

*The Interagency Process and Vietnam:
The American Experience*
Jeffrey Woods, Ph.D.
Arkansas Tech University

Moderator
Donald P. Wright, Ph.D.
US Army Combat Studies Institute

1430 – 1445   Break

Panel 3

1445 – 1615   Interagency Efforts at the National Level

*21st Century Security Challenges and the
Interagency Process: Historical Lessons about
Integrating Instruments of National Power*
Robert H. Dorff, Ph.D.
Strategic Studies Institute

1445 – 1615   Panel 3 (cont'd)

*The Independence of the International Red
Cross: The Value of Neutral and Impartial Action
Concurrent to the Interagency Process*
Mr. Geoffrey Loane
Head of Regional Delegation of the United
States and Canada for the International
Committee of the Red Cross (ICRC)

Moderator
Mr. Kelvin Crow
US Army Combat Studies Institute

1615 – 1630   Administrative Announcements

Coming in 2009

*A Different Kind of War: The United States Army Operation
Enduring Freedom (OEF) September 2001-September 2005*

By the CSI Contemporary Operations Study Team

CSI's study of Operation ENDURING FREEDOM (OEF) examines US Army operations in Afghanistan between September 2001 and September 2005. Focusing on the operational and tactical levels of war, this work follows the evolution of the Army mission in OEF as it moved from limited involvement in combat and humanitarian operations toward a more comprehensive full spectrum campaign directed at defeating insurgent forces and fostering stability in the new Afghanistan. Like *On Point II*, CSI's study of OEF is based on a large collection of primary sources including hundreds of oral interviews and unit documents.

0800 – 0830   **Registration**

0830 – 0845   **Administrative Announcements**

0845 – 1000   **Featured Speaker**

Richard Stewart, Ph.D.
Chief Historian
US Army Center of Military History

1000 – 1015   **Break**

Panel 4

1015 – 1145   **Interagency Case Studies**

*Approaching Iraq 2002 in the Light of Three
Previous Army Interagency Experiences:
Germany 1944-48, Japan 1944-48, and
Vietnam 1963-75*
Lieutenant General (Ret.) John H. Cushman

*A Mile Deep and an Inch Wide: Foreign Internal
Defense Campaigning in Dhofar, Oman, and
El Salvador*
Mr. Michael P. Noonan
Foreign Policy Research Institute

Moderator
Mr. John J. McGrath
US Army Combat Studies Institute

1145 – 1300   **Lunch**

Panel 5

1300 – 1430   **Post-War Germany**

*Role and Goal Alignment: The Military-NGO
Relationship in Post WWII Germany*
Major Tania M. Chacho, Ph.D.
United States Military Academy

416

1300 – 1430   **Panel 5 (cont'd)**

Between Catastrophe and Cooperation:
The US Army and the Refugee Crisis in
West Germany, 1945-50
Adam R. Seipp, Ph.D.
Texas A&M University

Moderator
Curtis King, Ph.D.
US Army Combat Studies Institute

1430 – 1445   **Break**

Panel 6

1445 – 1615   **The Interagency Process in Asia**

Post-Cold War Interagency Process in East Timor
Major Eric Nager
United States Army, Pacific

Joint Military-Civilian Civil Affairs
Operations in Vietnam
Nicholas J. Cull, Ph.D.
Director, Master of Public Diplomacy Program
University of Southern California

Moderator
Lieutenant Colonel Scott Farquhar
US Army Combat Studies Institute

1615 – 1630   Administrative Announcements

0800 – 0830   **Registration**

0830 – 0845   **Administrative Announcements**

Panel 7

0845 – 1015   **Interagency Process Within the United States**

Cold War Interagency Relations and the Struggle to Coordinate Psychological Strategy
Robert T. Davis II, Ph.D.
US Army Combat Studies Institute

"Our Most Reliable Friends": Army Officers and Tribal Leaders in Western Indian Territory, 1875-1889
Mr. William A. Dobak
US Army Center of Military History

Moderator
Ricardo A. Herrera, Ph.D.
US Army Combat Studies Institute

1015 – 1030   **Break**

Panel 8

1030 – 1200   **Military Governments and Courts**

Army Lawyers in the Interagency: Three Case Studies of Military Tribunals
Colonel Gary M. Bowman
US Army Center of Military History

Provincial Reconstruction Team (PRT): A Grass Roots Interagency Counter-Insurgency Methodology in Afghanistan
Lieutenant Colonel Lynda Granfield
Former PRT Commander for Jalalabad

Moderator
Mr. Robert Ramsey III
US Army Combat Studies Institute

1200 – 1300	**Lunch**

1300 – 1430   **Featured Speaker**

Mr. Mark T. Kimmitt
Assistant Secretary of State for
Political-Military Affairs
US Department of State

1430 – 1440   **Administrative Announcements**

1440 – 1500   **Concluding Remarks**

Dr. William G. Robertson
Director, US Army Combat Studies Institute

Appendix B
Conference Program

About the Presenters

Gary M. Bowman, Colonel, JA, US Army Reserve, is the Deputy Commander (IMA) at the US Army Center of Military History. He is a graduate of the Virginia Military Institute, and he earned his law degree and his Ph.D. in government at the University of Virginia. He is the author of *Highway Politics in Virginia* and numerous articles on legal, military, and political topics. He practices law in Roanoke, Virginia, is a member of the Consulting Faculty at the US Army Command & General Staff College, and has deployed four times during the Global War on Terror.

William B. Caldwell IV, Lieutenant General, US Army, serves as the Commander of the Combined Arms Center at Ft. Leavenworth, Kansas, the command that oversees the Command and General Staff College and 17 other schools, centers, and training programs located throughout the United States. The Combined Arms Center is also responsible for: development of the Army's doctrinal manuals, training of the Army's commissioned and noncommissioned officers, oversight of major collective training exercises, integration of battle command systems and concepts, and supervision of the Army's center for the collection and dissemination of lessons learned. His prior deployments and assignments include serving as Deputy Chief of Staff for Strategic Effects and spokesperson for the Multi National Force – Iraq; Commanding General of the 82d Airborne Division; Senior Military Assistant to the Deputy Secretary of Defense; Deputy Director for Operations for the United States Pacific Command; Assistant Division Commander, 25th Infantry Division; Executive Assistant to the Chairman of the Joint Chiefs of Staff; Commander, 1st Brigade, 10th Mountain Division; a White House Fellow, The White House; Politico-Military Officer in Haiti during Operation RESTORE/UPHOLD DEMOCRACY; Assistant Chief of Staff for Operations, 3d Brigade, 82d Airborne Division during Operations DESERT SHIELD and DESERT STORM; and Chief of Plans for the 82d Airborne Division during Operation JUST CAUSE in Panama. Lieutenant General Caldwell graduated from the United States Military Academy at West Point in 1976. He earned Master's Degrees from the United States Naval Postgraduate School and from the School for Advanced Military Studies at the United States Army Command and General Staff College. Lieutenant General Caldwell also attended the John F. Kennedy School of Government, Harvard University as a Senior Service College Fellow.

Tania M. Chacho, Ph.D., Major, US Army, is an Academy Professor and the Director of the Comparative Politics program in the Department of Social Sciences at the United States Military Academy, West Point. She serves in the US Army, with previous experience as a Military Intelligence Officer and then a Foreign Area Officer, specializing in Western Europe. Her last assignment was as a Special Advisor to the Supreme Allied Commander, Europe, working out of the Political Advisor's office in Mons, Belgium. Her past assignments include tours at Fort Bragg, Fort Huachuca, and Germany, and she has deployed to Bosnia and Kosovo. She has published articles on soldier motivation and a book chapter on European defense initiatives.

Nicholas J. Cull, Ph.D., is Professor of Public Diplomacy and Director of the Masters Program in Public Diplomacy at the University of Southern California. He took both his B.A. and Ph.D. at the University of Leeds. While a graduate student he studied at Princeton in the USA as a Harkness Fellow of the Commonwealth Fund of New York. His research and teaching interests are broad and inter-disciplinary, and focus on the the role of culture, information, news and propaganda in foreign policy. He is the author of *The Cold War and the United States Information Agency: American Propaganda and Public Diplomacy, 1945-1989* (Cambridge 2008). His first book, *Selling War* (Oxford University Press, New York, 1995), was a study of British information work in the United States before Pearl Harbor, and was named by Choice Magazine as one of the ten best academic books of that year. He is the co-editor (with David Culbert and David Welch) of *Propaganda and Mass Persuasion: A Historical Encyclopedia, 1500-Present* (2003) which was one of Booklist magazine's reference books of the year, and co-editor with David Carrasco of *Alambrista and the US-Mexico Border: Film, Music, and Stories of Undocumented Immigrants* (University of New Mexico Press, Albuquerque, 2004). He has published numerous articles on the theme of propaganda and media history. He is an active film historian who has been part of the movement to include film and other media within the mainstream of historical sources. He is President of the International Association for Media and History, a member of the Public Diplomacy Council and has worked closely with the British Council's Counterpoint Think Tank.

John H. Cushman, Lieutenant General (Ret.), US Army. General Cushman enlisted in the US Army in 1940 and in 1944 graduated from the United States Military Academy. He earned a Masters degree in Civil Engineering at the Massachusetts Institute of Technology, then attended the US Army Engineer School. From 1976 to 1978, General Cushman commanded I Corps (ROK/US) Group. He retired in 1978 and has since been a writer and consultant. He is the author of many books, papers, and articles on strategy, multiservice and multinational operations, warfare simulation, and military command and control most notably his 1993 pamphlet *Thoughts for Joint Commanders*. In 1994 he was named "Author of the Year" by the US Naval Institute *Proceedings*.

Robert T. Davis II, Ph.D., is a historian at the Combat Studies Institute. He received a B.A. in History from the University of Kansas in 1998, and an M.A. and Ph.D. from Ohio University in 2003 and 2008, respectively. He is the author of *The Challenge of Adaptation: The U.S. Army in the Aftermath of Conflict, 1953-2000* and has recently completed a study on military-media relations in the 20th century. He is currently revising his dissertation, "The Dilemma of NATO Strategy, 1949-1969," for publication.

William A. Dobak worked at the National Archives in Washington, DC, for nearly six years before joining the US Army Center of Military History in 2002 as a historian. He is the author of *Fort Riley and Its Neighbors: Military Money and Economic Development* (University of Oklahoma Press, 1998) and co-authored, with Thomas D. Phillips, *The Black Regulars, 1866-1898* (University of Oklahoma Press, 2001). His articles have appeared in *Kansas History*, *Montana*, *Prologue*, and the *Western Historical Quarterly*.

Robert H. Dorff, Ph.D., joined the Strategic Studies Institute in June 2007 as Research Professor of National Security Affairs. He previously served on the US Army War College faculty as a Visiting Professor (1994-96) and as Professor of National Security Policy and

Strategy in the Department of National Security and Strategy (1997-2004), where he also held the General Maxwell D. Taylor Chair (1999-2002) and served as Department Chairman (2001-2004). Dr. Dorff has been a Senior Advisor with Creative Associates International, Inc., in Washington, DC, and served as Executive Director of the Institute of Political Leadership in Raleigh, NC (2004-2006). He is the author or co-author of three books and numerous journal articles. Professor Dorff holds a B.A. in Political Science from Colorado College and an M.A. and Ph.D. in Political Science from the University of North Carolina-Chapel Hill.

Robert J. Felderman, Brigadier General, US Army, is Special Assistant to the Chief, National Guard Bureau, detailed as the United States Deputy Director of Plans, Policy & Strategy for North American Aerospace Defense Command and United States Northern Command, Peterson Air Force Base, Colorado. Prior to his current assignment, General Felderman served as the Operations Deputy Director for National Guard Matters at United States Northern Command. General Felderman was commissioned as a Second Lieutenant upon graduation from the Army Officer Candidate School at Fort Benning, Georgia in 1977. He was previously enlisted in both the Army and the Air National Guard, having achieved the rank of Sergeant. General Felderman is a Master Army Aviator, with over 2,200 flight hours throughout twenty-two years of aviation duty. He has commanded at the company, battalion and brigade level, and is branch qualified infantry, armor (cavalry), aviation, medical service corps and strategic plans. He is a graduate of the National War College, National Defense University, Washington, District of Columbia.

John T. Fishel, Ph.D., Professor Emeritus from the National Defense University, was Professor of National Security Policy and Research Director at the Center for Hemispheric Defense Studies of the National Defense University from December 1997 until August 2006. He is also Professor Emeritus (Adjunct) from the School of International Service of American University. He has specialized in security and defense policy as well as Latin American affairs throughout his career focusing on issues of national development and stability operations of various kinds. He has written extensively on civil military operations and peacekeeping and is the author (with Max G. Manwaring) of *Uncomfortable Wars Revisited* (2006), *Civil Military Operations in the New World* (1997) and the editor and co-author of *"The Savage Wars of Peace:" Toward a New Paradigm of Peace Operations* (1998), and with Walter E. Kretchik and Robert F. Baumann the author of *Invasion, Intervention, "Intervasion"* (1997) (a study of the Haiti peacekeeping operation of 1994-95). While he was on active duty as a Lieutenant Colonel in the US Army he served in the United States Southern Command where he was, successively, Chief of the Civic Action Branch of the Directorate of Policy, Strategy, and Plans (J5), Chief of Research and Assessments of the Small Wars Operations Research Directorate (SWORD), Chief of the Policy and Strategy Division of the J5, and Deputy Chief of the US Forces Liaison Group. Concurrent with the latter position he served as Special Assistant to the Commander, US Military Support Group-Panama and to the Commander, US Army-South. Dr. Fishel received his Ph.D. in Political and Administrative Development from Indiana University in 1971, his M.A. in Political Science, also from Indiana in 1967, and his A.B. from Dartmouth College in International Relations in 1964.

Lynda M. Granfield, Lieutenant Colonel, US Army, earned a B.S. from University of Nebraska and M.S. in International Relations from Troy State University in 2001. She has served over the past 15 years as an active duty civil affairs officer. Her earlier assignments include Assistance Chief of Staff, G-9, Civil Military Operations for the Southern European Task Force (SETAF) in Vicenza, Italy; Assistant S-5, ARCENT-Kuwait; Headquarters Company Commander, 3d Battalion, 1SWTG; Tactical Civil Affairs Team Leader, Delta Company, 96th Civil Affairs Battalion; and Civil Affairs Planner, DCSOPS, USASOC. Lieutenant Colonel Granfield has also served in Operations DESERT SHIELD/STORM, UPHOLD DEMOCRACY, JOINT ENDEAVOR, PROVIDE PROMISE, and CJTF-Liberia. Her most recent deployment was to Jalalabad City, Nangarhar Province, Afghanistan where she served as the Provincial Reconstruction Team (PRT) Commander from June 2005-May 2006.

Benjamin Grob-Fitzgibbon, Ph.D., is an Assistant Professor of History at the University of Arkansas. He received his doctoral degree from Duke University in May 2006, after which he taught as a Visiting Assistant Professor at Duke and a Lecturer at North Carolina State University before coming to the University of Arkansas in July 2007. In 2004, Dr. Grob-Fitzgibbon published his first book, *The Irish Experience during the Second World War: An Oral History* (Irish Academic Press), which was named a Book of the Year by the *Sunday Irish Independent*. His second book, based on his doctoral dissertation, was published in 2007 by Palgrave Macmillan, titled *Turning Points of the Irish Revolution: The British Government, Intelligence, and the Cost of Indifference, 1912-1921*. Dr. Grob-Fitzgibbon has also published articles in *The Historian* (2003), *Terrorism and Political Violence* (2004), *Hemisphere* (2004), *Peace and Change* (2005), *The Journal of Intelligence History* (2006), and *Perspectives on History* (2008), as well a chapter in the book *Defending the Homeland: Historical Perspectives on Radicalism, Terrorism, and State Responses* (West Virginia University Press, 2007). His current project, provisionally titled *Imperial Endgame: Britain's Dirty Wars and the End of Empire*, explores British counterinsurgency campaigns in Palestine, Malaya, Kenya, Cyprus, and Aden in the years 1945 to 1965.

Mark T. Kimmitt, Brigadier General (Ret.), US Army, was sworn in as Assistant Secretary of State for Political-Military Affairs on August 8, 2008. Prior to assuming this position, he was the Deputy Assistant Secretary of Defense for Middle East Policy, responsible for military policy development, planning, guidance and oversight for the region. Mr. Kimmitt served for over 30 years as an officer in a wide variety of command, operational, and policy positions with extensive operational experience abroad before retiring with the rank of Brigadier General in the United States Army in 2006. His assignments included Deputy Director of Strategy and Plans at United States Central Command and Deputy Director of Operations and Chief Military Spokesman for Coalition Forces in Iraq during Operation IRAQI FREEDOM. Mr. Kimmitt is a graduate of the United States Military Academy. He holds a Master's Degree (with Distinction) from Harvard Business School. He holds additional Master's Degrees from the School of Advanced Military Studies and the National Defense University, and earned a certification as a Chartered Financial Analyst (CFA) while serving as Assistant Professor of Finance and Economics in the Department of Social Sciences at the United States Military Academy.

Geoffrey Loane is the current Head of Regional Delegation of the United States and Canada for the International Committee of the Red Cross (ICRC). In this capacity he oversees ICRC visits to the detention facility in Guantanamo Bay, Cuba and is responsible for day to day working relationships with the United States Government. Mr. Loane has also worked in the Balkans, Middle East and spent more than a decade in the Horn of Africa during the major conflicts there. These include Ethiopia, Sudan, Somalia, and Rwanda.

He has also served as the Head of the Emergency Relief Unit of the ICRC in Geneva. He has published books on the unintended consequences of humanitarian assistance and has conducted extensive field research in assistance operations.

Eric Nager, Major, is the Deputy Historian for the US Army, Pacific Command. A member of the US Army Reserve, he is a graduate of the Combined Arms and Services Staff School and the Command and General Staff College (ILE). Prior to his current assignment he served in the J4 section of the US Pacific Command; as Company Commander of the 375th Transportation Group; in the Support Operations section of the 167th Corps Support Group; and as the executive officer for Service Battery in the 3/83rd Field Artillery Battalion. Commissioned in 1989 through the ROTC program at Washington University of St. Louis, Major Nager is a graduate of Principia College in Elsah, Illinois, from which he holds a Bachelor of Arts in business administration and history, with a minor in Latin American studies. He also earned an M.B.A .from the University of South Alabama in 1991, and a Master of Liberal Arts (A.L.M.) from Harvard in 1999. He has been an adjunct faculty member for Faulkner University and Huntingdon College, teaching primarily undergraduate business courses.

Michael P. Noonan is the managing director of the Foreign Policy Research Institute's Program on National Security. His current research focuses on civil-military relations, defense transformation, and the military's role in the war on terrorism. As a Captain in the US Army Reserve, in June 2007 he returned from a 14-month deployment to Texas, Kuwait, and Iraq, where he served on a Military Transition Team (MiTT) with an Iraqi light infantry battalion in and around the northern city of Tal`Afar. Among other professional affiliations, he is a member of the International Institute of Strategic Studies, a fellow of the Inter-University Seminar on Armed Forces and Society, a lifetime member of the Veterans of Foreign Wars, and has consulted for the Institute for Defense Analyses. Mr. Noonan is a doctoral candidate in political science from Loyola University, Chicago; his dissertation deals with foreign internal defense, civil-military relations, and political-military effectiveness. His writings have appeared in *Orbis*, *The American Interest*, *Parameters*, *National Security Studies Quarterly*, *FPRI Wire*, and *FPRI E-Notes*.

Edgar Raines, Ph.D., US Army Center of Military History, is a graduate of Southern Illinois University and the University of Wisconsin where he received his Ph.D. in history in 1976. He served as an assistant academic dean at Silver Lake College in Manitowoc, Wisconsin, and as a historian at the Office of Air Force History before joining the US Army Center of Military History in November 1980. He co-authored (with Major David R. Campbell) *The Army and the Joint Chiefs of Staff: Evolution of Army Ideas on the Command, Control, and Coordination of the U.S. Armed Forces, 1942–1985* (1986) and *Eyes of Artillery: The Origins of Modern U.S. Army Aviation in World War II* (2000). He is currently working on a manuscript tentatively titled "*The Rucksack War: U.S. Army Operational Logistics in Grenada, October–December 1983.*" Dr. Raines has written numerous unpublished special studies as well as several articles in military and social history.

William Glenn Robertson, Ph.D., is the Director of the US Army Combat Studies Institute, US Army Combined Arms Center, Fort Leavenworth, Kansas. A graduate of the University of Richmond, he received his M.A. and Ph.D. degrees in history from the University of Virginia. Before joining the Combat Studies Institute in 1981, the Suffolk, Virginia native taught military history for ten years at colleges and universities in three states. Beginning in 1983, he led the resurrection of the Staff Ride teaching technique at the Command and General Staff College. Among his publications are two books, *The Bermuda Hundred Campaign*, and *The Battle of Old Men and Young Boys*, the Bull Run chapter in *America's First Battles 1776-1965*, the monograph *Counterattack on the Naktong*, and the US Army's guide to *The Staff Ride*. He has published articles and book reviews in numerous journals and periodicals, to include *Military Review*, *Military Affairs*, *Civil War History*, *Journal of American History*, *Journal of Southern History*, *Civil War Times Illustrated*, and *Blue and Gray Magazine*. He is currently working on *River of Death: The Campaign of Chickamauga*, a book-length study of that campaign, and two smaller works, *The Blackwater Line, 1861-1865*, and *The Post of Albuquerque, 1846-1867*. His awards include Phi Beta Kappa (1966), Command and General Staff College Civilian Instructor of the Year (1993), and the Harry S. Truman Award of the Kansas City Civil War Roundtable (1995). He has been the Director of the US Army Combat Studies Institute since August 2008.

Adam R. Seipp, Ph.D., joined the Texas A&M University history department in 2005 after completing his Ph.D. at the University of North Carolina at Chapel Hill. His current research project focuses on the demobilization of European societies after the First World War. Dr. Seipp has published articles in *War and Society*, and *The Journal of Contemporary History*. He authored the "An Immeasurable Sacrifice of Blood and Treasure: Demobilization and the Politics of the Streets in Britain and Germany, 1917-1921" chapter in *The Street as Stage: Protest Marches and Public Rallies since the Nineteenth Century*, Matthias Reiss, ed., (London: Studies of the German Historical Institute, 2007).

Richard Stewart, Ph.D., is currently the Chief Historian of the US Army Center of Military History. He has served previously at the Center as the Chief of Histories Division from 1998-2006, as the Command Historian, US Army Special Operations Command, Fort Bragg, North Carolina from 1990-1998, and Historian, Center for Army Lessons Learned, Fort Leavenworth, Kansas from 1987-1990. He has a Bachelor's Degree in history from Stetson University (1972), a Master's degree in history from the University of Florida (1980) and a Ph.D. in History from Yale University (1986). Dr. Stewart also has a Master's of Science in National Security Strategy from the National War College (2006). A retired Colonel in military intelligence, USAR, with 30 years of commissioned service, he has deployed as a combat historian to Saudi Arabia, Somalia, Haiti, Bosnia, and Afghanistan.

Jeffrey Woods, Ph.D., is an Associate Professor of History at Arkansas Tech University. He received his doctoral degree from Ohio University, Contemporary History Institute in June 2000. He published his first book, *Black Struggle, Red Scare: Segregation and Anticommunism in the South, 1948-1968* (Louisiana State University Press) in 2004. His second book, *Richard Russell, Southern Nationalism, and American Foreign Policy* (Rowman and Littlefield) was published in 2007. He has also published articles in the *Arkansas Historical Quarterly* (1998), the *Arkansas Times* (2002), *American Presidential Campaigns and Elections* (2003), *Disasters, Accidents, and Crises* (2007), and *Passport* (2008).

Lawrence A. Yates, Ph.D., is a native of Kansas City, Missouri. He received his Ph.D. in history from the University of Kansas in 1981, after which he joined the Combat Studies Institute at Fort Leavenworth, Kansas. During his 24 years with CSI, he taught and wrote about US military interventions, contingency and stability operations, and unconventional warfare. In 1989, he was in Panama during Operation JUST CAUSE. He is the author of Leavenworth Paper No. 15, *Power Pack: US Intervention in the Dominican Republic, 1965-1966*; *The US Military's Experience in Stability Operations, 1789-2005*; *The US Military Intervention in Panama: Origins, Planning, and Crisis Management, June 1987-December 1989*; co-author of *My Clan Against the World: US and Coalition Forces in Somalia, 1992-1994*; and co-editor/contributor to *Block by Block: The Challenges of Urban Operations*. He is currently working on a history of Operation JUST CAUSE.

About the Moderators

Kelvin D. Crow earned a B.S. from the University of Missouri in 1976. He served as an infantry officer with assignments in the Berlin Brigade, HQ US Army, Europe, and as a Staff Ride Instructor in the US Army Command and General Staff College. He earned an M.A. from Oregon State University in 1988 and an M.M.A.S. from the Command and General Staff College in 1989. Mr. Crow has published several articles in the popular historical press and a book, *Fort Leavenworth: Three Centuries of Service* (Command History Office, Combined Arms Center and Fort Leavenworth, 2004). Since 2002 he has served as the Assistant Command Historian for the Combined Arms Center and Fort Leavenworth.

Charles D. Collins Jr. is an assistant professor and the Sioux Wars course author for the Staff Ride Team, Combat Studies Institute, Fort Leavenworth, Kansas. He received a B.A. in History from Southwest Missouri State University and an M.M.A.S. in History from the US Army Command and General Staff College. While on active duty, Mr. Collins served in various armor and cavalry assignments. He retired from the Army in 1996. Mr. Collins' published works include *The Corps of Discovery: Staff Ride Handbook for the Lewis and Clark Expedition*; *The Atlas of the Sioux Wars*; and numerous articles on a wide variety of military topics.

Scott Farquhar, Lieutenant Colonel, US Army, earned a B.A. from Providence College in 1985 and a Master of Arts degree in History from Kansas State University in 2007. An infantry officer, he has served in a variety of command and staff positions in the United States, Europe, the Balkans and Southwest Asia. His most recent deployment was to Iraq 2007-2008 where he served as senior advisor to an Iraqi infantry brigade.

Ricardo A. Herrera, Ph.D., joined the Combat Studies Institute's Staff Ride Team in January 2006. A 1998 Marquette University Ph.D. in US history, he received his B.A. in history from the University of California, Los Angeles in 1984. Formerly Director of Honors at Mount Union College and Department Chair at Texas Lutheran University, Dr. Herrera also served as a Cavalry and Armor officer. He is the author of several articles and chapters on American military history.

Curtis S. King, Ph.D., graduated from the US Military Academy (USMA) in 1982 with a B.S. in History and English Literature. He received an M.A. from the University of Pennsylvania in 1992 and then was an instructor at USMA for three years. In 1998, he became an associate professor, Combat Studies Institute, US Army Command and General Staff

College. While at CSI, Dr. King received his Ph.D. in Russian and Soviet history (1998) and spent a 6-month tour in Sarajevo, Bosnia (1999-2000) as a NATO historian. Dr. King retired from the Army in May 2002. In October 2002, he joined the Staff Ride Team, CSI, as a civilian associate professor and is an adjunct professor at Kansas State University.

John J. McGrath has worked for the US Army in various capacities since 1978. After retirement he worked for four years at the US Army Center of Military History in Washington, DC, as a historian and archivist. Mr. McGrath is a graduate of Boston College and holds an MA in history from the University of Massachusetts at Boston. He is the author of numerous articles and military history publications including *Theater Logistics in the Gulf War*, published by the Army Materiel Command in 1994 and several works with the Combat Studies Institute, which he joined in 2003. He was also the General Editor for *An Army at War: Change in the Midst of Conflict*, published by CSI in 2006. He is currently pursuing a Ph.D. from Kansas State University.

Robert D. Ramsey III retired from the US Army in 1993 after 24 years of service as an Infantry officer that included tours in Vietnam, Korea, and the Sinai. He earned an M.A. in history from Rice University. Mr. Ramsey taught military history for three years at the United States Military Academy and six years at the US Army Command and General Staff College. Mr. Ramsey is the author of Global War on Terrorism Occasional Paper 18, *Advising Indigenous Forces: American Advisors in Korea, Vietnam, and El Salvador*; Occasional Paper 19, *Advice for Advisors: Suggestions and Observations from Lawrence to the Present*; Occasional Paper 24, *Savage Wars of Peace: Case Studies of Pacification in the Philippines, 1900–1902*; and Occasional Paper 25, *A Masterpiece of Counterguerrilla Warfare: BG J. Franklin Bell in the Philippines, 1901–1902*.

Donald P. Wright, Ph.D., is the chief of CSI's Contemporary Operations Study Team and a co-author of the recently released *On Point II: Transition to the New Campaign—The US Army in Operation IRAQI FREEDOM, May 2003–January 2005*. Dr. Wright has also served in the Active and Reserve Components of the US Army as an infantry and military intelligence officer.